Pandemic Over the Centuries

M. H. Fulekar · Geetika Madan Patel · Ashita Rai

Pandemic Over the Centuries

Causes and Consequences in a Changing Environment

Springer

M. H. Fulekar
Parul University
Vadodara, Gujarat, India

Geetika Madan Patel
Parul University
Vadodara, Gujarat, India

Ashita Rai
Parul University
Vadodara, Gujarat, India

ISBN 978-3-031-94269-3 ISBN 978-3-031-94270-9 (eBook)
https://doi.org/10.1007/978-3-031-94270-9

Jointly published with Capital Publishing Company
The print edition is not for sale in Afghanistan, Bangladesh, Bhutan, India, Sri Lanka, the Maldives, Nepal and Pakistan. Customers from Afghanistan, Bangladesh, Bhutan, India, Sri Lanka, the Maldives, Nepal and Pakistan please order the print book from: Capital Publishing Company.

This Springer imprint is published by the registered company Springer Nature Switzerland AG
The registered company address is: Gewerbestrasse 11, 6330 Cham, Switzerland

In loving memory of my brothers
Yadav H. Fulekar and Dilip H. Fulekar
who lost their lives during COVID-19

—M. H. Fulekar

To my beloved parents
Dinesh K. Rai and Mamta Rai
for their unconditional love, support, and blessings

—Ashita Rai

Preface

Pandemics are an inevitable part of human existence and likely to become more prevalent in the coming years, given the rapidly growing global population, increasing population mobility, and the likely change in disease dynamics that will come with climate.

Pandemics are outbreaks of infectious diseases that spread globally, affecting millions of people. They are dynamic events, often occurring in waves and causing significant economic, social, and political disruption. Infectious agents such as bacteria, viruses can readily be transmitted from an infected person, animal, or reservoir to a susceptible host, either directly or indirectly through an intermediate vector or inanimate environment and trigger outbreaks having pandemic potential. These etiological agents can induce infection across all age groups, with varying degree of severity resulting into associated morbidity, co-morbidity, and mortality.

Pandemics can have unusual epidemiological patterns and large outbreaks can occur in a peculiar condition based on the behaviour of the etiological agents. Environment also plays significant role in their spread and cause major setback to humanity at large.

Historically, pandemics have decimated societies, determined outcomes of wars, wiped out millions of lives across the world. It is also reported that humanity has witnessed an outbreak with pandemic potential across every century. Plague, Spanish flu, Cholera have already afflicted the society unleashing its severity into the twenty-first century as witnessed in the ongoing COVID-19 pandemic. However, they have also paradoxically cleared the way for innovations and advances in sciences (including medicine and public health), economy, and political systems. The history of each global outbreak, the origin, cause, consequences, epidemiological patterns, preventive measures, therapeutic intervention, and vaccine developed to control the pandemic for building, understanding and acquiring in-depth knowledge could contribute significantly to the field of medical science.

The book highlights on the pandemics—Plague 1720–1722, Cholera 1817–1824, Spanish Flu 1918–1919, Corona Virus 2019–2023 which occurred across several centuries with specific origin, causes, epidemiological patterns, impact, prevention, control, and treatment interventions. This book will give the first-hand information on the epidemic outbreaks leading to pandemic throughout the world. The society will be enriched with the knowledge of pandemic and preventive measures to be taken thereof.

Vadodara, India

M. H. Fulekar
Geetika Madan Patel
Ashita Rai

About This Book

Pandemic is a global outbreak of new infectious disease that occurs through the transmission of emerging and re-emerging pathogens infecting host species. The impact or severity tends to be higher in pandemics because of the much larger number of people who lack pre-existing immunity to the new virus. The number of severe cases rises globally from endemic to pandemic when huge large number of populations are infected.

Pandemics can have unusual epidemiological patterns and large outbreaks can occur in a peculiar condition based on the behaviour of the virus. Manmade and natural disasters can intensify the risk of epidemics considerably; exclusive outbreaks with high case fatality rates are the result of pandemic. In the history population all over the world have sporadically been affected by devastating outbreak of pandemic. It is also reported that every century the outbreak of epidemic caused by specific virus spread up throughout the world as pandemic.

The book highlights pandemic occurred in centuries with specific epidemiological patterns, outbreak consequences scenario in the world. The book will give the first-hand information on the subject and enrich the society with the understanding, knowledge, and preventive measures to be taken for the pandemic.

Contents

1 Introduction to Pandemic 1
1 Introduction 1
2 Characteristics of Pandemic 2
2.1 Wider Geographic Extension 2
2.2 Disease Transmission 3
2.3 High Attack Rates 3
2.4 Minimal Population Immunity 3
2.5 Emerging Infectious Disease 3
2.6 Infectivity 4
2.7 Disease Severity 4
3 The Continuum of Pandemic Phases 4
3.1 Interpandemic Phase 4
3.2 Alert Phase 5
3.3 Pandemic Phase 5
3.3.1 Phase 1 5
3.3.2 Phase 2 6
3.3.3 Phase 3 6
3.3.4 Phase 4 6
3.3.5 Phase 5 7
3.3.6 Phase 6 7
3.3.7 Post-peak 7
3.3.8 Post-pandemic period 7
3.4 Pandemic Transition Phase 7
4 Recommended Actions Before, During and After a Pandemic 8
4.1 Planning and Coordination 8
4.2 Situation Monitoring and Assessment 8
4.3 Reducing Spread of Disease 8
4.4 Continuity of Health Care Provision 9
4.5 Communications 9

5 Prevention and Control of Pandemic ... 9
5.1 Testing and Diagnosis ... 10
5.2 Treatment ... 11
5.3 Development of New Antimicrobial Drugs ... 11
5.4 Vaccination ... 11
5.5 Surveillance ... 12
5.6 Reduction of Contact with Wildlife ... 12
5.7 Environment and Climate Change ... 13
5.8 Urban Planning ... 14
5.9 Food Security and Proper Nutrition ... 14
5.10 Social Inequalities and Poverty ... 15
6 Conclusion ... 15
References ... 16

2 Pandemic: Historical Perspective ... 23
1 Introduction ... 23
2 Timeline of Pandemic ... 24
2.1 The Plague ... 24
2.1.1 Athenian Plague ... 24
2.1.2 Antonine Plague ... 25
2.1.3 Justinian Plague ... 25
2.1.4 The Black Death ... 26
2.2 Cholera ... 26
2.2.1 The Seven Pandemics of Cholera ... 27
2.3 Influenza ... 29
2.3.1 Russian Flu ... 29
2.3.2 Spanish Flu ... 30
2.3.3 Asian Flu ... 30
2.4 HIV/AIDS ... 31
2.5 Severe Acute Respiratory Syndrome (SARS) ... 32
2.6 Swine Flu ... 33
2.7 COVID-19 ... 33
3 Conclusion ... 34
References ... 35

3 Climate Change: Zoonoses in Changing Environment ... 43
1 Introduction ... 43
2 Classification of Zoonoses ... 44
2.1 Classification Based on Etiological Agents ... 45
2.1.1 Bacterial Zoonoses ... 45
2.1.2 Viral Zoonoses ... 45
2.1.3 Mycotic Zoonoses ... 45
2.1.4 Protozoal Zoonoses ... 45
2.1.5 Parasitic Zoonoses ... 45
2.1.6 Rickettsial Zoonoses ... 45
2.1.7 Chlamydial Zoonoses ... 46

2.2 Classification Based on Transmission Cycle (Chomel, 2009) 46
2.2.1 Orthozoonoses (Direct Zoonoses) 46
2.2.2 Cyclozoonoses 46
2.2.3 Metazoonoses (Also Called Pherozoonoses) 46
2.2.4 Saprozoonoses 46
2.3 Classification Based on Reservoir Hosts (Tayde, n.d.) 47
2.3.1 Anthrapozoonoses 47
2.3.2 Zooanthroponoses 47
2.3.3 Amphixenoses 47
2.4 Classification Based on Environment (Resh & Cardé, 2009) 48
2.4.1 Synanthropic 48
2.4.2 Exoanthropic 48
3 Climatic Factors Affecting Zoonoses 48
3.1 Land Temperatures 48
3.2 Ocean Temperatures, Sea Level and Acidity 49
3.3 Rainfall Patterns 49
3.4 Soil Conditions 49
4 Anthropogenic Factors Affecting Zoonoses 50
4.1 Travel and Tourism 50
4.2 Trade 50
4.3 Pets 51
4.4 Agricultural Practices and Livestock Farming 51
4.5 Food Production and Distribution 51
4.6 Deforestation and Urbanization 52
4.7 Political Factors 52
5 Epidemiological Aspects of Zoonosis 52
6 Prevention and Control of Zoonoses 54
6.1 Reservoir Neutralization 54
6.2 Reducing Contact Potential 55
6.3 Increasing Host Resistance 56
6.4 Chemoprophylaxis 56
6.5 Immunization 56
7 Conclusion 56
References 57

4 Plague 63
1 Introduction 63
2 Three Pandemics of Plague: Historical Account 64
2.1 The Justinian Plague 64
2.2 'Black Death' of Europe 66
2.3 Third Pandemic of 1894 68
3 Types of Plague 69
3.1 Bubonic Plague 70

3.2 Septicemic Plague 70
3.3 Pneumonic Plague 71
4 Ecology and Transmission 72
5 Prevention and Control Strategies 73
5.1 Surveillance 73
5.2 Rodent Control 73
5.2.1 Environmental Sanitation 73
5.2.2 Physical Methods 74
5.2.3 Chemical Methods 74
5.2.4 Biological Methods 74
5.3 Vector Control 74
5.3.1 Personal Prophylactic Measures 74
5.3.2 Insufflation 75
5.3.3 Residual Insecticide Spray 75
5.3.4 Health Education and Community Participation 75
6 Diagnosis and Treatment 75
6.1 Clinical Diagnosis 76
6.2 Laboratory Testing (CDC, 2022) 76
6.3 Treatment (CDC, 2022) 77
6.4 Vaccines 78
7 Plague as Biowarfare Weapon 78
References 78

5 Spanish Flu 83
1 Introduction 83
2 The Great Influenza Pandemic of 1918 84
2.1 Origin and Transmission 84
2.2 Diagnosis and Symptomology 86
3 Spanish Flu: Environmental Perspectives 87
4 Prevention and Control of Spanish Flu 88
5 Influenza Diagnosis, Prevention and Control Strategies: Present Status 89
6 World Health Organization: Global Influenza Programme (GIP) and the Global Influenza Surveillance and Response System (GISRS) 90
7 Conclusion 91
References 91

6 Cholera 95
1 Introduction 95
2 Seven Pandemics of Cholera 96
2.1 First Cholera Pandemic 96
2.2 Second Cholera Pandemic 96
2.3 Third Cholera Pandemic 97
2.4 Fourth Cholera Pandemic 98
2.5 The Pandemic's Fifth Wave 98

2.6 Sixth Cholera Pandemic ... 98
2.7 Seventh Pandemic of Cholera ... 99
3 Cholera: Epidemiology ... 99
3.1 Global Distribution ... 99
3.2 *Vibrio Cholerae* Strain and Transmission of Cholera ... 100
3.3 Risk Factors ... 100
4 Cholera: Diagnosis and Clinical Manifestations ... 101
5 Cholera: Environmental Perspective ... 101
6 Prevention and Control of Cholera ... 102
6.1 Implementation/Reinforcement of Surveillance Systems ... 103
6.2 Water, Sanitation and Hygiene Promotion (WASH) ... 103
6.3 Deployment of Oral Cholera Vaccine (OCV) ... 103
6.4 Social Mobilization and Health Promotion ... 103
6.5 Collaboration ... 104
7 Treatment of Cholera ... 104
7.1 Oral Rehydration Therapy ... 105
7.2 Antibiotics/Antimicrobials ... 105
7.3 Nutritional Supplements ... 106
7.4 Vaccines ... 106
7.5 Probiotics ... 107
References ... 108

7 Coronavirus Disease (COVID-19) ... 115
1 Introduction ... 115
2 SARS-COV-2 Epidemiology ... 118
2.1 Origin of SARS-CoV-2 ... 118
2.2 Structure of SARS-CoV-2 ... 120
2.3 Classification of SARS-COV-2 ... 121
2.4 Variants of SARS-COV-2 ... 122
2.4.1 SARS-CoV-2 Variants Being Monitored (VBM) ... 123
2.4.2 SARS-CoV-2 Variants of Concern (VOCs) ... 123
2.4.3 Variant of Interest (VOI) ... 127
2.4.4 Variant of High Consequence (VOHC) ... 128
2.5 Transmission of SARS-COV-2 ... 128
2.5.1 Respiratory Transmission (Aerosol/Droplet) ... 128
2.5.2 Direct Contact and Fomites Transmission ... 129
2.5.3 Faecal-Oral Transmission ... 129
2.5.4 Transmission Through Surgical Operations ... 129
2.6 SARS-COV-2: Life Cycle ... 130
2.6.1 Entry of Virus ... 130
2.6.2 Translation of Viral Replication Machinery and Replication ... 130

2.6.3 Translation of Viral Structure Proteins, Virion Assembly and Release of Virus ... 131
3 SARS-CoV-2 Clinical Manifestations ... 131
3.1 Pulmonary Manifestation ... 133
3.2 Cardiovascular Manifestation ... 134
3.3 Gastrointestinal and Liver Related Manifestations ... 134
3.4 Haematological Manifestations ... 135
3.5 Neurological Manifestations ... 135
3.6 Renal Manifestations ... 136
3.7 Dermal Manifestations ... 136
3.8 Endocrine Organs ... 136
4 Mucormycosis and SARS-COV-2: An Epidemic Within Pandemic ... 137
4.1 Types of Mucormycosis ... 138
4.1.1 Rhino-Orbital Cerebral Mucormycosis (ROCM) ... 139
4.1.2 Pulmonary Mucormycosis ... 139
4.1.3 Cutaneous Mucormycosis ... 140
4.1.4 Gastrointestinal Mucormycosis ... 140
4.1.5 Disseminated Mucormycosis ... 140
4.2 Immunocompromised Conditions in COVID-19 Patients Favouring Mucormycosis ... 140
4.3 Clinical Identification and Diagnosis of Fungus ... 141
4.4 Risk Factors Associated with Mucormycosis ... 142
4.4.1 Diabetes Mellitus ... 142
4.4.2 Immunomodulator Use ... 142
4.4.3 Iron and Zinc Supplements ... 143
4.5 Prevention and Treatment of Mucormycosis ... 144
4.5.1 First-Line Therapy ... 144
4.5.2 Salvage Therapy ... 144
4.5.3 Surgical Treatment ... 145
5 Prevention of SARS-COV-2 ... 145
5.1 Administrative Management ... 145
5.2 Social Distancing ... 146
5.3 Cleaning and Disinfection ... 146
5.4 Personal Protective Measures ... 148
5.5 Quarantine/Home Isolation ... 148
6 Therapeutics for COVID-19 ... 149
6.1 Antiviral Treatments ... 149
6.2 Monoclonal (mAb) Antibodies ... 150
6.3 Long-Acting Antibody (LAAB) Treatment ... 150
7 Vaccination for SARS-COV-2 ... 150
7.1 Vaccine Development: Phases of Clinical Trials ... 155
7.1.1 Clinical Trial Phase I ... 156
7.1.2 Clinical Trial Phase II ... 156

7.1.3 Clinical Trial Phase III ... 156
7.1.4 Approval and Licensure ... 157
7.2 Vaccine Development Platforms ... 158
7.2.1 Component Viral Vaccines ... 159
7.2.2 Whole Virus Vaccines ... 162
8 SARS-CoV-2: Environmental Perspective ... 164
8.1 Role of Climatic Factors in COVID-19 ... 165
8.2 Impact on Water Quality ... 166
8.3 Impact of COVID-19 on Air Quality ... 167
8.4 Impact of COVID-19 on Biomedical Waste Management ... 168
8.5 Impact of Municipal Waste Management ... 168
8.6 Impact of COVID-19 on Energy Consumption ... 169
9 COVID-19: Indian Scenario ... 169
9.1 First Wave of Pandemic ... 170
9.1.1 The Unlocks ... 171
9.2 Second Wave of Pandemic ... 172
9.3 Third Wave of Pandemic ... 174
10 Conclusion ... 175
References ... 176

8 COVID-19 as an Occupational Disease ... 197
1 Introduction ... 197
1.1 COVID-19: Exposure in the Workforce ... 197
1.2 Occupational Health and Hygiene ... 198
1.3 Occupational Diseases ... 199
1.4 Coronavirus (COVID-19): Outbreak Risk to Other Diseases and Challenges ... 200
1.5 Modes of Entry ... 200
1.5.1 Inhalation ... 200
1.5.2 Direct Contact ... 200
1.5.3 Ingestion ... 201
2 Occupational Safety and Health Administration Classification Related to Risk of Worker Exposure to SARS-CoV-2 ... 201
3 COVID-19: Hierarchy of Hazard Controls ... 203
3.1 Elimination and Substitution ... 204
3.2 Engineering Controls ... 204
3.3 Administrative Controls ... 205
3.4 Personal Protective Equipment ... 205
4 Occupational Safety and Health Administration Guidelines ... 206
4.1 Employers ... 206
4.2 Health Workers ... 206
5 Conclusion ... 207
References ... 207

9 Migrants During Lockdown and Green Building Perspectives 211
1 Introduction 211
2 Covid-19: Lockdown Phases in India 212
2.1 Phase 1 (25th March to 14th April) 212
2.2 Phase 2 (15th April to 3rd May, 2020) 212
2.3 Phase 3 (4th to 17th May, 2020) 213
2.4 Phase 4 (18th to 31st May, 2020) 213
3 Lockdown Impact 213
4 Green Building for Migrant Workers 214
4.1 Green Building: Definition 215
4.2 Components of Green Building 216
4.2.1 Energy Efficiency and Renewable Energy 216
4.2.2 Water Efficiency 217
4.2.3 Environmentally Preferred Building Materials and Specification 218
4.2.4 Waste Reduction 218
4.2.5 Toxics Reduction 218
4.2.6 Indoor Air Quality 219
4.2.7 Smart Growth and Sustainable Development 219
5 Conclusion 220
References 220

Glossary 225

Index 233

About the Authors

M. H. Fulekar is a well-known Environmentalist and is listed among the top 2% of scientists in the world by Stanford University. He has worked as Professor, Department of Life Sciences, University of Mumbai from 2002 to 2011 and Senior Professor and Dean, School of Environment and Sustainable Development, Central University of Gujarat (2011–2019). At present he is working as Senior Professor and Joint Director (R&D), Research and Development Cell, Parul University, Gujarat, India. Professor Fulekar has 38 years of teaching and research experience. As an academician, he has developed various curriculums: M.Sc., M.Phil./Ph.D., Ph.D. Environment Science and M.Phil./Ph.D. in Nano Sciences. Under his guidance 30 Ph.Ds. 22 M.Phils. have been awarded. He has more than 350 research articles, five Indian patents and several state and national projects to his credit. He is a globally known author and has published 17 Books with reputed publishers. He has also rendered his expertise internationally to several countries on safety, health, and environment.

Geetika Madan Patel Vice President (Quality, Research and Health Sciences), Parul University is a proficient doctor and administrator having pursued her MBBS and MD in Community Medicine with Gold Medal from the illustrious Maharaja Sayajirao University of Baroda. She is also the Member of Board of Governance, Parul University and Medical Director of Parul Sevashram Hospital. She has played key role in the establishment of various medical institutions and hospitals in Parul University.

Ashita Rai is working as an Assistant Professor, Research and Development Cell, Parul University. She is a distinguished scholar, received gold medal with distinction in academic curriculum. She has been awarded her doctoral degree in Environment and Sustainable Development from Central University of Gujarat, Gandhinagar. She is a leading young scientist and has authored research articles and published two patents and filed three patents in the area of wastewater treatment, solid waste remediation and climate change. Currently, she is also writing books on (i) Rhizosphere: Green Technology (Springer) and (ii) Eco Nanotechnology for Wastewater Treatment and Purification (CRC Press, USA).

Chapter 1
Introduction to Pandemic

1 Introduction

COVID-19 has unleashed a worldwide shock wave with severe health, economic and social consequences. However, the world has faced several pandemics in past viz. plague, cholera, Spanish flu, severe acute respiratory syndrome coronavirus (SARS-CoV) and Middle East respiratory syndrome coronavirus (MERS-CoV) that have caused havoc among societies. The most fatal pandemic in recorded history was the Black Death (also known as The Plague) which killed an estimated 75–200 million people in the 14th century (DeLeo & Hinnebusch, 2005). Viral outbreaks do occur regularly but the severity of outcome may vary. Most of all these pandemics have originated through the cross-species transmission of microorganisms from animals to humans, overwhelmingly in the Old World (Jones et al., 2008). Genetic reassortment gives rise to novel viral strains causing new outbreaks with epidemic or pandemic potential against which mankind is not prepared to fight. The terms epidemic and pandemic relate to the occurrence of a health condition compared to its predicted rate as well as to its spread in geographic areas (Grennan, 2019). An epidemic is an outbreak that spreads to larger geographic areas where a pandemic is an epidemic that could significantly raise morbidity and mortality globally and cause major economic, social and political disturbance (Bonotti & Zech, 2021; Brooks, 2020). A pandemic is triggered by the combined impacts of spark risk (represents the pandemic's onset, and activities such as contact tracing, quarantine, and isolation are often mandate) and spread risk (when a pandemic takes off, measures used during the spark period are continued, and treatment—though essential—may be limited and its efficacy uncertain) (Das et al., 2021). Consequences of pandemics can be felt across every segment of the life. They result in substantial, widespread increase in mortality and morbidity, with significantly greater mortality implications on low- and middle-income countries (LMICs) (Walker et al., 2020); pandemics can generate financial damage through a variety of mechanisms, particularly short-term fiscal crises, and relatively long bad economic growth consequences (Shang et al.,

M. H. Fulekar et al., *Pandemic Over the Centuries*,
https://doi.org/10.1007/978-3-031-94270-9_1

2021). Individual behavioral changes, including fear-induced hostility to workplaces and other public gathering areas, are the leading cause of negative economic growth disruptions during pandemics (Tran et al., 2020). Further, mitigation methods can result in substantial societal and financial disruptions (Organisation for Economic Co-operation and Development [OECD], 2021). In nations with weak governance and a history of political fragility, pandemics can exacerbate political turmoil and burdens (OECD, 2020). In these circumstances, outbreak control tactics including quarantines often provoked unrest and animosity amongst governments and citizens (Wood et al., 2022).

There is a clear indication that during the past few decades, growing global travel and interconnectedness, urbanisation, changes in land use, and higher exploitation of the natural ecosystem have all raised the likelihood of pandemics (Jones et al., 2008; Morse, 1995). Such tendencies would intensify and continue to accelerate. Thus, to facilitate improved defence against significant disease outbreaks, pandemic preparedness and disease surveillance embedded in strong medical systems which include all citizens are essential. The imperative to identify and manage emerging breakouts that could spread to become pandemics, in addition to increase and sustain funding in preparation and medical systems, has received considerable legislative prominence (Smolinsky et al., 2003). Additionally, probabilistic simulation and analytical methods, such as probability of exceedance (EP) curves, are useful for assessing the likelihood of a pandemic and forecasting its potential effects (Chu, 2021; Shadabfar et al., 2021).

2 Characteristics of Pandemic

The term pandemic has some key features which help us better understand the concept by examining the similarities and differences among them.

2.1 Wider Geographic Extension

A pandemic cuts across international boundaries. This wide geographical reach is what makes pandemics lead to large-scale social disruption, economic loss, and general hardship. It is important to note that a once-declared epidemic can progress into pandemic status. While an epidemic is large, it is also generally contained or expected in its spread, while a pandemic is international and out of control.

2.2 *Disease Transmission*

Skin-to-skin contact and droplet spreading are the two direct methods of transmitting microorganisms from their natural repository to a host organism. Indirect methods include suspended air particles, inanimate objects, or animate intermediates (vectors) (Centers for Disease Control and Prevention [CDC], 2022).

2.3 *High Attack Rates*

An attack rate, also known as an incidence proportion, is a measurement of the number of fresh instances of a disease that emerge during a specific period from a populace that was existing at the onset of an infection (Moore & Bell, 2020). Pandemics tend to have higher attack rate in addition to wider geographical spread. For instance, Influenza occurs all over the world, with an annual global attack rate estimated at 5–10% in adults and 20–30% in children (World Health Organization [WHO], 2022).

2.4 *Minimal Population Immunity*

The expression of antibodies to an ailment in an individual confers immunity against that infection. New pathogenic strains lay stress upon the immune stress. This stress is heightened in an already immunocompromised individuals as it causes more harm. Number of cases and death toll increases exponentially during pandemics as evidenced in COVID-19 and previous pandemics as well.

2.5 *Emerging Infectious Disease*

These are diseases which have recently emerged among a community or whose occurrence or geographic distribution is rapidly growing or has an immediate potential threat of infection (Morse, 1995). The World Health Organization warned in its 2007 report that infectious diseases are emerging at a rate that has not been seen before. Since 1970s, about 40 infectious diseases have been discovered, including SARS, MERS, Ebola, chikungunya, avian flu, swine flu, Zika and most recently COVID-19, caused by a new coronavirus, SARS-CoV-2 (Baylor College of Medicine, n.d.).

2.6 *Infectivity*

This refers to diseases that are transmitted directly, is the capacity of an infectious agent to develop a new infection in a host organism. It is measured by the secondary attack rate (SAR), which is the proportion of susceptible individuals (i.e., household or close contacts) that develop the infection after exposure to a primary case (Barreto et al., 2006). The value of SAR is considerably high during pandemics. According to a research article SAR varies widely across countries with lowest reported rate as 4.6% and highest as 49.56% (Shah et al., 2020).

2.7 *Disease Severity*

It is commonly used to describe the effect that an illness has had on resources, complications and fatality. For example, the Pandemic Severity Assessment Framework (PSAF) of 1918 pandemic characterized it as one with very high transmissibility and clinical severity. The results help public health officials and health care professionals make timely and informed decisions, and to take appropriate actions. Ongoing COVID-19 can also be determined to have higher severity value owing to its unique symptoms, morbidity, co-morbidity, secondary infections like mucormycosis, and mortality rates.

3 The Continuum of Pandemic Phases

The World Health Organization (2017) has described continuum of phases in response to a pandemic from emerging viruses. The phases are applicable to the entire world and provide a global framework to aid countries in pandemic preparedness and response planning. The global phases: interpandemic, alert, pandemic, and transition as highlighted below (Ministero della Salute & Istituto Superiore di Sanità, 2020).

3.1 *Interpandemic Phase*

There are two phases to the interpandemic era. Phase 1 is when the virus's subtype is prevalent in animals although it may appear to pose minimal danger to human beings whereas in phase 2, a strain of the spreading animal virus presents a significant danger to mankind (Webby & Webster, 2003). The most effective method to minimize morbidity and death during this phase is vaccination, although an effective vaccine is unlikely to be widely accessible during the pandemic's onset. Therefore, it will

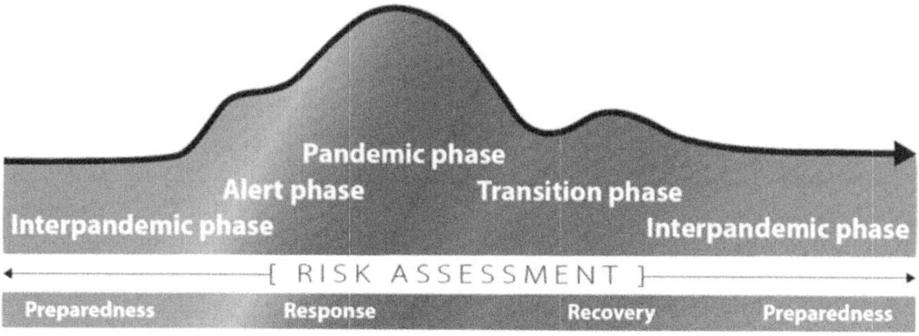

Fig. 1 Continuum of pandemic phases. *Source* WHO (2017)

be necessary to put forth other measures of control, such as pharmaceutical and non-pharmaceutical approaches (Duerr et al., 2007).

3.2 Alert Phase

Discovery of a novel, emergent disease in humans. During this phase, it is important to pay closer attention and conduct local, national, and international risk evaluation. De-escalate or shift initiatives towards less rigorous precautions, equivalent to those used during the interpandemic period, if evaluations demonstrate that the new disease is unlikely to have the potential to grow into a pandemic strain (Fig. 1).

3.3 Pandemic Phase

Time frame marked by the emerging pathogen's global transmission, which is tracked by monitoring on a worldwide scale. Relying mostly on serologic, epidemiologic, and clinical evidence, the transformation between interpandemic stage, the alert phase, as well as the pandemic phase might take place quickly or steadily. A six-stage categorization was formerly used by the World Health Organization (2009) to explain the progression of a pandemic influenza virus from the early human infections to a pandemic. It begins with a virus that largely affects animals and a few incidents where animals infect people, proceeds through a stage where the virus starts to transfer amongst people directly, and finishes with a stage where the viral infections in humans have spread globally. The six stages of the pandemic phase are as follows.

3.3.1 Phase 1

No reports of infections spreading amongst animals infecting humans.

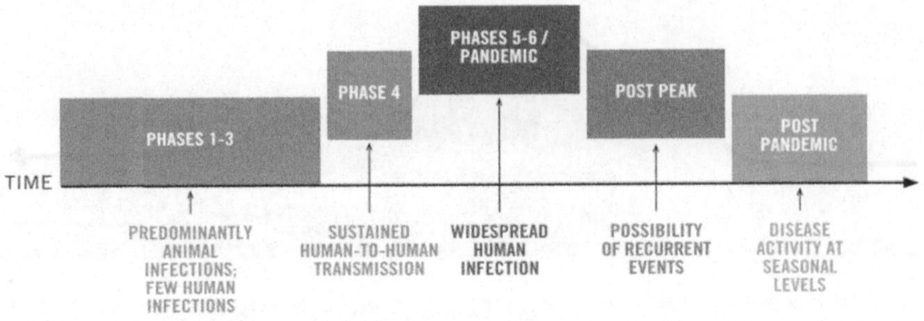

Fig. 2 Phases in pandemic. *Source* WHO (2017)

3.3.2 Phase 2

A virus that is spreading amongst either domestic or wild species is believed to have infected living beings, making it a possible pandemic hazard.

3.3.3 Phase 3

There have been isolated incidents or small clusters of illness in humans brought on by an animal or human-animal recombinant virus, but there has not been enough human-to-human spread to support widespread breakouts. Limited human-to-human transmission may happen in situations, such as when an infected individual comes into close contact with an unprotected caretaker. Furthermore, a small amount of transmission in such constrained conditions does not necessarily mean that the virus has attained the degree of human-to human transmissibility required to trigger a pandemic (Fig. 2).

3.3.4 Phase 4

The stage is identified by the confirmed spread of an animal or human-animal influenza recombinant virus from person-to-person that can lead to "community-level breakouts." A substantial increase in the likelihood of a pandemic is signalled by the capacity to produce continuous outbreaks in a region. In order to collectively evaluate the circumstances and determine if the execution of a prompt pandemic prevention effort is necessary, every nation which detects or has confirmed such an occurrence must immediately communicate with WHO. Phase 4 signals a considerable increase in the probability of a pandemic, but it does not imply that a pandemic is inevitable.

3.3.5 Phase 5

This phase is characterized by the virus spreading from person-to-person into at least two nations in one WHO area. Even if most nations are unaffected at this point, the proclamation of Phase 5 is a clear indication that a pandemic is approaching and that there is little time left to prepare, disseminate, and put the specified mitigation strategies into action.

3.3.6 Phase 6

In addition to the conditions listed in Phase 5, this phase is defined by community-level infections in at least one other nation in a different WHO area. The declaration of this phase might imply the onset of a pandemic on a worldwide scale.

3.3.7 Post-peak

In most nations with sufficient surveillance, pandemic infection rates may have fallen beneath maximum documented rates. The post-peak era indicates that pandemic incidence seems to be waning, but it is unknown if there may be further waves, in which nations ought to be ready.

3.3.8 Post-pandemic period

By the moment the time has passed, disease activity may have reverted to its usual levels. The pandemic pathogen is anticipated to act similarly to a seasonal pathogen. Maintaining surveillance and updating pandemic preparedness and response strategies as necessary are crucial at this point. It might be necessary to go through a rigorous evaluation and recuperation procedure.

3.4 Pandemic Transition Phase

De-escalation of efforts, depending on nation risk evaluation, may take place as the threat declines internationally, with a drop in crisis response at the national scale as well as a transition towards recovery efforts.

4 Recommended Actions Before, During and After a Pandemic

The WHO (2009) has identified five elements of preparedness and response, that are discussed as follows.

4.1 Planning and Coordination

The purpose of planning and coordinating operations is to provide direction and collaboration among sectors. A key element of national emergency preparedness approaches is the integration of pandemic readiness.

4.2 Situation Monitoring and Assessment

Information about a pandemic's potential must be collected, evaluated, and communicated before it commences. Once a pandemic has begun, it needs to be watched carefully for severity as well as other factors. To ascertain whether the probability of a pandemic is increasing, it will be essential to monitor the pathogens, how communicable it is to humans, as well as how infection is spread among communities. It is essential to gather information on pathogens, mutations that take place and affect biological characteristics, in addition to quickly investigate and assess situations. If a pandemic breaks out, evaluating the effectiveness of the prevention strategies will be essential.

4.3 Reducing Spread of Disease

The key to stopping the spread of disease will be to construct increased "social distance" amongst people. In addition to administering antivirals, other drugs, and vaccines, safeguards will need to be taken at the individual/household, social, and international scales. A few examples of interventions conducted at the individual or family levels are risk information, basic hygiene, self-protection, personal care, and contact quarantine. A change in public behaviour, the involvement of multiple sectors, the mobilisation of resources, effective communication, and media backing are all essential for these initiatives. Limitations on overseas travel would impact commerce and transportation while also preventing the pandemic from propagating to uninfected countries. Nations must achieve a balance between reducing threats to the general public's health and preventing unwanted disruptions in international transportation and commerce. There are several ways to treat or prevent the infection utilising medicinal medications. Effective treatment and prevention of secondary or pre-existing ailments will frequently also play a significant role in reducing the overall impact of morbidity and mortality.

4.4 Continuity of Health Care Provision

During a pandemic, medical centres are required to provide healthcare simultaneously managing the influx of sick people. Monitoring for surge abilities in healthcare centres will help decide how much the existing medical system can be upgraded to meet the additional caseloads. Medical centres must provide effective evaluation and disease management protocols in order to protect medical staff, patients and visitors.

4.5 Communications

Relevant information must be communicated to the general public, collaborators, and stakeholders, both before and during an outbreak, to help them decide wisely, take appropriate actions to protect the public's well-being and safety. This is crucial for effective risk mitigation. Planning, trust, transparency, making early announcements, and listening are the five major components which communications should abide by, in accordance with the WHO's planning framework for outbreak communications (Fig. 3).

5 Prevention and Control of Pandemic

Several variables, including humanitarian crises, armed conflicts, weakened medical systems, inadequate monitoring, and limited laboratory diagnosis, have contributed to a rise in the danger of pandemic pathogens over the past few years (Santoro et al., 2016). Combating outbreaks continues to be a top priority for governments everywhere in the world in order to safeguard public health. Additionally, it plays a crucial role in ensuring that people are well-protected from worldwide transmission of these infections. As a result, the following steps and actions can help in the prevention and management of infections at the local, national and international levels.

<table>
<tr><th></th><th>ESTIMATED PROBABILITY OF PANDEMIC</th><th>DESCRIPTION</th><th>MAIN ACTIONS IN AFFECTED COUNTRIES</th><th>MAIN ACTIONS IN NOT-YET AFFECTED COUNTRIES</th></tr>
<tr><td>PHASE 1</td><td rowspan="3">Uncertain</td><td>No animal influenza virus circulating among animals has been reported to cause infection in humans.</td><td colspan="2" rowspan="3">Producing, implementing, exercising, and harmonizing national pandemic influenza preparedness and response plans with national emergency preparedness and response plans.</td></tr>
<tr><td>PHASE 2</td><td>An animal influenza virus circulating in domesticated or wild animals is known to have caused infection in humans and is therefore considered a specific potential pandemic threat.</td></tr>
<tr><td>PHASE 3</td><td>An animal or human-animal influenza reassortant virus has caused sporadic cases or small clusters of disease in people, but has not resulted in human-to-human transmission sufficient to sustain community-level outbreaks.</td></tr>
<tr><td>PHASE 4</td><td>Medium to high</td><td>Human-to-human transmission of an animal or human-animal influenza reassortant virus able to sustain community-level outbreaks has been verified.</td><td>Rapid containment.</td><td>Readiness for pandemic response.</td></tr>
<tr><td>PHASE 5</td><td>High to certain</td><td>The same identified virus has caused sustained community-level outbreaks in at least two countries in one WHO region.</td><td rowspan="2">Pandemic response: each country to implement actions as called for in their national plans.</td><td rowspan="2">Readiness for imminent response.</td></tr>
<tr><td>PHASE 6</td><td>Pandemic in progress</td><td>In addition to the criteria defined in Phase 5, the same virus has caused sustained community-level outbreaks in at least one other country in another WHO region.</td></tr>
<tr><td colspan="2">POST-PEAK PERIOD</td><td>Levels of pandemic influenza in most countries with adequate surveillance have dropped below peak levels.</td><td>Evaluation of response, recovery, preparation for possible second wave.</td><td rowspan="3">-</td></tr>
<tr><td colspan="2">POSSIBLE NEW WAVE</td><td>Level of pandemic influenza activity in most countries with adequate surveillance is rising again.</td><td>Response</td></tr>
<tr><td colspan="2">POST-PANDEMIC PERIOD</td><td>Levels of influenza have returned to the levels seen for seasonal influenza in most countries with adequate surveillance.</td><td>Evaluation of response; revision of plans; recovery.</td></tr>
</table>

Fig. 3 Summary table for recommended actions. *Source* WHO (2009)

5.1 Testing and Diagnosis

The most crucial measures for the effectiveness of communicable diseases control programmes are indeed the testing of pathogens and the diagnosis of affected individuals within a certain population. This action reduces the spread of the disease and serves as the first step in treating sick people. By identifying the individuals or communities most impacted by a specific virus, mass-scale testing makes it possible to identify the regions wherein the issue is most severe and more effectively target control and preventative strategies to these specific populations and locales. By identifying and monitoring viral strains can quantify current infections and give vital information about viral etiology (Lot et al., 2004). Additionally, testing is essential for monitoring persistent communicable diseases and containing outbreaks (Banoo et al., 2007; Kozel & Burnham-Marusich, 2017; Mabey et al., 2004; Walensky et al., 2007).

5.2 Treatment

Treatment of infectious diseases significantly enhances the life quality for those who are afflicted by limiting negative effects on social and economic spheres and preventing potential co-infection related problems for the patient (da Silva et al., 2014). Additionally, treating certain diseases at the community level requires that infections are properly treated (Cohen et al., 2011). In conclusion, treating infections lessens the incidence of infectious diseases and lowers the risk of developing new infections.

5.3 Development of New Antimicrobial Drugs

Antimicrobial treatments are among the most important instruments that civilization requires to treat a wide range of infectious diseases. Due to their widespread use across the globe, antibiotics play a significant role among antimicrobial medications. Conversely, the development of multi-resistant microbes has been caused by the uncontrolled and frequently unethical utilization of antibiotics in both humans and other animals, as well as by the disposal of these medications in environments. One of the largest issues currently challenging the study of contagious diseases is antimicrobial resistance. To treat infections caused by multidrug-resistant bacteria, it is vital to create new antimicrobial medications (Bronzwaer et al., 2002; Coates et al., 2002; Marston et al., 2016; Morgan et al., 2011; Piddock, 2012; Tenover, 2006). However, the COVID-19 outbreak showed that this resource needs to be used very cautiously. Whenever employed in varied settings to combat novel infectious pathogens, antimicrobials that are typically used to treat certain infections can be beneficial or even harmful (Kandeel et al., 2021; Mallhi et al., 2021; Martinez, 2020; Senanayake, 2020).

5.4 Vaccination

Vaccines are biological agents that produce a resistance mechanism towards pathogens. Every year, vaccinations avert 2.5 million deaths. Vaccines not just shield the vaccinated but also the general populace. Through the phenomenon known as "herd immunity," which is attributed to the presence of many people immunized against a specific pathogen, mass vaccination limits the susceptibility of unvaccinated people to pathogens and stops or slows down the chain of disease transmission. Governments ought to put a lot of effort into raising vaccination coverage by running vaccination campaigns and setting up frameworks to empower vaccination, like supplying vaccines at no cost to the general public and guaranteeing quick access to vaccination services.

5.5 Surveillance

Surveillance is among the most successful ways of controlling infectious disease outbreaks and containing infectious strains. Strong epidemiological surveillance systems enable the tracking of pathogen circulation in people, non-human animals, and human-animal interfaces, ensuring the rapid recognition of ailments with the possibility of spreading epidemically and pandemically. Additionally, the method for identifying abrupt spikes in cases of a specific disease is epidemiologic surveillance. Genomic sequence technologies are being employed in the quick and highly precise diagnosis and follow-up of illnesses. Such tools also enable precise and practical real-time surveillance of outbreaks (Ellwanger et al., 2017, 2019; Geoghegan & Holmes, 2017; Grady & Loman, 2018; Holmes et al., 2018). Outbreaks and pathogens disregard social and political limitations as well as national boundaries. Infectious disease outbreaks may turn into major international emergencies due to inadequacies in the surveillance systems. In order to effectively combat infectious diseases, global leaders must cooperate and practice medical policy, consideration for local and global ethical concerns, and strong international collaborations (Bedford et al., 2019). The COVID-19 pandemic serves as an excellent example of the value of epidemiological surveillance as well as early detection and management to a new potential pandemic pathogen.

5.6 Reduction of Contact with Wildlife

About 58–75% of infectious diseases that impact humans are caused by microbes that are housed in non-human species. In other words, the preponderance of infectious diseases that affect humans are transmitted by zoonoses (Ellwanger et al., 2020; Jones et al., 2008; Taylor et al., 2001; Woolhouse & Gowtage-Sequeria, 2005). Since it enables the introduction of new pathogens into the human population, frequent and close contact with wild animals presents potential for the emergence of fresh infectious diseases among humans (Nava et al., 2017; Wolfe et al., 2005). Zoonotic spillover name is given to intricate process of pathogen transmission across species. Such a process is influenced by environmental factors, pathogen genetics and host genetics. The likelihood of cross-species transmission and the development of new infections in humans are both increased by the frequency and severity of human-animal interaction. Pandemics or outbreaks might develop if the pathogen that has been transferred into the human population is exposed to conditions that are conducive to its reproduction and propagation (Ellwanger & Chies, 2018a; Morse, 1995; Plowright et al., 2017). In the cases of MERS-COV, SARS-COV, and SARS-COV-2, this phenomenon has already been observed. Therefore, the best strategy to decrease these occurrences is to restrict human contact with animals in the wild, particularly mammals and birds. Limiting the trade in wildlife and items generated from wildlife is a strong and reasonable method to accomplish this (Kock et al., 2020;

Li et al., 2020; Zhang et al., 2020). Additionally, it is critical to educate people who keep exotic pets and those who deal closely with animals (e.g. biologists, veterinarians and animal keepers) on the dangers of zoonosis and emerging infections in order to reduce the likelihood of infestations and spillover incidents (Chomel et al., 2007; Rabinowitz & Conti, 2013).

5.7 *Environment and Climate Change*

The abundance, diversity and circulation of microorganisms in the environment are impacted by human behaviour and interactions with non-human species. In a similar manner, microbes affects many different components of ecology, including prey-predator dynamics, animal behaviour, dispersal and frequency. Such linkages consequently have an impact on the epidemiology of pathogens and infectious ailments. Consequently, maintaining biodiversity is linked to fewer changes in the ecology of hosts and associated pathogens and, consequently, to a lower likelihood of the genesis of infectious diseases (French & Holmes, 2020; Gibb et al., 2020; Keesing et al., 2010; Mendoza et al., 2020; Mills, 2006). Habitat loss, mining, intensive agricultural use, as well as other ecosystem-degrading practices also enhance human interaction with many wild species, which facilitates spillover events. These practises also expand human exposure to disease vectors like mosquitoes (Ellwanger et al., 2020).

Adequate sanitary conditions, which primarily include access to treated water, rubbish collection, and sewage system and treatment, should also be taken into account while preventing infection. Inaccessibility to potable water, inadequate sewage treatment facilities, as well as poor sanitation, are linked to a number of diseases (Adane et al., 2017; Freeman et al., 2017; Mara et al., 2010). Many public health issues could be reduced by good sanitation (Prüss-Ustün et al., 2017). Although using personal hygiene habits to avoid infections is a well-established technique in high-income nations, this is still not the case in many LMICs due to a lack of awareness or shortage of adequate infrastructure for the provision of clean water and sewage collection. Governments must provide acceptable living standards for the populace, with good sanitation, in combination to investing in training as how to combat diseases linked to poor sanitation, in order to lessen the infections related to a lack of basic hygiene. To prevent the propagation of viral infections, several basic steps are essential (Inaida et al., 2016).

Numerous abiotic elements, including temperature variations and wind and rain patterns, have an impact on host-pathogen interactions and vector dynamics. As a result, the genesis and transmission of infectious diseases are significantly influenced by climate change and extreme weather phenomena. Increased temperatures and severe weather conditions encourage forest fires, which contribute to habitat loss and endanger a number of species (Ellwanger & Chies, 2018b; French & Holmes, 2020; Khan et al., 2019; Wu et al., 2016).

5.8 Urban Planning

Agglomeration of people, unplanned urbanization, and de-urbanization contribute to the development and transmission of pathogens because they are linked to persistent human-to-human pathogen transmission, poor sanitation, increased interaction with wildlife, litter buildup, and the growth of disease vectors (Dalziel et al., 2018; Eskew & Olival, 2018; Gubler, 2011; Herfst et al., 2017; Neiderud, 2015; Tian et al., 2018). In order to effectively control infectious diseases, urban planning, that places a high priority on people's access to health infrastructure, is important because it establishes the prerequisites for healthier urban living and prevents the resettlement of people in areas with inadequate sanitation. It is crucial to remember that human mobility patterns in addition to population size might influence the transmission of infectious illnesses. Consider the COVID-19 pandemic as an illustration. After the virus spread quickly over the world, several steps were taken to slow or stop its transmission, using non-pharmaceutical treatments like limiting human mobility, closing schools, and instituting social isolation policies. Notably, non-pharmaceutical measures have considerably slowed the virus's spread in several nations (Candido et al., 2020; Flaxman et al., 2020; Lai et al., 2020; Pan et al., 2020).

5.9 Food Security and Proper Nutrition

The incidence and development of infectious diseases are greatly influenced by the nutritional condition in a number of ways. For instance, protein deficiencies and malnutrition promote disease aetiology and make people more susceptible to infections (Cegielski & McMurray, 2004; Ibrahim et al., 2017; Katona & Katona-Apte, 2008; Schaible & Kaufmann, 2007). Micronutrient deficiency has a significant impact on how the immune system reacts to infectious illnesses. For instance, zinc deficiency was linked to higher infection morbidity, particularly chronic respiratory diseases (Bahl et al., 1998; Bhaskaram, 2002; Ellwanger & Chies, 2011; Walker & Black, 2004). These instances demonstrate how good nutrition affects both public health and the infectious disease prevention.

Food security, which includes not only having access to food with proven nutritional and microbiological quality but also having enough finances for people and families to buy these meals adequately, consistently and sustainably, is crucial for good nutrition (Glamann et al., 2017; Schmidhuber & Tubiello, 2007). Most populations rely on eating raw or undercooked meat to get enough protein, which increases the danger of new infectious diseases spreading to humans. In other words, avoiding human exposure to potentially novel infections from wild animal meat and biofluids depends greatly on food security (Friant et al., 2020).

5.10 Social Inequalities and Poverty

Infectious diseases are directly and very visibly impacted by social inequality and poverty. Families in LMICs frequently have poor ventilation and shelter (Lönnroth et al., 2009; Lygizos et al., 2013). Furthermore, those living in poverty frequently lack access to clean water and other basic sanitation facilities, which fosters the spread of infectious diseases (Alvar et al., 2006; Cifuentes & Rodriguez, 2005; Echazú et al., 2015; Hotez et al., 2005). The spread of these diseases is correlated with low income, limited education, unemployment, inequality, brutality, as well as other poverty-related issues (Haacker & Birungi, 2018; Kalichman et al., 2006; Leddy et al., 2019). In addition to living in environments more conducive to the transmission of SARS-CoV-2 while also being more vulnerable to infection-related fatality, the poor are more vulnerable to the COVID-19 pandemic (Ahmed et al., 2020). Considering as a whole, reducing social inequalities is a reliable method of infection control and prevention at the macro level.

6 Conclusion

Pandemics can cause massive morbidity and mortality, as well as social, political and economic upheaval. The Black Death, the Spanish flu, and the current COVID-19 pandemics are just a few of the notable pandemics that have affected the planet. Since a pandemic is primarily a geographical phenomenon, it encompasses a wide range of infections and public health issues, each having its own severity, frequency and other disease-related characteristics. Every scenario requires a different approach for preparation and response. The large array of pandemic possibilities is driven by the enormous diversity of viruses and how they interact with humans. Numerous characteristics of pathogens vary, including the mechanism and dynamics of disease transmission, the severity and particularity of associated morbidity. These and other factors affect how quickly cases are identified and treated and also whether an infection spreads. Therefore, the scope of possible health, economic, and sociopolitical consequences, as well as the necessary resources, skills and countermeasures, vary substantially among diseases with pandemic potential. There are numerous reasons why preparedness for a pandemic is challenging. Consequently, ethical and global health imperatives strongly demand building capacity to recognise and respond to pandemic dangers, especially in countries with little resilience and a significant risk of spark and spread.

In the event of a public health emergency, there is no single ideal course of action; rather, approaches must be modified in accordance with the local environment, the intensity of the pandemic and its character. It is challenging to develop pandemic situational awareness and calls for cooperation among bureaucracies, the public and private sectors, and disciplines with varying levels of education and norms (including epidemiology, clinical medicine, logistics, and disaster response). A vital necessity is, however, for an adequately large and skilled health team (including doctors, pharmacists, epidemiologists, veterinarians, lab technicians, and many others), backed by suitable institutional frameworks. We will be able to recognize any population that may be affected by a pathogen with a high pandemic threat if there are more qualified healthcare professionals working in the field. Researchers need to fill in significant information gap regarding pandemic preparedness and response in LMICs. Investment and assistance flows primarily devoted to pandemic prevention and preparedness need to be carefully tracked in order to identify gaps and adjust help supplies for maximum efficacy. Addressing these knowledge gaps can improve pandemic preparedness by promoting evidence-based decision-making and supporting initiatives to avoid and minimise pandemics.

References

Adane, M., Mengistie, B., Kloos, H., Medhin, G., & Mulat, W. (2017). Sanitation facilities, hygienic conditions, and prevalence of acute diarrhea among under-five children in slums of Addis Ababa, Ethiopia: Baseline survey of a longitudinal study. *PLoS ONE, 12*(8), Article e0182783.

Ahmed, F., Ahmed, N. E., Pissarides, C., & Stiglitz, J. (2020). Why inequality could spread COVID-19. *The Lancet Public Health, 5*(5), Article e240.

Alvar, J., Yactayo, S., & Bern, C. (2006). Leishmaniasis and poverty. *Trends in Parasitology, 22*(12), 552–557.

Bahl, R., Bhandari, N., Hambidge, K. M., & Bhan, M. K. (1998). Plasma zinc as a predictor of diarrheal and respiratory morbidity in children in an urban slum setting. *The American Journal of Clinical Nutrition, 68*(2), 414S-417S.

Banoo, S., Bell, D., Bossuyt, P., Herring, A., Mabey, D., Poole, F., & Ramsay, A. (2007). Evaluation of diagnostic tests for infectious diseases: general principles. *Nature Reviews Microbiology, 5*(11), S21–S31.

Barreto, M. L., Teixeira, M. G., & Carmo, E. H. (2006). Infectious diseases epidemiology. *Journal of Epidemiology and Community Health, 60*(3), 192–195. https://doi.org/10.1136/jech.2003.011593

Baylor College of Medicine. (n.d.). Emerging infectious diseases. Retrieved from https://www.bcm.edu/departments/molecular-virology-and-microbiology/emerging-infections-and-biodefense/emerging-infectious-diseases

Bedford, J., Farrar, J., Ihekweazu, C., Kang, G., Koopmans, M., & Nkengasong, J. (2019). A new twenty-first century science for effective epidemic response. *Nature, 575*(7781), 130–136.

Bhaskaram, P. (2002). Micronutrient malnutrition, infection, and immunity: An overview. *Nutrition Reviews, 60*(suppl_5), S40–S45.

Bonotti, M., & Zech, S. T. (2021). The human, economic, social, and political costs of COVID-19. *Recovering Civility during COVID-19*, 1–36. https://doi.org/10.1007/978-981-33-6706-7_1

Bronzwaer, S. L., Cars, O., Buchholz, U., Mölstad, S., Goettsch, W., Veldhuijzen, I. K., & Degener, J. E. (2002). The relationship between antimicrobial use and antimicrobial resistance in Europe. *Emerging Infectious Diseases, 8*(3), 278.

Brooks, O. T. (2020). What if they threw a pandemic and nobody came? *Journal of the National Medical Association, 112*(2), 115–116. https://doi.org/10.1016/j.jnma.2020.04.005

Candido, D. S., Claro, I. M., De Jesus, J. G., Souza, W. M., Moreira, F. R., Dellicour, S., & Faria, N. R. (2020). Evolution and epidemic spread of SARS-CoV-2 in Brazil. *Science, 369*(6508), 1255–1260.

Cegielski, J. P., & McMurray, D. N. (2004). The relationship between malnutrition and tuberculosis: Evidence from studies in humans and experimental animals. *The International Journal of Tuberculosis and Lung Disease, 8*(3), 286–298.

Centers for Disease Control and Prevention [CDC]. (2016). Pandemic Severity Assessment Framework (PSAF). Retrieved from https://www.cdc.gov/flu/pandemic-resources/national-strategy/severity-assessment-framework.html

Centers for Disease Control and Prevention [CDC]. (2022). How Are Diseases Transmitted? Retrieved from https://www.cdc.gov/csels/dsepd/ss1978/lesson1/section10.html

Chomel, B. B., Belotto, A., & Meslin, F. X. (2007). Wildlife, exotic pets, and emerging zoonoses. *Emerging infectious diseases, 13*(1), 6–11. https://doi.org/10.3201/eid1301.060480

Chu, J. (2021). A statistical analysis of the novel coronavirus (COVID-19) in Italy and Spain. *PLoS ONE, 16*(3), Article e0249037.

Cifuentes, E., & Rodriguez, S. (2005). Urban sprawl, water insecurity, and enteric diseases in children from Mexico City. *EcoHealth, 2*(1), 70–75.

Coates, A., Hu, Y., Bax, R., & Page, C. (2002). The future challenges facing the development of new antimicrobial drugs. *Nature Reviews Drug Discovery, 1*(11), 895–910.

Cohen, M. S., Chen, Y. Q., McCauley, M., Gamble, T., Hosseinipour, M. C., Kumarasamy, N., & Fleming, T. R. (2011). Prevention of HIV-1 infection with early antiretroviral therapy. *New England Journal of Medicine, 365*(6), 493–505.

Da Silva, G. K., Vianna, P., Veit, T. D., Crovella, S., Catamo, E., Cordero, E. A. A., & Chies, J. A. B. (2014). Influence of HLA-G polymorphisms in human immunodeficiency virus infection and hepatitis C virus co-infection in Brazilian and Italian individuals. *Infection, Genetics and Evolution, 21*, 418–423.

Dalziel, B. D., Kissler, S., Gog, J. R., Viboud, C., Bjørnstad, O. N., Metcalf, C. J. E., & Grenfell, B. T. (2018). Urbanization and humidity shape the intensity of influenza epidemics in US cities. *Science, 362*(6410), 75–79.

Das, G., Jain, S. P., Maheswaran, D., Slotegraaf, R. J., & Srinivasan, R. (2021). Pandemics and marketing: Insights, impacts, and research opportunities. *Journal of the Academy of Marketing Science, 49*(5), 835–854.

DeLeo, F. R., & Hinnebusch, B. J. (2005). A plague upon the phagocytes. *Nature Medicine, 11*(9), 927–928.

Duerr, H.P., Brockmann, S.O., Piechotowski, I., Schwehm, M., & Eichner, M. (2007). Influenza pandemic intervention planning using InfluSim: Pharmaceutical and non-pharmaceutical interventions. *BMC Infectious Diseases, 7*(1), 1–13.

Echazú, A., Bonanno, D., Juarez, M., Cajal, S. P., Heredia, V., Caropresi, S., & Krolewiecki, A. J. (2015). Effect of poor access to water and sanitation as risk factors for soil-transmitted helminth infection: Selectiveness by the infective route. *PLoS Neglected Tropical Diseases, 9*(9), Article e0004111.

Ellwanger, J. H., & Chies, J. A. (2018a). Wind: A neglected factor in the spread of infectious diseases. *The Lancet Planetary Health, 2*(11), Article e475.

Ellwanger, J. H., & Chies, J. A. B. (2011). Zinc and immunity: The influence of zinc on the immune response and infection resistance. *Archives of Biochemistry and Biophysics, 512*(1), 37–44.

Ellwanger, J. H., & Chies, J. A. B. (2018b). Zoonotic spillover and emerging viral diseases—Time to intensify zoonoses surveillance in Brazil. *Brazilian Journal of Infectious Diseases, 22*, 76–78.

Ellwanger, J. H., de Lima Kaminski, V., & Chies, J. A. (2019). Emerging infectious disease prevention: Where should we invest our resources and efforts? *Journal of Infection and Public Health, 12*(3), 313–316.

Ellwanger, J. H., Kaminski, V. D. L., & Chies, J. A. B. (2017). How to detect new viral outbreaks or epidemics? We need to survey the circulation of viruses in humans and other animals using fast, sensible, cheap, and broad-spectrum methodologies. *Brazilian Journal of Infectious Diseases, 21*, 211–212.

Ellwanger, J. H., Kulmann-Leal, B., Kaminski, V. L., Valverde-Villegas, J., Veiga, A. B. G., Spilki, F. R., & Chies, J. A. B. (2020). Beyond diversity loss and climate change: Impacts of Amazon

deforestation on infectious diseases and public health. *Anais da Academia Brasileira de Ciências, 92*.
Eskew, E. A., & Olival, K. J. (2018). De-urbanization and zoonotic disease risk. *Ecohealth, 15*(4), 707–712.
Flaxman, S., Mishra, S., Gandy, A., Unwin, H. J. T., Mellan, T. A., Coupland, H., & Bhatt, S. (2020). Estimating the effects of non-pharmaceutical interventions on COVID-19 in Europe. *Nature, 584*(7820), 257–261.
Freeman, M. C., Garn, J. V., Sclar, G. D., Boisson, S., Medlicott, K., Alexander, K. T., & Clasen, T. F. (2017). The impact of sanitation on infectious disease and nutritional status: A systematic review and meta-analysis. *International Journal of Hygiene and Environmental Health, 220*(6), 928–949.
French, R. K., & Holmes, E. C. (2020). An ecosystems perspective on virus evolution and emergence. *Trends in Microbiology, 28*(3), 165–175.
Friant, S., Ayambem, W. A., Alobi, A. O., Ifebueme, N. M., Otukpa, O. M., Ogar, D. A., & Rothman, J. M. (2020). Eating bushmeat improves food security in a biodiversity and infectious disease "hotspot." *EcoHealth, 17*(1), 125–138.
Gardy, J. L., & Loman, N. J. (2018). Towards a genomics-informed, real-time, global pathogen surveillance system. *Nature Reviews Genetics, 19*(1), 9–20.
Geoghegan, J. L., & Holmes, E. C. (2017). Predicting virus emergence amid evolutionary noise. *Open Biology, 7*(10), Article 170189.
Gibb, R., Redding, D. W., Chin, K. Q., Donnelly, C. A., Blackburn, T. M., Newbold, T., & Jones, K. E. (2020). Zoonotic host diversity increases in human-dominated ecosystems. *Nature, 584*(7821), 398–402.
Glamann, J., Hanspach, J., Abson, D. J., Collier, N., & Fischer, J. (2017). The intersection of food security and biodiversity conservation: A review. *Regional Environmental Change, 17*(5), 1303–1313.
Grennan, D. (2019). What is a pandemic? *JAMA, 321*(9), 910–910.
Gubler, D. J. (2011). Dengue, urbanization and globalization: The unholy trinity of the 21st century. *Tropical Medicine and Health, 39*(4 Supplement), S3–S11.
Haacker, M., & Birungi, C. (2018). Poverty as a barrier to antiretroviral therapy access for people living with HIV/AIDS in Kenya. *African Journal of AIDS Research, 17*(2), 145–152.
Herfst, S., Böhringer, M., Karo, B., Lawrence, P., Lewis, N. S., Mina, M. J., & Menge, C. (2017). Drivers of airborne human-to-human pathogen transmission. *Current Opinion in Virology, 22*, 22–29.
Holmes, E. C., Rambaut, A., & Andersen, K. G. (2018). Pandemics: Spend on surveillance, not prediction. *Nature, 558*, 180–182.
Hotez, P. J., Bethony, J., Bottazzi, M. E., Brooker, S., & Buss, P. (2005). Hookworm: The great infection of mankind. *PLoS Medicine, 2*(3), Article e67.
Ibrahim, M. K., Zambruni, M., Melby, C. L., & Melby, P. C. (2017). Impact of childhood malnutrition on host defense and infection. *Clinical Microbiology Reviews, 30*(4), 919–971.
Inaida, S., Shobugawa, Y., Matsuno, S., Saito, R., & Suzuki, H. (2016). Delayed norovirus epidemic in the 2009–2010 season in Japan: Potential relationship with intensive hand sanitizer use for pandemic influenza. *Epidemiology & Infection, 144*(12), 2561–2567.
Jones, K. E., Patel, N. G., Levy, M. A., Storeygard, A., Balk, D., Gittleman, J. L., & Daszak, P. (2008). Global trends in emerging infectious diseases. *Nature, 451*(7181), 990–993.
Kalichman, S. C., Simbayi, L. C., Kagee, A., Toefy, Y., Jooste, S., Cain, D., & Cherry, C. (2006). Associations of poverty, substance use, and HIV transmission risk behaviors in three South African communities. *Social Science & Medicine, 62*(7), 1641–1649.
Kandeel, M., Abdelrahman, A. H., Oh-Hashi, K., Ibrahim, A., Venugopala, K. N., Morsy, M. A., & Ibrahim, M. A. (2021). Repurposing of FDA-approved antivirals, antibiotics, anthelmintics, antioxidants, and cell protectives against SARS-CoV-2 papain-like protease. *Journal of Biomolecular Structure and Dynamics, 39*(14), 5129–5136.

Katona, P., & Katona-Apte, J. (2008). The interaction between nutrition and infection. *Clinical Infectious Diseases, 46*(10), 1582–1588.

Keesing, F., Belden, L. K., Daszak, P., Dobson, A., Harvell, C. D., Holt, R. D., & Ostfeld, R. S. (2010). Impacts of biodiversity on the emergence and transmission of infectious diseases. *Nature, 468*(7324), 647–652.

Khan, M. D., Thi Vu, H. H., Lai, Q. T., & Ahn, J. W. (2019). Aggravation of human diseases and climate change nexus. *International Journal of Environmental Research and Public Health, 16*(15), 2799.

Kock, R. A., Karesh, W. B., Veas, F., Velavan, T. P., Simons, D., Mboera, L. E., & Zumla, A. (2020). 2019-nCoV in context: Lessons learned? *The Lancet Planetary Health, 4*(3), e87–e88.

Kozel, T. R., & Burnham-Marusich, A. R. (2017). Point-of-care testing for infectious diseases: Past, present, and future. *Journal of Clinical Microbiology, 55*(8), 2313–2320.

Lai, S., Ruktanonchai, N. W., Zhou, L., Prosper, O., Luo, W., Floyd, J. R., & Tatem, A. J. (2020). Effect of non-pharmaceutical interventions to contain COVID-19 in China. *Nature, 585*(7825), 410–413.

Leddy, A. M., Turan, J. M., Johnson, M. O., Neilands, T. B., Kempf, M. C., Konkle-Parker, D., & Turan, B. (2019). Poverty stigma is associated with suboptimal HIV care and treatment outcomes among women living with HIV in the US. *AIDS (London, England), 33*(8), 1379.

Li, X., Giorgi, E. E., Marichannegowda, M. H., Foley, B., Xiao, C., Kong, X. P., & Gao, F. (2020). Emergence of SARS-CoV-2 through recombination and strong purifying selection. *Science Advances, 6*(27), eabb9153.

Lönnroth, K., Jaramillo, E., Williams, B. G., Dye, C., & Raviglione, M. (2009). Drivers of tuberculosis epidemics: The role of risk factors and social determinants. *Social Science & Medicine, 68*(12), 2240–2246.

Lot, F., Semaille, C., Cazein, F., Barin, F., Pinget, R., Pillonel, J., & Desenclos, J. C. (2004). Preliminary results from the new HIV surveillance system in France. *Eurosurveillance, 9*(10), 7–8.

Lygizos, M., Shenoi, S. V., Brooks, R. P., Bhushan, A., Brust, J., Zelterman, D., & Friedland, G. H. (2013). Natural ventilation reduces high TB transmission risk in traditional homes in rural KwaZulu-Natal, South Africa. *BMC Infectious Diseases, 13*(1), 1–8.

Mabey, D., Peeling, R. W., Ustianowski, A., & Perkins, M. D. (2004). Diagnostics for the developing world. *Nature Reviews Microbiology, 2*(3), 231–240.

Mallhi, T. H., Khan, Y. H., Alotaibi, N. H., Alzarea, A. I., Alanazi, A. S., Qasim, S., & Tanveer, N. (2021). Drug repurposing for COVID-19: A potential threat of self-medication and controlling measures. *Postgraduate Medical Journal, 97*(1153), 742–743.

Mara, D., Lane, J., Scott, B., & Trouba, D. (2010). Sanitation and health. *PLoS Medicine, 7*(11), Article e1000363.

Marston, H. D., Dixon, D. M., Knisely, J. M., Palmore, T. N., & Fauci, A. S. (2016). Antimicrobial resistance. *JAMA*, 316(11), 1193–1204.

Martinez, M. A. (2020). Clinical trials of repurposed antivirals for SARS-CoV-2. *Antimicrobial Agents and Chemotherapy, 64*(9), e01101-e1120.

Mendoza, H., Rubio, A. V., García-Peña, G. E., Suzán, G., & Simonetti, J. A. (2020). Does land-use change increase the abundance of zoonotic reservoirs? Rodents say yes. *European Journal of Wildlife Research, 66*(1), 1–5.

Mills, J. N. (2006). Biodiversity loss and emerging infectious disease: An example from the rodent-borne hemorrhagic fevers. *Biodiversity, 7*(1), 9–17.

Ministero della Salute, Istituto Superiore di Sanità. (2020). Prevention and response to COVID-19: Evolution of strategy and planning in the transition phase for the autumnwinter season. English version. Rome.

Moore, C., & Bell, D. (2020). *Attack rate*. Retrieved from https://radiopaedia.org/articles/attack-rate

Morgan, D. J., Okeke, I. N., Laxminarayan, R., Perencevich, E. N., & Weisenberg, S. (2011). Non-prescription antimicrobial use worldwide: A systematic review. *The Lancet Infectious Diseases, 11*(9), 692–701.

Morse, S. S. (1995). Factors in the emergence of infectious diseases. *Emerging Infectious Diseases, 1*(1), 7–15. https://doi.org/10.3201/eid0101.950102

Nava, A., Shimabukuro, J. S., Chmura, A. A., & Luz, S. L. B. (2017). The impact of global environmental changes on infectious disease emergence with a focus on risks for Brazil. *ILAR Journal, 58*(3), 393–400.

Neiderud, C. J. (2015). How urbanization affects the epidemiology of emerging infectious diseases. *Infection Ecology & Epidemiology, 5*(1), 27060.

Organisation for Economic Co-operation and Development [OECD]. (2021). *OECD policy responses to coronavirus (COVID-19) responding to the COVID-19 and pandemic protection gap in insurance.* Retrieved from https://www.oecd.org/coronavirus/policy-responses/responding-to-the-covid-19-and-pandemic-protection-gap-in-insurance-35e74736/

Organisation for Economic Co-operation and Development [OECD]. (2020). *OECD policy responses to coronavirus (COVID-19) the territorial impact of COVID-19: Managing the crisis across levels of government.* Retrieved from https://www.oecd.org/coronavirus/policy-responses/the-territorial-impact-of-covid-19-managing-the-crisis-across-levels-of-government-d3e314e1/

Pan, A., Liu, L., Wang, C., Guo, H., Hao, X., Wang, Q., & Wu, T. (2020). Association of public health interventions with the epidemiology of the COVID-19 outbreak in Wuhan, China. *JAMA, 323*(19), 1915–1923.

Piddock, L. J. (2012). The crisis of no new antibiotics—What is the way forward? *The Lancet Infectious Diseases, 12*(3), 249–253.

Plowright, R. K., Parrish, C. R., McCallum, H., Hudson, P. J., Ko, A. I., Graham, A. L., & Lloyd-Smith, J. O. (2017). Pathways to zoonotic spillover. *Nature Reviews Microbiology, 15*(8), 502–510.

Prüss-Ustün, A., Wolf, J., Corvalán, C., Neville, T., Bos, R., & Neira, M. (2017). Diseases due to unhealthy environments: An updated estimate of the global burden of disease attributable to environmental determinants of health. *Journal of Public Health, 39*(3), 464–475.

Rabinowitz, P., & Conti, L. (2013). One Health and emerging infectious diseases: clinical perspectives. *Current topics in microbiology and immunology*, 365, 17–29. https://doi.org/10.1007/82_2012_263

Santoro, A., Abu Rmeileh, N., Khader, A., Seita, A., & McKee, M. (2016). Primary healthcare reform in the United Nations relief and works agency for Palestine refugees in the near east. *EMHJ-Eastern Mediterranean Health Journal, 22*(6), 417–421.

Schaible, U. E., & Kaufmann, S. H. E. (2007). Malnutrition and infection: Complex mechanisms and global impacts. *PLoS Medicine, 4*(5), Article e115.

Schmidhuber, J., & Tubiello, F. N. (2007). Global food security under climate change. *Proceedings of the National Academy of Sciences, 104*(50), 19703–19708.

Senanayake, S. L. (2020). Drug repurposing strategies for COVID-19. *Future Drug Discovery, 2*(2).

Shadabfar, M., Mahsuli, M., Khoojine, A. S., & Hosseini, V. R. (2021). Time-variant reliability-based prediction of COVID-19 spread using extended SEIVR model and Monte Carlo sampling. *Results in Physics, 26*, Article 104364.

Shah, K., Saxena, D., & Mavalankar, D. (2020). Secondary attack rate of COVID-19 in household contacts: A systematic review. *QJM: An International Journal of Medicine, 113*(12), 841–850.

Shang, Y., Li, H., & Zhang, R. (2021). Effects of pandemic outbreak on economies: Evidence from business history context. *Frontiers in Public Health, 9*, Article 632043.

Smolinski, M. S., Hamburg, M. A., & Lederberg, J. (Eds.). (2003). *Microbial threats to health: Emergence, detection, and response.* National Academies Press (US).

Taylor, L. H., Latham, S. M., & Woolhouse, M. E. (2001). Risk factors for human disease emergence. *Philosophical Transactions of the Royal Society of London. Series B: Biological Sciences, 356*(1411), 983–989.

Tenover, F. C. (2006). Mechanisms of antimicrobial resistance in bacteria. *The American Journal of Medicine, 119*(6), S3–S10.

Tian, H., Hu, S., Cazelles, B., Chowell, G., Gao, L., Laine, M., & Stenseth, N. C. (2018). Urbanization prolongs hantavirus epidemics in cities. *Proceedings of the National Academy of Sciences, 115*(18), 4707–4712.

Tran, B. X., Nguyen, H. T., Le, H. T., Latkin, C. A., Pham, H. Q., Vu, L. G., & Ho, R. C. (2020). Impact of COVID-19 on economic well-being and quality of life of the Vietnamese during the national social distancing. *Frontiers in Psychology, 11*, Article 565153.

Walensky, R. P., Freedberg, K. A., Weinstein, M. C., & Paltiel, A. D. (2007). Cost-effectiveness of HIV testing and treatment in the United States. *Clinical Infectious Diseases*, *45*(Supplement_4), S248–S254.

Walker, C. F., & Black, R. E. (2004). Zinc and the risk for infectious disease. *Annual Review of Nutrition, 24*(1), 255–275.

Walker, P. G., Whittaker, C., Watson, O. J., Baguelin, M., Winskill, P., Hamlet, A., & Ghani, A. C. (2020). The impact of COVID-19 and strategies for mitigation and suppression in low-and middle-income countries. *Science, 369*(6502), 413–422.

Webby, R. J., & Webster, R. G. (2003). Are we ready for pandemic influenza? *Science, 302*(5650), 1519–1522.

Wolfe, N. D., Daszak, P., Kilpatrick, A. M., & Burke, D. S. (2005). Bushmeat hunting, deforestation, and prediction of zoonotic disease. *Emerging Infectious Diseases, 11*(12), 1822.

Wood, R., Reinhardt, G. Y., RezaeeDaryakenari, B., & Windsor, L. C. (2022). Resisting lockdown: The influence of COVID-19 restrictions on social unrest. *International Studies Quarterly, 66*(2), sqac015.

Woolhouse, M. E., & Gowtage-Sequeria, S. (2005). Host range and emerging and reemerging pathogens. *Emerging Infectious Diseases, 11*(12), 1842.

World Health Organization [WHO]. (2009). Pandemic influenza preparedness and response: A WHO guidance document.

World Health Organization [WHO]. (2017). Pandemic influenza risk management: A WHO guide to inform and harmonize national and international pandemic preparedness and response (No. WHO/WHE/IHM/GIP/2017.1). World Health Organization.

World Health Organization [WHO]. (2022). *Influenza*. Retrieved from https://www.who.int/teams/health-product-policy-and-standards/standards-and-specifications/vaccine-standardization/influenza#:~:text=Influenza%20occurs%20all%20over%20the,20%20%E2%80%93%2030%25%20in%20children

Wu, X., Lu, Y., Zhou, S., Chen, L., & Xu, B. (2016). Impact of climate change on human infectious diseases: Empirical evidence and human adaptation. *Environment International, 86*, 14–23.

Zhang, T., Wu, Q., & Zhang, Z. (2020). Probable pangolin origin of SARS-CoV-2 associated with the COVID-19 outbreak. *Current Biology, 30*(7), 1346–1351.

Chapter 2
Pandemic: Historical Perspective

1 Introduction

Pandemics have catastrophically struck humanity across the recorded human history; these are not a new phenomenon. The word "*Pandemic*" stems from Greek word *pandēmos* (pan 'all' + dēmos 'people') (Memish et al., 2020). Past pandemics which have affected mankind include plague, Spanish flu, cholera, HIV resulting in mass deaths, toppling of governments, in addition to economic and psychological consequences (Huremović, 2019). Currently, COVID-19 pandemic has unleased its severity ever since 2019, throughout which we have experienced extreme suffering, fatality and instability in regular living. These catastrophic events have occurred and are now more probable to appear for a variety of reasons (Joi, 2020). The transition from nomadic to farming communities, expanded trade between communities, increased interactions between humans and animals, expanded cities and trade territories, tourism and climate change have contributed towards the transmission of infectious pathogens in the human species (Pike et al., 2010). Additionally, one of the main factors raising the likelihood of a future pandemic is antibiotic resistance and potential employment of microbes as biological weapons (Clark & Pazdernik, 2016).

This chapter reviews major pandemics that have afflicted humankind throughout history such as plague, cholera, influenza and coronavirus. Managing any surging pandemic and minimizing the emergence of a potentially new one requires detailed knowledge and an understanding of the mechanisms behind the emergence of such pandemics (Al Ahdab, 2021). Our capability to control the circumstance in the event of a pandemic is also impacted by the lack of health personnel and the poorly equipped health systems (Haldane et al., 2021). On the brighter side, past pandemics have helped in enhancing preparedness and strenghthen mitigation measures to curb the effects of pandemics. Unfortunately, since pathogens may easily spread through international commerce and travel, pathogens continue to pose a danger to people's health (Findlater & Bogoch, 2018). Therefore, in order to manage such pathogens, in

M. H. Fulekar et al., *Pandemic Over the Centuries*,
https://doi.org/10.1007/978-3-031-94270-9_2

addition to identification and detection of pathogens, global surveillance systems are essential (Lee et al., 2020). World Health Organization constantly monitors disease activity on a global scale through a network of surveillance centres located in countries worldwide (Rogers, 2022). To further combat and prevent the spread of contagious diseases in the human population, both pharmaceutical and non-pharmaceutical strategies must be adopted (Iezadi et al., 2020).

2 Timeline of Pandemic

Pandemics recorded across history are discussed below.

2.1 *The Plague*

Plague is transmitted by the flea-borne bacteria *Yersinia pestis*, which is accountable for human plague pandemics (the Justinian plague and Black Death) (Barbieri et al., 2020). *Yersinia pestis* is a rod-shaped, gram-negative, coccobacillus facultative anaerobic bacteria that is spread by fleas linked mostly to rats and other mammal (Lin, 2015). By drawing blood from an infested rat, fleas receive the bacterium. In the flea's digestive tract, bacteria rapidly proliferate and obstruct the alimentary. Further, the fleas spread bacteria to new rodent hosts via regurgitating blood that has been coagulated (Bland et al., 2018). Depending on the mode of infection, the bubonic, septicemic and pneumonic types of plague emerge (Sebbane and Lemaître, 2021). The bubonic variant is the most prevalent and is transmitted through flea bites. Clinical signs include flu-like symptoms including fever, cold, headache, body aches, weakness, vomiting, and nausea, followed by painful lymph node hypertrophy. Almost 50–90% of cases involving bubonic plague are deadly (McEvedy, 1988). Septicemic plague is uncommon (10–25% of cases) and characterised by a bloodstream infection that progresses without lymphadenopathy (Kwit et al., 2015). Individuals with the septicemic type of plague have a higher death rate than those suffering from the bubonic form (Bennett et al., 2019). Pneumonic plague arises whenever the bacterium affects the lungs, either through infectious respiratory secretions or as a consequence of bubonic plague. This variety has severe clinical impact and therefore is lethal if left untreated (Anderson et al., 2009).

2.1.1 Athenian Plague

The Athenian plague of 430 BC was marked by rashes, headaches and conjunctivitis, which progressed to abdominal pains and hemoptysis, ultimately resulting in death in several victims within 7–8 days. The precise figure is uncertain, although it is thought that the plague killed approximately 25% of Athenians and neighbouring countrymen (Littman, 2009). It is considered that battlefield population is one of the primary reasons. The diversity of reported symptoms is so extensive that this outbreak may have been caused by numerous contagious diseases. The pathogen which caused this outbreak remains unknown (Horgan, 2016).

2.1.2 Antonine Plague

The Antonine plague lasted from 165 to 180 AD. During their eastern operations (now Iraq), warriors carried the disease, which spread rapidly throughout the entire Roman empire down to western Germany. Marcus Aurelius, the emperor at that time, was believed to be among the 5 million people who succumbed to the pandemic (Sabbatani & Fiorino, 2009). The Antonine plague, which wiped off 33% of the Roman empire, was not confined to a specific place, as the empire's massive political network linked several provinces over a broad territory (Scheidel, 2002). According to Galen, the signs included rashes, hemorrhagic pustules, bloody diarrhoea, fever and occasionally hemoptysis (Fears, 2004). According to modern research, it was presumably smallpox (Elliott, 2016).

2.1.3 Justinian Plague

Egypt was the origin of the Justinian plague, which spread all through the eastern Roman empire and its allies (Rosen, 2007). The detection of the disease responsible for the deaths of victims of past pandemics typically relies on ancient DNA techniques designed to extract DNA directly from skeletal remains (Spyrou et al., 2019). This outbreak was triggered by *Yersinia pestis*, as verified by DNA isolated from the dental pulp of bones found in grave from that era (Arning & Wilson, 2020). The symptoms included fever, cough and breathlessness in pneumonic plague, and blisters in the groyne or axilla in bubonic plague (Retief & Cilliers, 2006). Several investigations suggest hallucinations, accompanied with fever and exhaustion (Skwarecki, 2016). Occasionally, sore throat or diarrhoea were the first signs. It was spread from infectious rats to people by flea bites, and also spread through human-to-human transmission (Raoult et al., 2013). Across its military and commercial channels, the better evolved system of the Roman Empire favoured the transmission of the Justinian plague. The plague did not impact the less structured barbarian civilizations outside the limits of Rome (Habicht et al., 2020). During 541 and 544 AD, the plague is thought to have killed 100 million people in the Roman Empire, mainly in capital Constantinople (Rosen, 2007). The high death rate induced by the plague may have resulted towards the empire's decadence and eventual collapse.

2.1.4 The Black Death

The Black Death, often referred as Plague, was a bubonic plague pandemic that ravaged Europe, Asia and Africa during 1346–1353 wreaking havoc in its aftermath; certain projections predict that it killed more than 50% of Europe's population (Benedictow, 2004; Cantor, 2001). It is regarded to be the deadliest pandemic in recorded existence, with an estimated 200 million fatalities (Feehan & Apostolopoulos, 2021). The pandemic was induced by an extinct strain of the bacterium *Y. pestis* that has been transmitted by fleas biting infested rats (Gage & Kosoy, 2005). The corpses were laid to rest in mass graves (Byrne, 2004). Considering the number of dead, workers became more difficult to find, resulting in worker compensation and the abolition of feudalism in Europe (Moore, 2021). According to investigations, surviving workers got accessibility to meat and bread of superior quality. Also, the shortage of cheap labour may have also led to technical advancement (Herlihy, 1997). There was no appropriate treatment for plague at the time. During the Black Death, early institutionalized solutions to infection control were initiated (Makris et al., 2000). Security personnel enforced a hygienic barrier across commuting networks and entrances to towns (Shepard, 2015). In camps (and later in dedicated plague hospitals), segregation amongst healthy and sick individuals was maintained (called lazarettos) (Crawshaw, 2016). Port cities restricted ships from plague-infected regions from docking. Ships suspected of carrying the plague were quarantined, passengers and staff were confined in lazarettos, while vessels were fumigated and held for 40 days (Tognotti, 2013).

2.2 *Cholera*

Cholera is an acute, potentially serious condition of the digestive tract that is triggered by *Vibrio cholerae* (Kumar et al., 2020). *V. cholerae* is a coma-shaped, gram-negative, facultative anaerobe with a flagellum and pili at one of its cell poles (Bitton, 2014). Bacterium colonise the small intestine and generate toxin, that causes a quick and enormous loss of bodily fluids resulting in dehydration, hypovolemic shock and death (Muanprasat & Chatsudthipong, 2013). On the basis of their main lipopolysaccharide O antigens, *V. cholerae* strains are categorised into about 206 serogroups, among which serogroups O1 (containing of two biotypes known as classical and El Tor) and O139 are responsible for epidemic cholera (Chowdhury et al., 2017). Cholera bacterium is transmitted by water and people are affected by consuming or cooking food with polluted water (Baker-Austin et al., 2018). In most cases, the infection is minor or asymptomatic, and the pathogens are excreted in 1 or 2 weeks via faeces (Daniels & Shafaie, 2000).

2.2.1 The Seven Pandemics of Cholera

The very first pandemic of cholera, often referred as Asiatic cholera, started in 1817 (Howard-Jones, 1974). It began in India and spread to Nepal, Indonesia, China, Japan, the Middle East, and some regions of Russia (Awofeso & Aldbak, 2018). This pandemic evolved during a period of growing urbanization owing to technological advancements in communication. Certainly, the introduction of steamships and railroads facilitated a huge reduction in commute time and an expansion of commerce (Johnson, 1997). Preventive health techniques at that time were basically identical to those employed during the Black Death (Booth, 1991). Infected individuals were quarantined in lazarettos. Cholera-infected regions were restricted from entering ports. Those who had contact with infected individuals or travelled from a region where cholera was prevalent were confined (Maglen, 2002) (Fig. 1).

The first pandemic of cholera was followed by six further outbreaks that occurred between 1829 and 1923 and spread across various continents, particularly Europe and the USA (Lee & Dodgson, 2017). Notably, the first six pandemics were spread through polluted water by the classic subtype of Vibrio cholerae serogroup O1 (Safa et al., 2010). The second cholera pandemic spread to the British Isles. During 1854 the cholera outbreak in Soho (London), physician John Snow employed epidemiological tools for the first time to locate the outbreak's source. He recounted the progression of the pandemic and its geographical distribution throughout the city and found public drinking water pumps in these places and concluded that water was indeed the cause of contamination. Furthermore, he recommended practical steps to prevent transmission by removing pump handles in urban areas in which the outbreak had started (Koch & Denike, 2010). During the fifth epidemic that ravaged South America, Robert Koch in 1884 found the cholera bacillus and realized the importance of clean water in limiting its propagation (Lipp et al., 2002). Only in 1959 was the toxin causative for the illness identified (Lippi & Gotuzzo, 2014). In 1961, the seventh pandemic of cholera started. It began in Indonesia, extended to more nations and continents which later on became endemic in a number of world zones. This pandemic was the longest and most widespread in terms of geographic distribution (Lee, 2001). El Tor biotype strain of *V. cholerae* gained virulence genes from the surroundings and triggered outbreaks in Zimbabwe (2008), Haiti (2010), Sierra Leone (2012), Mexico (2013), South Sudan and Ghana (2014) and Yemen (2016) on a recurring basis (Balasubramanian et al., 2021; Faruque et al., 2007). El Tor biotype strain of *V. cholerae* continues to cause global outbreaks, the most recent of which occurred in Somalia in March 2021 (Challa et al., 2022). The first six cholera pandemics were estimated to cause millions of death across all continents (World Health Organization [WHO], 2022). The seventh cholera pandemic predicted to cause roughly 2.86 million cases annually, with approximately 95,000 deaths annually (Ali et al., 2015). In 2019, WHO received reports of 923,037 incidents and 1,911 fatalities from 31 countries (WHO, 2021a). Cholera cannot be completely eradicated since Its pathogen inhabits water environments naturally. Poor living conditions, such as a lack of clean water, inadequate hygiene, poor housing, and ineffective sewer systems, contribute to the recurrence of cholera (Luby et al., 2020). Re-emergence

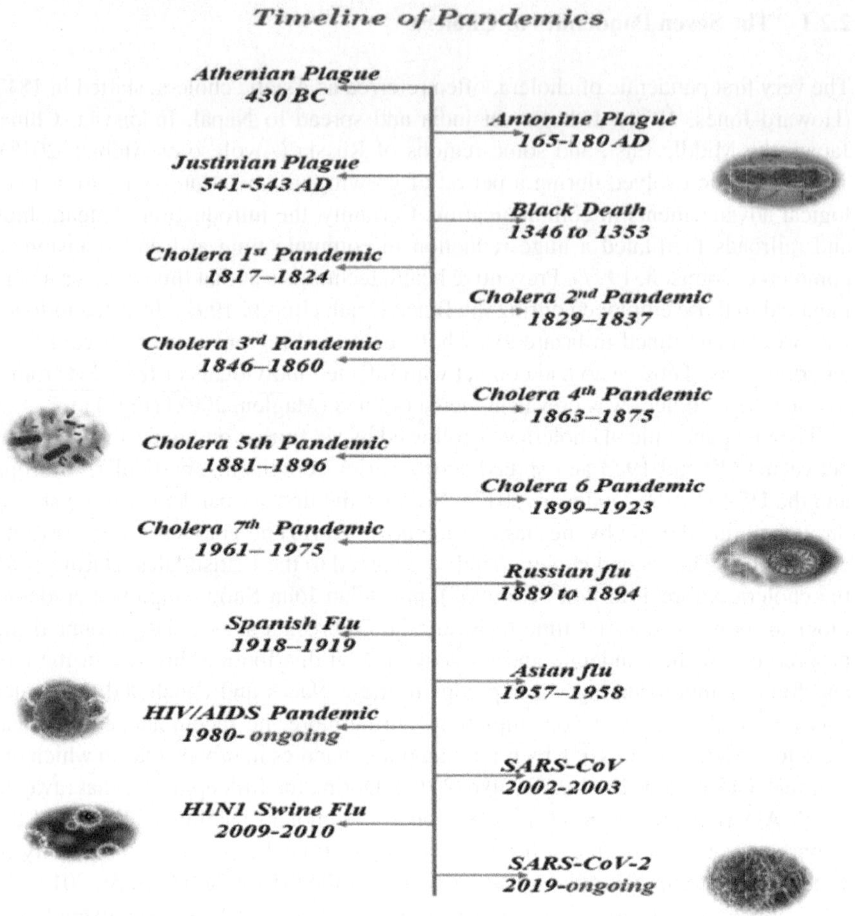

Fig. 1 Timeline of pandemics

of the infection is also possible in the aftermath of catastrophic events such as earthquakes, that impair accessibility to clean water supplies. Cholera episodes can be anticipated through real-time monitoring of marine regions, climate variations, and an epidemiological surveillance programme (Chowdhury et al., 2017). Also, implementing public health measures to guarantee appropriate cleanliness and safe water supply could avoid the infection (Bartram & Cairncross, 2010).

2.3 *Influenza*

Influenza viruses are members of the family Orthomyxoviridae. They are enclosed, single-stranded, negative-sense RNA viruse. The 7 or 8 RNA segments encode at least 10 structural and non-structural proteins (Brázda et al., 2021). Hemagglutinin (HA), neuraminidase (NA), two matrix proteins, and a nucleoprotein are the structural proteins in the genome of virion (Hurwitz et al., 1985). Influenza viruses can be differentiated in categories A, B, C and D. Influenza A and B viruses are accountable for outbreaks in tropical regions and periodic epidemics in temperate regions, but only influenza A viruses have pandemic potential (Webster et al., 1992). In fact, influenza A virus is endemic in multiple species, notably humans, birds and pigs. Thus, gene reassortments between human and animal influenza A viruses can result in a novel virus subtype that is potentially dangerous to individuals (Medina & García-Sastre, 2011).

In seasonal epidemics, the influenza virus causes between 3 and 5 million instances of clinical complications and about 500,000 fatalities globally (Ginsberg et al., 2009). The vast majority of seasonal influenza infections are asymptomatic or cause relatively moderate or classic influenza illness, defined by 4 or 5 days of fever, cough, cold, headache, muscular soreness, fatigue, and occasionally upper respiratory ailments. Symptoms can be significant, particularly in babies, the elderly, and those with chronic illnesses like diabetes and cardiac/pulmonary problems. Pneumonia, which is connected with subsequent bacterial infection, is one of the most severe implications (Nypaver et al., 2021).

2.3.1 Russian Flu

It is indeed difficult to establish when influenza virus started to cause disease or trigger a pandemic, even though many researchers believe that the initial influenza pandemic probably happened in 1510 (Dunning et al., 2020). Around 1889 and 1893, the Russian flu became the first well-described pandemic (Chakarova & Kichukova, 2021). Depending on serologic and epidemiological studies, this pandemic was likely caused by the A/H3N8 virus (Rice, 2018). It took the pathogen only 4 months to traverse the globe (Smith, 1995). The pandemic virus resurfaced annually for 3 years, causing an estimated 1 million deaths across Europe (Honigsbaum, 2013). The highest incident percentages was amongst individuals aged 1–60 years, whereas newborns and elderly had lower rates. In comparison, the death rate followed a J-shaped curve, with the highest rate among babies and those above 20 years (Valtat et al., 2011).

2.3.2 Spanish Flu

The A/H1N1 virus that caused the Spanish flu 25 years later to Russian flu is thought to have evolved through genetic adaptation of a pre-existing avian influenza virus to a new human host (Schnitzler & Schnitzler, 2009). Prior to its detection, the pathogen spread discreetly across the globe, and its place of genesis was unknown. Examination of samples with formalin-fixed, paraffin-embedded, and permafrost-frozen from that time revealed that the strain was an A/H1N1 influenza virus (Anhlan et al., 2011). Within 9 months, at least three separate waves of the 1918–1919 influenza pandemic occurred. The initial wave of the 1918 influenza pandemic began during the spring and summer months and resulted in high suffering and low fatality. In the summer-fall of 1918 and the winter of 1918–1919, the second and third waves caused considerable fatality (Chowell et al., 2007). The most prevalent clinical manifestation was acute aggressive bronchopneumonia with epithelial necrosis, microvasculitis/vascular necrosis, haemorrhages, edoema, and severe tissue damage to the lung, often accompanied by secondary bacteria invasion with *Streptococcus pneumonia*, *Streptococcus pyogenes*, *Staphylococcus aureus* and *Haemophilus influenza* (Morens & Fauci, 2007). It is hypothesised that the 1918 influenza virus had a heightened ability to travel to and damage bronchial and bronchiolar epithelial cells, allowing bacteria to break the mucociliary barrier and cause deadly bacterial pneumonia (Humphreys, 2018). The majority of fatalities occurred in many days to weeks (median 7–10 days) following the onset of symptoms (Tuite et al., 2010). The pandemic influenza of 1918–1919 caused around 500 million illnesses and 50 million deaths globally (Nickol & Kindrachuk, 2019). In major cities of the western world, health officials implemented many containment techniques to limit the transmission of the disease, e.g. closing of schools, churches and theatres, as well as suspension of public meetings. Individual interventions, such as pulmonary hygiene and social isolation, were recommended for by doctors (Jones, 2007). Due to World War I, unfortunately, these steps were enacted very late and incoherently, making it impossible to establish travel restrictions and border controls. The movement of armed forces and the poor living conditions of soldiers during the European trench warfare helped the disease to spread (Humphries, 2014).

2.3.3 Asian Flu

The novel subtype A/H2N2 that triggered the 1957–1958 pandemic (Asian flu) originated from the 1918 virus through the acquisition of three new avian gene segments (HA, NA and PB1 polymerase) via reassortment, whereas the circulating A/H1N1 was completely replaced (Cox & Subbarao, 2000). Multiple waves of persistent infection of the 1957–1959 pandemic virus began in December 1957 and continued for several years (Potter, 2001). Morbidity was greatest among children, while mortality was greatest in elderly (Chin et al., 1960). The global death toll from the influenza pandemic of 1957–1959 was anticipated to be between 1 and 2 million, depending on the additional fatality from breathing problems (Viboud et al., 2016). Rapid development of bronchial epithelial necrosis, preservation of the basal layer, limited inflammatory response, and indications of fast healing were observed in autopsy histopathology studies (Dybing et al., 2000). Perhaps as a result of the extensive use of antibiotics, secondary bacterial pneumonia was a relatively critical reason for death. In fatal instances, the proportion of antibiotic-resistant strains was substantially higher than in situations where the patient survived (Oswald et al., 1958). Mechanical ventilation systems were accessible in intensive care units (ICU) to assist patients with hypoxemia (Oshitani et al., 2008). Country-specific containment measures merely slowed the emergence of the disease by a few weeks (Forbes, 2004). However, inadequate vaccination coverage and vaccine efficacy of 53–60%; it failed to have a considerable impact on the pandemic. Besides, in 1952, the WHO established a global influenza surveillance network that supplied data regarding the appearance and transmission of the influenza virus (Kitler et al., 2002).

2.4 HIV/AIDS

Human Immune Deficiency virus infection/Acquired Immune Deficiency Syndrome (HIV/AIDS) is a worldwide concern that spreads gradually over decades, multiple continents, and diverse people, raising new threats with each successive stage and in each new group it impacts (Defo, 2014). It began in early 1980s in the USA, prompting widespread public concern as HIV eventually advanced to AIDS and fatality (Broder, 2010). The early spread of HIV was characterised by its predominance among the hetrosexual population and its high death rate, which led to pronounced social exclusion and discrimination (WHO, 2001). The 40–90% of persons infected may appear with an acute retroviral syndrome 2–6 weeks after HIV-1 exposure; symptoms include fever, weight loss, exhaustion, sore throat, lymphadenopathy, myalgia, headache, nausea and diarrhoea. In addition to thrombocytopenia, leukopenia, aseptic meningitis, mononucleosis-like symptoms, and secondary infection, numerous other nonspecific signs and manifestations might be observed (Kahn & Walker, 1998; Remien et al., 2009) (Table 1).

HIV affects around 40 million individuals worldwide (prevalence rate: 0.79%) and has killed nearly the same number since 1981. It results in approximately 1 million fatalities annually (Cohen et al., 2008). In several Sub-Saharan African nations (Botswana, Lesotho and Swaziland), the prevalence of HIV exceeds 25% (De Wagt & Connolly, 2005). In the USA, over 1.2 million individuals are diagnosed with HIV and 12,000 die annually (HIV.GOV, 2022).

As a pandemic that spreads rather gradually, HIV has garnered substantial public health concern from both international and national organizations and medicines (Berkman et al., 2005). HIV is now a chronic illness that may be treated with medication, owing to the advancements in treatment (protease inhibitors and antiretrovirals) (Voshavar, 2019).

UNAIDS has declared a 90-90-90 treatment target for HIV as of 2014, which entails that 90% of those infected should be diagnosed, 90% among those diagnosed

Table 1 Major pandemics, causative pathogen/host and death toll

Disease	Pathogen/host	Estimated death toll
Athenian plague	–	75,000–100,000
Antonine plague	–	5 Million
Justinian plague	*Yersinia pestis*/rat flea	30–50 Million
The Black Death	*Yersinia pestis*/rat flea	200 Million
Seven cholera pandemics	*Vibrio cholerae*	Over 100 Million to date
Russian flu	A/H3N8 virus/avian origin	1 Million
Spanish flu	A/H1N1 virus/avian origin	40–50 Million
Avian flu	A/H2N2 virus/avian origin	1.1 Million
HIV/AIDS	HIV/chimpanzees	40.1 Million
SARS-CoV	Coronavirus/bats, civets	774
H1N1 swine flu	H1N1 virus/pigs	151,700–575,400
SARS-CoV-2	Coronavirus/bats	6.57 Million (as of 14/10/2022)

must undergo ART, and 90% of those undergoing treatment must establish viral suppression (UNAIDS, n.d.).

2.5 Severe Acute Respiratory Syndrome (SARS)

The severe acute respiratory syndrome coronavirus (SARS-CoV), which is a member of the Coronaviridae family and the Coronavirus genus, was responsible for a worldwide pandemic that lasted from 2002 to 2003 (Acter et al., 2020). This pandemic migrated from China's Guangdong province to 29 nations in North America, South America, Europe and Asia (Lam et al., 2003). It depends upon bats, palm civets and raccoon dogs as intermediate hosts.

Exposure with SARS-CoV generated a condition resembling influenza with chills, exhaustion, and high fever. Nausea, vomiting, and diarrhoea are less frequent complications. The 20–30% infected persons had atypical pneumonia, characterised by difficulty breathing and poor oxygen exchanges in the alveoli, necessitating ICU or ventilators. The majority of these patients also had diarrhoea accompanied with active viral shedding. Respiratory failure was the leading cause of death among SARS-CoV-infected people (Cheng et al., 2013; Daga et al., 2019). SARS-CoV's low infectiousness and prolonged incubation (viral load peaks 6–11 days after symptom start) allowed for the implementation of a number of control measures to prevent its spread (Petersen et al., 2020). It only resulted in 774 deaths (Shi & Hu, 2008). Case detection and quarantine, coupled with contact tracking and monitoring, were successful in containing the worldwide concern and eliminating the virus in around 7 months (Chan-Yeung & Xu, 2003).

2.6 *Swine Flu*

The 2009 swine flu promptly impacted one-tenth of the population, but it was not as lethal (Lees et al., 2011). The infection had the same characteristics as the Spanish flu, affecting younger generation significantly due to their robust immune systems (Klemm et al., 2016). Swine flu, caused by the H1N1 virus that emerged in Mexico, primarily impacted adolescent and young adults (Fitzgerald, 2009). Due to international commerce and travel, the virus spread to 122 nations in 6 weeks and had three episodes in the spring, summer and fall (Rajoura et al., 2011). Victims were suffering with high fever, coughing, sore throat, myalgia, breathlessness, and occasionally diarrhea and vomiting. Respiratory failure was the typical reason for death in extreme situations (Jilani et al., 2018). According to the CDC, the virus infected as many as 1.4 billion individuals worldwide in 1 year and killed between 151,700 and 575,400 (Centers for Disease Control and Prevention [CDC], 2019).

2.7 *COVID-19*

Early in December 2019, a group of people in Wuhan (China) were diagnosed with a typical pneumonia. It was later determined that the infection, which is known as COVID-19, is apparently caused by a novel coronavirus named SARS-CoV-2 (Chu et al., 2020). Most probably, the animal repositories were bats. Pangolins were hypothesised to be the animal carriers which spread the virus to humans, however the intermediate host, if some, has yet to be conclusively found (Jo et al., 2021). The 80% of laboratory-confirmed patients presented with mild symptoms. The 14–19% of patients are hospitalised, and 3–5% of cases necessitate transfer to the intensive care unit, most frequently due to hypoxemic respiratory failure. The 29–91% of these patients require invasive respiratory support (Shang et al., 2020). Further, 60–90% of hospitalised patients have comorbidities including high blood pressure, diabetes, cardiovascular disease, chronic lung illness, chronic renal disease, cancer and chronic liver disease (Ejaz et al., 2020). The 40% of SARS-CoV-2 infections may be asymptomatic, or they may generate a variety of disorders ranging from minor symptoms to severe life-threatening condition (Perico et al., 2021). Fever, dry cough, breathlessness, exhaustion, myalgia, nausea/vomiting or diarrhoea, headache, lethargy, rhinorrhea, anosmia and ageusia are the most prevalent symptoms of infection. Pneumonia, ARDS, acute liver injury, heart injury, prothrombotic coagulopathy, acute renal injury, and neurologic symptoms can be frequently seen among hospitalised patients (Alimohamadi et al., 2020). Patients who are critical may potentially develop a cytokine storm and a macrophage activation syndrome (Tang et al., 2020). Overall, hospitalised patients with COVID-19 have a fatality rate of 15–20% (Macedo, Gonçalves and Febra, 2021). COVID-19 has spread across the world within months, resulting in over 621,797,133 cases reported and 6,545,561 deaths as of October 17, 2022.

Most nations have implemented a variety of preventative measures, including individual behavior (physical distance, washing hands, usage of face masks and cough protocol), detection of clusters (case identification, contact tracing, and isolation), regulatory actions (school closure, workplace closure, stay-at-home order, public transport closure and limitation, limits on the number of gatherings and company capacity), closures of internal and external borders (Chiu et al., 2020). The purpose of these public health interventions is to halt and flatten the pandemic curve, thereby minimizing an overburdening of the medical system and safeguarding those at highest risk of a tragic outcome prior to the development of reliable and safe vaccinations and therapies (Schneiders et al., 2022). First vaccination based on new platform consisting of mRNA encoding the virus spike protein (developed by Pfizer/BioNTech) was authorised for Emergency Use by the US FDA in less than a year (as of December 11, 2020) (Food and Drug Administration [FDA], 2021). Several other vaccines based on inactivated whole virus, recombinant non-replicative viral vector, recombinant spike protein and virus-like particles have also been developed as of September, 2022 (WHO, 2021b).

3 Conclusion

Pandemics had a profound impact on the mindset of the human race all across the history. On one hand, they have impacted the economy, society, and psychological health while also causing millions of fatalities. On the other hand, they resulted in important scientific advancements such as the establishment of quarantine, the construction of vaccinations and new treatment techniques. The ongoing COVID-19 pandemic demonstrated that we are still ill-equipped to identify new disease pandemics, anticipate their onset, or stop their transmission. To lessen the severe load on medical systems, there is an unrealized necessity raise knowledge and practise pandemic readiness. Additionally, it serves as a vivid reflection of the gap in access to healthcare between those who can afford it versus those who cannot, particularly in nations without national healthcare. During emergency preparedness of infectious disease pandemics, national health policy agencies should give priority to extensive testing, contact tracking, quarantine and vaccine research. Newer pandemics need to be prevented from spreading, and a unified response needs to be developed, according to future consideration.

References

Acter, T., Uddin, N., Das, J., Akhter, A., Choudhury, T. R., & Kim, S. (2020). Evolution of severe acute respiratory syndrome coronavirus 2 (SARS-CoV-2) as coronavirus disease 2019 (COVID-19) pandemic: A global health emergency. *Science of the Total Environment, 730*, Article 138996.

Al Ahdab, S. (2021). A cross-sectional survey of knowledge, attitude and practice (KAP) towards COVID-19 pandemic among the Syrian residents. *BMC Public Health, 21*, 1–7.

Ali, M., Nelson, A. R., Lopez, A. L., & Sack, D. A. (2015). Updated global burden of cholera in endemic countries. *PLoS Neglected Tropical Diseases, 9*(6), Article e0003832.

Alimohamadi, Y., Sepandi, M., Taghdir, M., & Hosamirudsari, H. (2020). Determine the most common clinical symptoms in COVID-19 patients: A systematic review and meta-analysis. *Journal of Preventive Medicine and Hygiene, 61*(3), E304–E312. https://doi.org/10.15167/2421-4248/jpmh2020.61.3.1530

Anderson, D. M., Ciletti, N. A., Lee-Lewis, H., Elli, D., Segal, J., DeBord, K. L., Overheim, K. A., Tretiakova, M., Brubaker, R. R., & Schneewind, O. (2009). Pneumonic plague pathogenesis and immunity in Brown Norway rats. *The American Journal of Pathology, 174*(3), 910–921. https://doi.org/10.2353/ajpath.2009.071168

Anhlan, D., Grundmann, N., Makalowski, W., Ludwig, S., & Scholtissek, C. (2011). Origin of the 1918 pandemic H1N1 influenza A virus as studied by codon usage patterns and phylogenetic analysis. *RNA* (New York, N.Y.), *17*(1), 64–73. https://doi.org/10.1261/rna.2395211

Arning, N., & Wilson, D. J. (2020). The past, present and future of ancient bacterial DNA. *Microbial genomics, 6*(7), mgen000384.

Awofeso, N., & Aldbak, K. (2018). Cholera, migration, and global health—A critical review. *International Journal of Travel Medicine and Global Health, 6*(3), 92–99.

Baker-Austin, C., Oliver, J.D., Alam, M., Ali, A., Waldor, M.K., Qadri, F., & Martinez-Urtaza, J. (2018). Vibrio spp. infections. *Nature Reviews Disease Primers, 4*(1), 1–19.

Balasubramanian, D., Murcia, S., Ogbunugafor, C. B., Gavilan, R., & Almagro-Moreno, S. (2021). Cholera dynamics: Lessons from an epidemic. *Journal of Medical Microbiology, 70*(2), Article 001298.

Barbieri, R., Signoli, M., Chevé, D., Costedoat, C., Tzortzis, S., Aboudharam, G., Raoult, D., & Drancourt, M. (2020). *Yersinia pestis*: The natural history of plague. *Clinical Microbiology Reviews, 34*(1), e00044-19. https://doi.org/10.1128/CMR.00044-19

Bartram, J., & Cairncross, S. (2010). Hygiene, sanitation, and water: Forgotten foundations of health. *PLoS Medicine, 7*(11), Article e1000367.

Benedictow, O. J. (2004). *The black death, 1346–1353: The complete history*. Boydell & Brewer.

Bennett, J. E., Dolin, R., & Blaser, M. J. (2019). *Mandell, Douglas, and Bennett's principles and practice of infectious diseases e-book*. Elsevier Health Sciences.

Berkman, A., Garcia, J., Muñoz-Laboy, M., Paiva, V., & Parker, R. (2005). A critical analysis of the Brazilian response to HIV/AIDS: Lessons learned for controlling and mitigating the epidemic in developing countries. *American Journal of Public Health, 95*(7), 1162–1172.

Bitton, G. (2014). *Microbiology of drinking water: Production and distribution*. John Wiley & Sons.

Bland, D. M., Jarrett, C. O., Bosio, C. F., & Hinnebusch, B. J. (2018). Infectious blood source alters early foregut infection and regurgitative transmission of *Yersinia pestis* by rodent fleas. *PLoS Pathogens, 14*(1), Article e1006859. https://doi.org/10.1371/journal.ppat.1006859

Booth, C. (1991). Plague and cholera. *International Journal of STD & AIDS, 2*(1_suppl), 3–7.

Brázda, V., Porubiaková, O., Cantara, A., Bohálová, N., Coufal, J., Bartas, M., & Mergny, J. L. (2021). G-quadruplexes in H1N1 influenza genomes. *BMC Genomics, 22*(1), 1–11.

Broder, S. (2010). The development of antiretroviral therapy and its impact on the HIV-1/AIDS pandemic. *Antiviral research, 85*(1), 1–18. https://doi.org/10.1016/j.antiviral.2009.10.002

Byrne, J. P. (2004). *The Black Death*. Greenwood Press.

Cantor, N.F. (2001). *In the wake of the plague: The black death and the world it made*. Simon and Schuster.

Centers for Disease Control and Prevention [CDC]. (2019). Summary of progress since 2009. Retrieved from https://www.cdc.gov/flu/pandemic-resources/h1n1-summary.htm

Chakarova, B., & Kichukova, K. (2021). Ancient and newly emerged diseases affecting the course of human history. *Trakia Journal of Sciences, 19*(2), 91–105.

Challa, J. M., Getachew, T., Debella, A., Merid, M., Atnafe, G., Eyeberu, A., Birhanu, A., & Regassa, L. D. (2022). Inadequate hand washing, lack of clean drinking water and latrines as major determinants of cholera outbreak in Somali Region, Ethiopia in 2019. *Frontiers in Public Health, 10*, Article 845057. https://doi.org/10.3389/fpubh.2022.845057

Chan-Yeung, M., & Xu, R. H. (2003). SARS: Epidemiology. *Respirology* (Carlton, Vic.), *8*(Suppl 1), S9–S14. https://doi.org/10.1046/j.1440-1843.2003.00518.x

Cheng, V. C., Chan, J. F., To, K. K., & Yuen, K. Y. (2013). Clinical management and infection control of SARS: Lessons learned. *Antiviral Research, 100*(2), 407–419.

Chin, T. D., Foley, J. F., Doto, I. L., Gravelle, C. R., & Weston, J. (1960). Morbidity and mortality characteristics of Asian strain influenza. *Public Health Reports, 75*(2), 149.

Chiu, N. C., Chi, H., Tai, Y. L., Peng, C. C., Tseng, C. Y., Chen, C. C., Tan, B. F., & Lin, C. Y. (2020). Impact of wearing masks, hand hygiene, and social distancing on influenza, enterovirus, and all-cause pneumonia during the coronavirus pandemic: Retrospective national epidemiological surveillance study. *Journal of Medical Internet Research, 22*(8), Article e21257. https://doi.org/10.2196/21257

Chowdhury, F. R., Nur, Z., Hassan, N., von Seidlein, L., & Dunachie, S. (2017). Pandemics, pathogenicity and changing molecular epidemiology of cholera in the era of global warming. *Annals of Clinical Microbiology and Antimicrobials, 16*(1), 1–6.

Chowell, G., Ammon, C. E., Hengartner, N. W., & Hyman, J. M. (2007). Estimating the reproduction number from the initial phase of the Spanish flu pandemic waves in Geneva, Switzerland. *Mathematical Biosciences & Engineering, 4*(3), 457.

Chu, H., Chan, J. F. W., Yuen, T. T. T., Shuai, H., Yuan, S., Wang, Y., & Yuen, K. Y. (2020). Comparative tropism, replication kinetics, and cell damage profiling of SARS-CoV-2 and SARS-CoV with implications for clinical manifestations, transmissibility, and laboratory studies of COVID-19: An observational study. *The Lancet Microbe, 1*(1), e14–e23.

Clark, D. P., & Pazdernik, N. J. (2016). Biological warfare: Infectious disease and bioterrorism. *Biotechnology,* 687–719. https://doi.org/10.1016/B978-0-12-385015-7.00022-3

Cohen, M. S., Hellmann, N., Levy, J. A., DeCock, K., & Lange, J. (2008). The spread, treatment, and prevention of HIV-1: Evolution of a global pandemic. *The Journal of Clinical Investigation, 118*(4), 1244–1254.

Cox, N. J., & Subbarao, K. (2000). Global epidemiology of influenza: past and present. *Annual review of medicine, 51*, 407–421. https://doi.org/10.1146/annurev.med.51.1.407

Crawshaw, J. L. S. (2016). *Plague hospitals: Public health for the city in early Modern Venice.* Routledge.

Daga, M. K., Kumar, N., Aarthi, J., Mawari, G., Garg, S., & Rohatgi, I. (2019). From SARS-CoV to coronavirus disease 2019 (COVID-19)—A brief review. *Journal of Advanced Research in Medicine, 6*(4), 1–9. E-ISSN: 2349-7181 & P-ISSN: 2394-7047.

Daniels, N. A., & Shafaie, A. (2000). A review of pathogenic Vibrio infections for clinicians. *Infections in Medicine, 17*(10), 665–685.

De Wagt, A., & Connolly, M. (2005). Orphans and the impact of HIV/AIDS in sub-Saharan Africa. *Food Nutrition and Agriculture, 34*, 24.

Defo, B. K. (2014). Demographic, epidemiological, and health transitions: Are they relevant to population health patterns in Africa? *Global Health Action, 7*(1), 22443.

Dunning, J., Thwaites, R. S., & Openshaw, P. J. (2020). Seasonal and pandemic influenza: 100 years of progress, still much to learn. *Mucosal Immunology, 13*(4), 566–573.

Dybing, J. K., Schultz-Cherry, S., Swayne, D. E., Suarez, D. L., & Perdue, M. L. (2000). Distinct pathogenesis of Hong Kong-origin H5N1 viruses in mice compared to that of other highly pathogenic H5 avian influenza viruses. *Journal of Virology, 74*(3), 1443–1450.

Ejaz, H., Alsrhani, A., Zafar, A., Javed, H., Junaid, K., Abdalla, A. E., Abosalif, K., Ahmed, Z., & Younas, S. (2020). COVID-19 and comorbidities: Deleterious impact on infected patients. *Journal of Infection and Public Health, 13*(12), 1833–1839. https://doi.org/10.1016/j.jiph.2020.07.014

Elliott, C. P. (2016). The Antonine Plague, climate change and local violence in Roman Egypt. *Past & Present, 231*(1), 3–31.

Faruque, S. M., Tam, V. C., Chowdhury, N., Diraphat, P., Dziejman, M., Heidelberg, J. F., & Nair, G. B. (2007). Genomic analysis of the Mozambique strain of Vibrio cholerae O1 reveals the origin of El Tor strains carrying classical CTX prophage. *Proceedings of the National Academy of Sciences, 104*(12), 5151–5156.

Fears, J. R. (2004). The plague under Marcus Aurelius and the decline and fall of the Roman Empire. *Infectious Disease Clinics, 18*(1), 65–77.

Feehan, J., & Apostolopoulos, V. (2021). Is COVID-19 the worst pandemic? *Maturitas, 149*, 56–58.

Findlater, A., & Bogoch, I. I. (2018). Human mobility and the global spread of infectious diseases: A focus on air travel. *Trends in Parasitology, 34*(9), 772–783. https://doi.org/10.1016/j.pt.2018.07.004

Fitzgerald, D. A. (2009). Human swine influenza A [H1N1]: Practical advice for clinicians early in the pandemic. *Paediatric Respiratory Reviews, 10*(3), 154–158.

Food and Drug Administration [FDA]. (2021). FDA Approves First COVID-19 Vaccine. Retrieved from https://www.fda.gov/news-events/press-announcements/fda-approves-first-covid-19-vaccine

Forbes, K. J. (2004). The Asian flu and Russian virus: The international transmission of crises in firm-level data. *Journal of International Economics, 63*(1), 59–92.

Gage, K. L., & Kosoy, M. Y. (2005). Natural history of plague: perspectives from more than a century of research. *Annu. Rev. Entomol., 50*, 505–528. https://doi.org/10.1146/annurev.ento.50.071803.130337

Ginsberg, J., Mohebbi, M. H., Patel, R. S., Brammer, L., Smolinski, M. S., & Brilliant, L. (2009). Detecting influenza epidemics using search engine query data. *Nature, 457*(7232), 1012–1014.

Habicht, M. E., Pate, F. D., Varotto, E., & Galassi, F. M. (2020). Epidemics and pandemics in the history of humankind and how governments dealt with them: A review from the Bronze Age to the Early Modern Age. *Rivista Trimestrale di Scienza dell'Amministrazione, 30*, 1–32.

Haldane, V., De Foo, C., Abdalla, S. M., Jung, A. S., Tan, M., Wu, S., & Legido-Quigley, H. (2021). Health systems resilience in managing the COVID-19 pandemic: Lessons from 28 countries. *Nature Medicine, 27*(6), 964–980.

Herlihy, D. (1997). *The Black Death and the transformation of the West*. Harvard University Press.

HIV.GOV. (2022). US statistics. Retrieved from https://www.hiv.gov/hiv-basics/overview/data-and-trends/statistics

Honigsbaum, M. (2013). *A history of the great influenza pandemics: Death, panic and hysteria, 1830–1920* (Vol. 30). IB Tauris.

Horgan, J. (2016). *The plague at Athens, 430-427 BCE*. Ancient History Encyclopedia.

Howard-Jones, N. (1974). Cholera nomenclature and nosology: A historical note. *Bulletin of the World Health Organization, 51*(3), 317.

Humphreys, M. (2018). The influenza of 1918: Evolutionary perspectives in a historical context. *Evolution, Medicine, and Public Health, 2018*(1), 219–229.

Humphries, M. O. (2014). Paths of infection: The First World War and the origins of the 1918 influenza pandemic. *War in History, 21*(1), 55–81.

Huremović, D. (2019). Brief history of pandemics (pandemics throughout history). *Psychiatry of Pandemics: A Mental Health Response to Infection Outbreak*, 7–35. https://doi.org/10.1007/978-3-030-15346-5_2

Hurwitz, J. L., Hackett, C. J., McAndrew, E. C., & Gerhard, W. (1985). Murine TH response to influenza virus: Recognition of hemagglutinin, neuraminidase, matrix, and nucleoproteins. *The Journal of Immunology, 134*(3), 1994–1998.

Iezadi, S., Azami-Aghdash, S., Ghiasi, A., Rezapour, A., Pourasghari, H., Pashazadeh, F., & Gholipour, K. (2020). Effectiveness of the non-pharmaceutical public health interventions against COVID-19: A protocol of a systematic review and realist review. *PloS One, 15*(9), Article e0239554. https://doi.org/10.1371/journal.pone.0239554

Jilani, T. N., Jamil, R. T., & Siddiqui, A. H. (2018). H1N1 influenza. incom.

Jo, W. K., de Oliveira-Filho, E. F., Rasche, A., Greenwood, A. D., Osterrieder, K., & Drexler, J. F. (2021). Potential zoonotic sources of SARS-CoV-2 infections. *Transboundary and Emerging Diseases, 68*(4), 1824–1834.

Johnson, D. G. (1997). Agriculture and the wealth of nations. *The American Economic Review, 87*(2), 1–12.

Joi, P. (2020). 5 reasons why pandemics like COVID-19 are becoming more likely. Retrieved from https://www.gavi.org/vaccineswork/5-reasons-why-pandemics-like-covid-19-are-becoming-more-likely

Jones, E. W. (2007). *Influenza 1918: Disease, death, and struggle in winnipeg* (Vol. 31). University of Toronto Press.

Kahn, J. O., & Walker, B. D. (1998). Acute human immunodeficiency virus type 1 infection. *New England Journal of Medicine, 339*(1), 33–39.

Kitler, M. E., Gavinio, P., & Lavanchy, D. (2002). Influenza and the work of the world health organization. *Vaccine, 20*, S5–S14.

Klemm, C., Das, E., & Hartmann, T. (2016). Swine flu and hype: A systematic review of media dramatization of the H1N1 influenza pandemic. *Journal of Risk Research, 19*(1), 1–20.

Koch, T., & Denike, K. (2010). Essential, illustrative, or… just propaganda? Rethinking John Snow's Broad Street map. *Cartographica: The International Journal for Geographic Information and Geovisualization, 45*(1), 19–31.

Kumar, A., Das, B., & Kumar, N. (2020). Vibrio pathogenicity island-1: The master determinant of cholera pathogenesis. *Frontiers in Cellular And Infection Microbiology, 10*, Article 561296.

Kwit, N., Nelson, C., Kugeler, K., Petersen, J., Plante, L., Yaglom, H., & Mead, P. (2015). Human plague—United States, 2015. *MMWR Morb Mortal Wkly Rep, 64*(33), 918–9.

Lam, W. K., Zhong, N. S., Tan, W. C. (2003). Overview on SARS in Asia and the world. *Respirology, 8*(Suppl 1), S2–S5. https://doi.org/10.1046/j.1440-1843.2003.00516.x

Lee, H. A., Kung, H. H., Lee, Y. J., Chao, J. C., Udayasankaran, J. G., Fan, H. C., Ng, K. K., Chang, Y. K., Kijsanayotin, B., Marcelo, A. B., & Hsu, C. Y. (2020). Global infectious disease surveillance and case tracking system for COVID-19: Development study. *JMIR Medical Informatics, 8*(12), Article e20567. https://doi.org/10.2196/20567

Lee, K. (2001). The global dimensions of cholera. *Global Change and Human Health, 2*(1), 6–17.

Lee, K., & Dodgson, R. (2017). Globalization and cholera: Implications for global governance. *Global Health*, 407–430. Routledge.

Lees, C. H., Avery, C., Asherin, R., Rainbow, J., Danila, R., Smelser, C., & Lynfield, R. (2011). Pandemic (H1N1) 2009–associated deaths detected by unexplained death and medical examiner surveillance. *Emerging Infectious Diseases, 17*(8), 1479.

Lin, C. N. (2015). Impacts on human health caused by zoonoses. *Biological Toxins and Bioterrorism, 1*, 211.

Lipp, E. K., Huq, A., & Colwell, R. R. (2002). Effects of global climate on infectious disease: The cholera model. *Clinical Microbiology Reviews, 15*(4), 757–770.

Lippi, D., & Gotuzzo, E. (2014). The greatest steps towards the discovery of Vibrio cholerae. *Clinical Microbiology and Infection, 20*(3), 191–195.

Littman, R. J. (2009). The plague of Athens: Epidemiology and paleopathology. *The Mount Sinai Journal of Medicine, 76*(5), 456–467. https://doi.org/10.1002/msj.20137

Luby, S. P., Davis, J., Brown, R. R., Gorelick, S. M., & Wong, T. H. (2020). Broad approaches to cholera control in Asia: Water, sanitation and handwashing. *Vaccine, 38*, A110–A117.

Macedo, A., Gonçalves, N., & Febra, C. (2021). COVID-19 fatality rates in hospitalized patients: Systematic review and meta-analysis. *Annals of Epidemiology, 57*, 14–21. https://doi.org/10.1016/j.annepidem.2021.02.012

Maglen, K. (2002). 'The first line of defence': British quarantine and the port sanitary authorities in the nineteenth century. *Social History of Medicine, 15*(3), 413–428.

Makris, A. T., Morgan, L., Gaber, D. J., Richter, A., & Rubino, J. R. (2000). Effect of a comprehensive infection control program on the incidence of infections in long-term care facilities. *American Journal of Infection Control, 28*(1), 3–7.

McEvedy, C. (1988). The bubonic plague. *Scientific American, 258*(2), 118–123. https://doi.org/10.1038/scientificamerican0288-118

Medina, R. A., & García-Sastre, A. (2011). Influenza A viruses: New research developments. *Nature Reviews Microbiology, 9*(8), 590–603.

Memish, Z. A., Ebrahim, S. H., Kattan, R. F., Alharthy, A., Alqahtani, S. A., & Karakitsos, D. (2020). Leadership to prevent COVID-19: Is it the most important mitigation factor? *Travel Medicine and Infectious Disease, 38*, Article 101925. https://doi.org/10.1016/j.tmaid.2020.101925

Moore, K. (2021). "Ther cam a Privee Theef, Men Clepeth Deeth": A tale of two plagues and of altered perspectives. *History, 7*(1), 71.

Morens, D. M., & Fauci, A. S. (2007). The 1918 influenza pandemic: Insights for the 21st century. *The Journal of Infectious Diseases, 195*(7), 1018–1028.

Muanprasat, C., & Chatsudthipong, V. (2013). Cholera: pathophysiology and emerging therapeutic targets. *Future medicinal chemistry, 5*(7), 781–798. https://doi.org/10.4155/fmc.13.42

Nickol, M. E., & Kindrachuk, J. (2019). A year of terror and a century of reflection: Perspectives on the great influenza pandemic of 1918–1919. *BMC Infectious Diseases, 19*(1), 1–10.

Nypaver, C., Dehlinger, C., & Carter, C. (2021). Influenza and influenza vaccine: A review. *Journal of Midwifery & Women's Health, 66*(1), 45–53.

Oshitani, H., Kamigaki, T., & Suzuki, A. (2008). Major issues and challenges of influenza pandemic preparedness in developing countries. *Emerging Infectious Diseases, 14*(6), 875.

Oswald, N. C., Shooter, R. A., & Curwen, M. P. (1958). Pneumonia complicating Asian influenza. *British Medical Journal, 2*(5108), 1305.

Perico, L., Benigni, A., Casiraghi, F., Ng, L. F., Renia, L., & Remuzzi, G. (2021). Immunity, endothelial injury and complement-induced coagulopathy in COVID-19. *Nature Reviews Nephrology, 17*(1), 46–64.

Petersen, E., Koopmans, M., Go, U., Hamer, D. H., Petrosillo, N., Castelli, F., & Simonsen, L. (2020). Comparing SARS-CoV-2 with SARS-CoV and influenza pandemics. *The Lancet Infectious Diseases, 20*(9), e238–e244.

Pike, B. L., Saylors, K. E., Fair, J. N., Lebreton, M., Tamoufe, U., Djoko, C. F., Rimoin, A. W., & Wolfe, N. D. (2010). The origin and prevention of pandemics. *Clinical Infectious Diseases: An Official Publication of the Infectious Diseases Society of America, 50*(12), 1636–1640. https://doi.org/10.1086/652860

Potter, C. W. (2001). A history of influenza. *Journal of Applied Microbiology, 91*(4), 572–579.

Rajoura, O. P., Roy, R., Agarwal, P., & Kannan, A. T. (2011). A study of the Swine Flu (H1N1) epidemic among health care providers of a medical college hospital of Delhi. *Indian Journal of Community Medicine: Official Publication of Indian Association of Preventive & Social Medicine, 36*(3), 187–190. https://doi.org/10.4103/0970-0218.86518

Raoult, D., Mouffok, N., Bitam, I., Piarroux, R., & Drancourt, M. (2013). Plague: History and contemporary analysis. *Journal of Infection, 66*(1), 18–26.

Remien, R. H., Higgins, J. A., Correale, J., Bauermeister, J., Dubrow, R., Bradley, M., Steward, W. T., Seal, D. W., Sikkema, K. J., Kerndt, P. R., Mayer, K. H., Truong, H. M., Casey, C. Y., Ehrhardt, A. A., & Morin, S. F. (2009). Lack of understanding of acute HIV infection among newly-infected persons-implications for prevention and public health: The NIMH multisite acute HIV infection study: II. *AIDS and Behavior, 13*(6), 1046–1053. https://doi.org/10.1007/s10461-009-9581-7

Retief, F. P., & Cilliers, L. (2006). The epidemic of Justinian (AD 542): A prelude to the Middle Ages. *Acta Theologica, 26*(2), 115–127.

Rice, G. W. (2018). Influenza in New Zealand before 1918: A preliminary report. *American Journal of Epidemiology, 187*(12), 2524–2529.

Rogers, K. (2022). *Pandemic. Encyclopedia Britannica*. https://www.britannica.com/science/pandemic

Rosen, W. (2007). *Justinian's flea: The first great plague and the end of the Roman Empire*. Penguin.

Sabbatani, S., & Fiorino, S. (2009). The Antonine Plague and the decline of the Roman Empire. *Le Infezioni in Medicina, 17*(4), 261–275.

Safa, A., Nair, G. B., & Kong, R. Y. (2010). Evolution of new variants of Vibrio cholerae O1. *Trends in Microbiology, 18*(1), 46–54.

Scheidel, W. (2002). A model of demographic and economic change in Roman Egypt after the Antonine plague. *Journal of Roman Archaeology, 15*, 97–114.

Schneiders, M. L., Naemiratch, B., Cheah, P. K., Cuman, G., Poomchaichote, T., Ruangkajorn, S., Stoppa, S., Osterrieder, A., Cheah, P. K., Ongkili, D., Pan-Ngum, W., Mackworth-Young, C., & Cheah, P. Y. (2022). The impact of COVID-19 non-pharmaceutical interventions on the lived experiences of people living in Thailand, Malaysia, Italy and the United Kingdom: A cross-country qualitative study. *PloS One, 17*(1), Article e0262421. https://doi.org/10.1371/journal.pone.0262421

Schnitzler, S. U., & Schnitzler, P. (2009). An update on swine-origin influenza virus A/H1N1: A review. *Virus Genes, 39*(3), 279–292.

Sebbane, F., & Lemaître, N. (2021). Antibiotic therapy of Plague: A review. *Biomolecules, 11*(5), 724. https://doi.org/10.3390/biom11050724

Shang, Y., Pan, C., Yang, X., Zhong, M., Shang, X., Wu, Z., Yu, Z., Zhang, W., Zhong, Q., Zheng, X., Sang, L., Jiang, L., Zhang, J., Xiong, W., Liu, J., & Chen, D. (2020). Management of critically ill patients with COVID-19 in ICU: Statement from front-line intensive care experts in Wuhan, China. *Annals of Intensive Care, 10*(1), 73. https://doi.org/10.1186/s13613-020-00689-1

Shepard, W. (2015). *Ghost cities of China: The story of cities without people in the world's most populated country*. Bloomsbury Publishing.

Shi, Z., & Hu, Z. (2008). A review of studies on animal reservoirs of the SARS coronavirus. *Virus Research, 133*(1), 74–87. https://doi.org/10.1016/j.virusres.2007.03.012

Skwarecki, B. (2016). *Outbreak!: 50 tales of epidemics that terrorized the world*. Simon and Schuster.

Smith, F. B. (1995). The Russian Influenza in the United Kingdom, 1889–1894. *Social History of Medicine, 8*(1), 55–73.

Spyrou, M. A., Bos, K. I., Herbig, A., & Krause, J. (2019). Ancient pathogen genomics as an emerging tool for infectious disease research. *Nature Reviews. Genetics, 20*(6), 323–340. https://doi.org/10.1038/s41576-019-0119-1

Tang, Y., Liu, J., Zhang, D., Xu, Z., Ji, J., & Wen, C. (2020). Cytokine storm in COVID-19: The current evidence and treatment strategies. *Frontiers in Immunology, 11*, 1708. https://doi.org/10.3389/fimmu.2020.01708

Tognotti, E. (2013). Lessons from the history of quarantine, from plague to influenza A. *Emerging Infectious Diseases, 19*(2), 254.

Tuite, A. R., Greer, A. L., Whelan, M., Winter, A. L., Lee, B., Yan, P., & Fisman, D. N. (2010). Estimated epidemiologic parameters and morbidity associated with pandemic H1N1 influenza. *Cmaj, 182*(2), 131–136.

UNAIDS. (n.d.). 90-90-90: An ambitious treatment target to help end the AIDS epidemic. Retrieved from https://www.unaids.org/en/resources/909090

Valtat, S., Cori, A., Carrat, F., & Valleron, A. J. (2011). Age distribution of cases and deaths during the 1889 influenza pandemic. *Vaccine, 29*, B6–B10.

Viboud, C., Simonsen, L., Fuentes, R., Flores, J., Miller, M. A., & Chowell, G. (2016). Global mortality impact of the 1957–1959 influenza pandemic. *The Journal of Infectious Diseases, 213*(5), 738–745.

Voshavar, C. (2019). Protease inhibitors for the treatment of HIV/AIDS: Recent advances and future challenges. *Current Topics in Medicinal Chemistry, 19*(18), 1571–1598.

Webster, R. G., Bean, W. J., Gorman, O. T., Chambers, T. M., & Kawaoka, Y. (1992). Evolution and ecology of influenza A viruses. *Microbiological Reviews, 56*(1), 152–179.

World Health Organization [WHO]. (2001). *HIV/AIDS in Asia and the Pacific Region*. WHO Regional Office for the Western Pacific.

World Health Organization [WHO]. (2021a). Cholera 2020. Weekly epidemiological record. Retrieved from https://apps.who.int/iris/bitstream/handle/10665/345267/WER9637-eng-fre.pdf

World Health Organization [WHO]. (2021b). The different types of COVID-19 vaccines. Retrieved from https://www.who.int/news-room/feature-stories/detail/the-race-for-a-covid-19-vaccine-explained

World Health Organization [WHO]. (2022). Cholera. Retrieved from https://www.who.int/news-room/fact-sheets/detail/cholera#:~:text=History,and%20the%20Americas%20in%201991

Chapter 3
Climate Change: Zoonoses in Changing Environment

1 Introduction

The United Nations Framework Convention on Climate Change [UNFCCC], defines climate change as "a change of climate which is attributed directly or indirectly to human activity that alters the composition of global atmosphere and which is in addition to natural climate variability observed over comparable time period". Climate change is a serious threat to the entire world, with numerous significant side-effects including social unrest, humanitarian crises, financial distress and ecological damage. To reduce these repercussions of anthropogenic climate change, immediate action is necessary (Ogunseitan, 2011).

A substantial knowledge base exists regarding how climate change may affect human health, notably the genesis and spread of infectious diseases (Hambling et al., 2011). The detrimental aspects of climate change on human health, both directly and indirectly, is highlighted in numerous research. For instance, climate change modifies the ecology of zoonotic disease infections and their vectors. The rising prevalence of West Nile and Usutu viruses and the emergence of novel vector species, like certain mosquitoes and ticks, in Europe and perhaps other regions of the world seem to be indicators of it (Myaing, 2011). Recognizing the link amongst zoonotic disease and climate change is crucial considering that it will exacerbate an already substantial impact that infections currently place on economic systems as well as population health. In order to prepare for or take measures to avoid these scenarios, officials must be able to analyse, foresee, and evaluate the susceptibility of public health to climatic changes (Mills et al., 2010).

Zoonosis pertains to the diseases communicated by animals to humans, sometimes directly (through physical human-animal contact or animal products) and indirectly (via an intermediary carrier like an arthropod or insect) (Pappas, 2011). Zoonotic infections might be parasitic, bacterial, viral, or even involve exotic entities (Tulchinsky & Varavikova, 2014). Alteration in disease epidemiology brought

M. H. Fulekar et al., *Pandemic Over the Centuries*,
https://doi.org/10.1007/978-3-031-94270-9_3

on by climate change are linked to shifts in habitats, human vulnerability, and over-exposure to causal organism. Additionally, it has an impact on the spatial range of intermediate hosts, such as migrating birds, rodents and insects (Ghazali et al., 2018). Climate change can alter human sociodemographics and cause a mass exodus of people and domestic animals from dry, hot regions to places wherein conditions are more favourable (El-Sayed & Awad, 2018). In addition, there are widespread alterations to the traditional distribution pattern of wildlife, flies and rodents (Walsh et al., 2018). The thawing of an ice layer that has been present for hundreds of years is among the most serious underrated consequences of global warming. In addition to causing numerous cities and river deltas to vanish, melting ice also has the potential to revive organisms that have been trapped in permafrost, increasing the risk of zoonotic disease episodes (Resnick, 2019). The harm that climate change and global warming do to people's health is widely ignored. According to estimates, alterations in environmental conditions are responsible for 36% among infants and young child mortality worldwide and 34% of all childhood illnesses. Climate change has a significant impact on outbreaks of cholera, severe malaria, diarrhea and dengue, among other diseases (Thompson et al., 2012).

Although there is a significant burden of zoonoses globally in terms of short- and protracted death and morbidity as well as emergence/reemergence and social economic, environmental and diplomatic relationships, there is still only a small amount of research and community healthcare funding dedicated to these illnesses (Akritidis, 2011; Christou, 2011; Pappas, Cascio & Rodriguez-Morales, 2012). In order to anticipate the projected impacts of climate-related variations in the coming years on the ecology of pathogens, their distribution, carriers, reservoirs and the last hosts, it may be helpful to consider the epidemiological linkages among climate variations, ecological disruption, exposure to vectors and transmission of diseases (Greer et al., 2008; Medone et al., 2015; Teklehaimanot et al., 2004).

2 Classification of Zoonoses

Zoonoses are divided into groups based on the reservoirs, aetiological agents, transmission cycle and environmental factors (Rahman et al., 2020; Schaechter, 2009).

2.1 Classification Based on Etiological Agents

2.1.1 Bacterial Zoonoses

Anthrax (*Bacillus anthracis*), Brucellosis (*Brucella* sp.), Bovine Tuberculosis (*Mycobacterium bovis*), Camphylobacteriosis (*Campylobacter* sp.), Listeriosis (*Listeria monocytogenesis*), Salmonellosis (*Salmonella* sp.), Leptospirosis (*Leptospira interrogans*).

2.1.2 Viral Zoonoses

Rabies virus (*Lyssavirus genera*), Influenza viruses (*Orthomyxovirus genera*), Yellow fever (Yellow fever virus), Encephalitis (West Nile virus), Chikungunya (Chikungunya virus).

2.1.3 Mycotic Zoonoses

Dermatophytosis (*Trichophyton verrucosum, Arthroderma vanbreuseghemii* etc.), Basidiobolo mycosis (*Basidiobolus ranarum*), Histoplasmosis (*Histoplasma capsulatum*), Sporotrichosis (*Sporothrix schenckii*).

2.1.4 Protozoal Zoonoses

Toxoplasmosis (*Toxoplasma gondii*), Sarcosporidiosis (*Sarcocystis hominis* and *S. suihomini*), Trypanosomiasis (*Trypanosoma brucei*).

2.1.5 Parasitic Zoonoses

Toxocariasis (*Toxocara canis*), Amebiasis (*Entamoeba histolytica*), Cryptosporidiosis (*Cryptosporidium parvum*).

2.1.6 Rickettsial Zoonoses

Human ehrlichosis (*Ehrlichia chaffeensis, E. muris muris*), Rickettsiosis (*Rickettsia aeschliman-nii*), Typhus fever (*Rickettsia prowazekii*).

2.1.7 Chlamydial Zoonoses

Human psittacosis (*Chlamydophila psittaci*), Enzootic ovine abortion (*Chlamydophila abortus*).

2.2 *Classification Based on Transmission Cycle (Chomel, 2009)*

2.2.1 Orthozoonoses (Direct Zoonoses)

Direct touch, contact with a fomite, or use of a mechanical vector are the three main ways that disease is transferred from an infected to a vulnerable host organism. Species like dogs or foxes with rabies or cattle, small ruminants, or swine with brucellosis, could be the source of a direct zoonoses in environment.

2.2.2 Cyclozoonoses

To complete the cycle of an infectious agent's transmission, cyclozoonoses need more than one type of vertebrate host, but no invertebrate recipient. Human taeniases and pentastomid infections are two examples. Cestodiases make up the majority of the relatively few cyclozoonoses.

2.2.3 Metazoonoses (Also Called Pherozoonoses)

Zoonoses that need both invertebrates and vertebrates to complete their life cycle in order to spread. There is usually a short incubation phase in the invertebrate host preceding transmission to a vertebrate host in pherozoonoses, wherein the pathogen replicates or simply matures. Examples include rickettsial infections, Lyme borreliosis, plague and arbovirus infections.

2.2.4 Saprozoonoses

These zoonoses have an inanimate developmental reservoir in addition to a vertebrate host. Nonanimal resources, like organic material, comprising food, soil and plants, are regarded as the developmental reservoir. Direct transmission in this group of zoonoses is typically uncommon or nonexistent. Examples include listeriosis and histoplasmosis (Fig. 1).

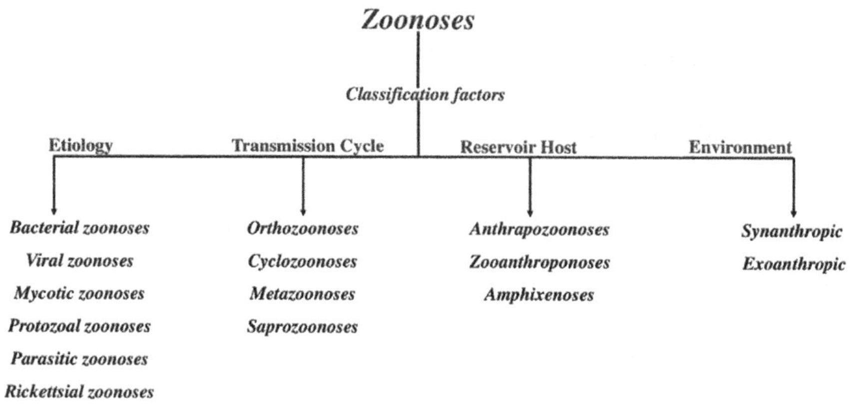

Fig. 1 Types of zoonoses based on classifying factors

2.3 Classification Based on Reservoir Hosts (Tayde, n.d.)

2.3.1 Anthrapozoonoses

Such infections affect both domesticated and wild animals, and they develop naturally without the involvement of humans. Animals can infect humans under rare conditions, such as through food or work-related contact. Leptospirosis, tularemia, rift valley fever, hydatidosis and rabies are a few examples.

2.3.2 Zooanthroponoses

These illnesses typically spread from people to other vertebrate animals. Examples include tuberculosis in humans and amoebias.

2.3.3 Amphixenoses

Both humans and animals are susceptible to contracting the infectious disease. It consists of Staphylococcosis, Salmonellosis and Streptococcosis.

2.4 Classification Based on Environment (Resh & Cardé, 2009)

2.4.1 Synanthropic

Domestic animals are the only species where the pathogens can spread. Examples include listeriosis, ringworm and bovine tuberculosis. These diseases are spread either through percutaneous contact or aerogenous conjunctival contact.

2.4.2 Exoanthropic

The life-cycle of pathogens, such as tick-borne encephalitis or arboviruses, occurs in wild or non-domestic animals. A blood-feeding vector is typically used in the infection transmission from animal to human.

3 Climatic Factors Affecting Zoonoses

The alteration in reservoir and vector dynamics is indicative of the impact of geoclimatic changes on zoonoses transmission. Climate change generates new ecological niches for vectors, influencing the temporal and spatial dispersal of disease (Lafferty, 2009). The spread and infestation of zoonotic pathogens are influenced by the following factors.

3.1 Land Temperatures

Anthropogenic greenhouse gas (GHG) emissions are primarily responsible for the exponential pace of global warming. The Intergovernmental Panel on Climate Change [IPCC] has estimated temperature increases averaging 0.2 °C every decade, with an estimated mean temperature rise of 1.8–4 °C by the end of the twenty-first century (Team et al., 2014). For vector-borne zoonotic pathogens, it is anticipated that vector spread and consequently disease risk will enhance. The most susceptible vectors to temperature variation in the environment are arthropods.

Life cycles of ectothermic insects like mosquitoes, ticks and sandflies rely on the temperature. If the temperature fluctuates between 14–18 °C and 35–40 °C, risk of transmission is most likely to occur. At 30–32 °C, vector densities ought to be at their highest (Githeko et al., 2000). Mosquitoes are directly impacted by temperature. It causes an increase in activity and reproduction, which causes blood feeds and also rapid digestion (Martin et al., 2008). Mosquito-borne infections typically reach maturity sooner. The development of mosquito larvae is accelerated by warmer water, which also increases the capacity of all vectors (Reiter, 2008). The first chikungunya infection spread in a temperate zone was caused by the abundance of the vector *Aedes albopictus* in Italy (Rezza et al., 2007). Tick development and over-winter survival rates both rise in response to higher temperatures. The impact of global warming can also be demonstrated in leishmaniasis transmission via sandflies as vectors. Because sandflies feed more frequently and are more active in hotter weather, transmission is enhanced. Leishmania parasite growth is accelerated by rising temperatures (Ready, 2010).

3.2 Ocean Temperatures, Sea Level and Acidity

Sea level rise is a significant concern since it results from melting glaciers and polar ice caps, as well as rising sea temperatures. Sea level rise will increase the risk of coastal floods and water-borne zoonoses. Oceanic carbon intake results in a pH decline, which threatens marine ecosystems (Team et al., 2014).

3.3 Rainfall Patterns

Precipitation has an indirect impact on vectors. Several fertile breeding habitats for mosquitoes are developed by changes in precipitation. After rain, the foliage becomes dense, which gives vectors places to rest and find shelter (Githeko et al., 2000). Rift valley fever (RVF) outbreaks are linked to situations after a lot of rain. The most significant mosquito species are *Aedes* spp., and they transmit viruses by transovarial transmission. They breed in floodwaters, and when it rains heavily, eggs are laid. Even during times of drought, these eggs are still viable and will hatch when the climate is more favourable. If there are vertebrate reservoirs around, outbreaks happen due to enhanced vector capacity caused by heavy rains and larval maturation. The outbreak can then spread through secondary vectors including *Culex* and *Anopheles* spp. There may be a connection between West Nile and chikungunya infections and instances of heavy rain (Chevalier et al., 2004). Since there are fewer mosquito predators during droughts, there are more vectors following, and the density of reservoir hosts around drainage holes makes disease transmission easier (Wang et al., 2010).

3.4 Soil Conditions

Tick development is influenced by soil moisture, and mortality is correlated with arid climates and soil evaporation (Randolph, 2008). Hyalomma ticks, the carriers of Crimean-Congo hemorrhagic fever, are better suited than other ticks to survive in arid environments. *Bacillus anthracis* spore distribution is influenced by soil composition.

Humus-rich, high-calcium, and alkaline (pH > 6.1) soil properties are ideal for spore persistence (McPhaden et al., 2006).

4 Anthropogenic Factors Affecting Zoonoses

The origin, persistence and spread of zoonotic illnesses are all influenced by a variety of anthropogenic influences.

4.1 Travel and Tourism

Intercontinental migration makes it easier for pathogens to spread. As seen by the SARS-COV-2 pandemic. Air and land transit connectivity makes disease management challenging (Wilder-Smith, 2006). Eco-adventure travel and tourism to more exotic locations are both growing in appeal. In Asia and Africa, volunteering is also prevalent. Non-immune travellers are thereby exposed to endemic infections. Recreational zoonoses are a risk associated with adventure sports. As animals excrete the bacterium in their urine, water activities like canoeing, kayaking and river rafting have been linked to leptospirosis outbreaks (Sejvar et al., 2003).

4.2 Trade

As a result of globalisation, items can now be transported by air, rail, sea and land to almost any location on the earth. Used tyre trading served as a haven for mosquitoes that can thrive in urban settings and in small containers (*Aedes aegypti* and *Aedes albopictus*). This enabled a transition from the ancient sylvatic transmission of chikungunya to the quick invasion of the Asian tiger mosquito, resulting in widespread human epidemics (Gould & Higgs, 2009).

Multiple exotic species are brought together in Asia's live animal markets, producing a breeding ground for the spread of interspecies diseases. Civet cats were discovered to harbour coronaviruses that were 99% identical to the strain responsible for the SARS coronavirus (SARS-CoV) outbreak in Guangdong Province. Civet cats had a part in the transmission to humans that began in these trading markets, despite the fact that they are not the natural reservoirs (Poon et al., 2004).

4.3 Pets

Zoonotic disease transmission is linked to close interaction between pet animals and their caretakers. Salmonellosis in humans is linked to owning exotic reptiles as pets. The domestic dog serving as the primary reservoir host has contributed to the recent rise of zoonotic leishmaniasis in Europe. In the Mediterranean region, *Leishmania infantum* induces cutaneous and visceral leishmaniasis and can spread through travel and the procurement of infected dogs (Ready, 2010). The primary reservoir for Bartonella species is domestic cats. Because the reservoirs are not likely to be removed, diseases like Cat Scratch Disease, *Bacillary angiomatosis* and Peliosis are likely to remain and even worsen (Chomel et al., 2006).

4.4 Agricultural Practices and Livestock Farming

The rapid expansion of the human population increases the need for food, land used for farming and causes ecosystems to be disrupted. Dams and canals constructed for agricultural purposes could serve as new mosquito vector breeding grounds. Animal husbandry on a massive scale is expanding to meet consumer demand (Cutler et al., 2010). The intensive farming of numerous animals or different species makes it easier for diseases to spread across species boundaries. Animal farming enables the exponential growth of novel diseases. One of the earliest cases of Methicillin-resistant *Staphylococcus aureus* (MRSA) associated with pig rearing was described in the Netherlands (Voss et al., 2005). By conventional procedures, it is impossible to identify these swine ST398 strains, that are present throughout Europe. Farming is now thought to increase the risk of MRSA in livestock (LaMRSA). The use of antibiotics in livestock production to promote growth and treat infection exerts a selection pressure that causes the establishment of antimicrobial resistance strains in livestock and poultry (Schlundt et al., 2004).

4.5 Food Production and Distribution

Large-scale food production, food processing and food distribution are all contributing to an increase in food-borne zoonotic diseases. Undercooked meat is linked to infections of enterohaemorrhagic *E. coli* O157:H7 (Schlundt et al., 2004).

4.6 Deforestation and Urbanization

Through the "dilution effect," the loss of biodiversity brought on by deforestation affects how zoonotic illnesses are spread. According to this logic, the disease is less severe in places with great biodiversity because more animals support disease vectors. The burden of disease is greater when there are fewer species present. Clear-cut logging does not raise the danger of zoonoses, but selective logging may do so if the forest's biodiversity is preserved (Wolfe et al., 2005).

The natural equilibrium between various species is disrupted by deforestation, which changes ecosystems. For example, water puddles in deforested areas have less acidity and are ideal mosquito breeding places, creating new biological niches for some vectors (Vora, 2008). The frequency of interactions between people and wildlife grows as a result of population growth and the incursion of human communities into natural ecosystems, which makes it easier for zoonotic pathogens to spread. This has been demonstrated in Malaysia by the spread of *Plasmodium knowlesi* from macaques to people (Singh et al., 2004). Additionally, urbanisation draws immigrant inhabitants. There is a chance that new diseases or vectors could be imported due to human migration for jobs. Also, people who go to foreign countries lack immunity to diseases that may exist there. Emerging infections spread more readily as a result of the expansion of road and rail networks. New highways expose rural, immune-deficient people to emerging diseases (Wolfe et al., 2005).

4.7 Political Factors

Civil conflict, war and political unrest result in a breakdown of the global health facilities and control mechanisms, which may cause an increase in infectious diseases generally, for instance through inadequate vector management.

5 Epidemiological Aspects of Zoonosis

Currently, zoonoses are regarded as one of the biggest hazards to public health across the globe. According to Bueno-Mar et al. (2015), around 60% of all pathogenic strains are zoonotic and roughly 75% of newly emerging infections that impact mankind are diseases of animal origin. Arthropods are most likely the principal zoonoses' possible vectors due to their abundance, great adaptability, resilience and coevolution to various infections, high levels of synanthropism in many groups and challenges in implementing efficient population control measures. Even while ticks, flies, sandflies, cockroaches, bugs and fleas are all potent carriers of viruses, parasites and bacteria, mosquitoes are without a doubt the most significant carriers of pathogens in domesticated animals, while ticks are the most significant carriers of infections in humans. In recent years, tropical and temperate countries have seen an increase in the occurrence of mosquito-borne diseases such malaria, equine encephalitis and West Nile (Hernández-Triana et al., 2014; Neira et al., 2014; Prow et al., 2014).

Different issues with zoonoses are significant in industrialised and underdeveloped nations. Chagas disease, also known as American trypanosomiasis, is one example of a vector-borne disease that is currently manageable with therapeutical, preventive and vector control tools and yet has a shocking occurrences in Central and South America (Finol & Roschman-González, 2014; Gonzalez-Britez et al., 2014; Garca-Alzate et al., 2014). Several overlooked illnesses in Africa and Asia, such as leishmaniasis or African trypanosomiasis, have a significant local effect on human communities (Darvishi et al., 2015; Dhiman, 2014). In the USA, Salmonella, *Escherichia coli* O157:H7 and Cryptosporidium are frequently implicated in disease outbreaks. The practice of keeping prairie dogs as pets is becoming more popular in USA. These animals introduce infection from the wild into homes; in 2003, a monkey pox outbreak was connected to exposure to and contact with people who had recently bought prairie dogs. Later research revealed that exposure to the Gambian rat was how the pets caught the uncommon virus (Merianos, 2007; Metzgar et al., 2010). *Rattus* spp. commensal rats are common in both rural and urban regions, mostly in homes and other structures. They are everywhere and constantly interact with people. Commensal rats are reservoirs for numerous zoonoses, including leptospirosis and bartonellosis, which are urgent public health priority, in addition to having negative economic effects from contaminating food storage facilities and inflicting infrastructural damage (Himsworth et al., 2014; Obiegala et al., 2019; Rothenburger et al., 2019).

Effective zoonotic disease surveillance, prevention and control are crucial yet provide a considerable barrier for many nations (Meslin et al., 2000). The lack of data collection on these ailments, especially in poor nations, is a significant obstacle to knowing and comprehending the epidemiological context and impact of such diseases. Furthermore, given the ecological contexts in which zoonotic diseases emerge, other ecoepidemiological variables need to be included in the analysis, but regrettably, many health organisations, particularly in developing countries, still do not take them into account. It is essential to comprehend zoonotic illnesses as a complex problem in order to stop them from spreading in terms of geographical and social predominance. Pathogen-related variables and human behavior related factors are two general categories for influencing their growth. These elements are interconnected, as changes in human behaviour affect pathogen ecology and life-cycles in multiple ways (Cascio et al., 2011).

6 Prevention and Control of Zoonoses

Immediate safeguarding of humans, the decrease or elimination of the illness in the animal reservoir and preventive actions are all included in the three-tiered approach to the prevention and control of zoonotic diseases (Chomel, 2009). It places a special, frequently heavy load on public health services, especially in environments with little resources. For many zoonotic illnesses, medical and veterinary health organisations have a significant stake in disease surveillance and control operations due to the serious effects zoonotic diseases can have on both animals and people. Partnership between agencies is essential, but it takes time and requires careful planning and effective activity coordination. In many real-world scenarios, where resources are scarce, institutional cultures and preferences vary, legal permissions distinguish and numerous other considerations can hinder the creation of formal structures required to ensure the successful execution of disease control and prevention programmes, reaching this degree of collaboration can be challenging (Shiferaw et al., 2017). Zoonoses must be prevented and controlled by the harmonization of national and international laws and regulations as well as interagency collaboration under the sponsorship of international organizations like Food and Agricultural Organization, World Animal Health Organization and the WHO (Chomel, 2009).

Following actions can aid in zoonoses' prevention and management.

6.1 Reservoir Neutralization

The contaminated reservoir host serves as the principal site of zoonotic infection. Other sources of infection gradually decrease in prominence or eliminate when infection in the reservoirs can be controlled or eliminated. Three techniques are utilised to neutralize the reservoir, viz. eliminating the diseased person; rendering the infected person "non shedder;" and altering the surroundings. As happened with human smallpox vaccination, increasing host resistance can potentially neutralise a reservoir if the number of sick animals is decreased (Martin, 2007).

By testing and killing those who test positive for the disease, infection can be eradicated from the herd. This approach has proven to be most successful with agents that spread through direct contact and involve a small number of reservoir species. Its use is restricted in many nations because of the exorbitant expense to government and business (Schellenberg et al., 2003). Mass treatment is a second strategy for shrinking the reservoir by getting rid of contaminated host. Mass treatment, which is typically limited to a local setting, involves treating all possibly infectious animals or people without first doing any tests to identify those who are actually ill. Treatment without prior screening may lower the cost factor by 2–6 times in some circumstances, especially with endemic diseases in the less developed countries wherein diagnostic facilities may be limited (Dubal et al., 2014). Nevertheless, dangers of mass treatment include the emergence of infectious agents with heightened resistance and unfavourable side effects, especially if administered incorrectly (Acha & Szyfres, 2001). Environmental manipulation is a reservoir neutralisation technique meant to disrupt the transmission chain between portal of exit of the diseased (shedder) host and the susceptible host by lowering the agent's survival in the vector of vehicle (food, water, soil, plants) (Christian et al., 2021). Anywhere the agent may be detected other than the vertebrate host is considered the environment of concern. The fact that this measure is local and only effective in the vicinity of where it is implemented is a constraint. The number of mosquitoes has been reduced with some effectiveness utilising biological control methods such as natural predators or vector-borne illnesses. Examples include introducing Gambusia fish to eat mosquito larvae and using *Marisa cornuarietis*, a competitive snail, to oust *Australorbis* spp. snails, an intermediate host of Schistosomes, from an area (Jones et al., 2006).

6.2 Reducing Contact Potential

A fundamental premise in limiting direct transmission of a pathogens from an infected individual to a susceptible individual is to decrease the possibility of contact (Centers for Disease Control and Prevention [CDC], 2012). Two groups are taken into account in disease control, viz. identified diseased and the potentially vulnerable (Barreto et al., 2006). Three techniques used are: population control, quarantine of potential sick individuals and case isolation and treatment. While quarantine is meant to keep the agent out and be utilized for asymptomatic people, isolation is meant to keep the agent in and is used for symptomatic people (Qureshi, 2017). Isolating an infected, clinically unwell animal has two benefits: it lowers the likelihood that it may come into touch with a susceptible person, and makes treatment and disinfection easier (Weese, 2004). Due to major flaws and frequent failure, this strategy relies on early and accurate diagnosis from efficient disease control programmes. These programmes are based on isolating and treating cases to stop the spread of infectious disease agents (Reta et al., 2020). Additionally, if shedding happens during the incubation phase or if healthy carriers are shedding, the strategy is typically ineffective (Widders et al., 2020). The second plague pandemic marked the beginning of the practise of placing under quarantine for the people who appeared to be healthy but may have come into contact with an infectious source (WHO, 2020). To be effective, quarantine must be maintained for the duration of the disease's longest incubation period to check on suspected infected individuals for the development of apparent symptoms (Bikbov & Bikbov, 2021).

6.3 *Increasing Host Resistance*

Although preventing infection is ideal, there are numerous situations in which improving host resistance may merely reduce the severity of sickness without raising resistance to infection (Schneider & Ayres, 2008). Genetic selection for resistance and stress reduction through improved nutrition or better shelter are common practices in veterinary medicine (Parmar et al., 2017). Maintaining animals at a healthy nutritional level improves both their capacity to fight-off infection and their capacity to react appropriately to vaccinations (Katona & Katona-Apte, 2008). By improving nutrition and shelter, we can lessen stress, which serves as a means of reducing the effects of infections by improving the capacity for survival in the affected area (Madzingira, 2018).

6.4 *Chemoprophylaxis*

It aims to stop the spread of infection or at the very least lessen the severity of the illness (McBride, 2010). It is a passive method of building host resistance as opposed to immunisation, lasting only as long as the drug does (Weinstein, 1955). Furthermore, the duration of the active reaction to vaccination can be months or even a life-time. When workers in laboratories unintentionally come into contact with an agent known to be drug-sensitive, chemoprophylaxis is utilised (McBride, 2010). In contrast to mass treatment, which is intended to eradicate infection, this is done to avoid infection. Chemoprophylaxis is only used when there are no other more effective ways to protect the host.

6.5 *Immunization*

Vaccines are used to both limit the spread of infectious agents by building up an immune population and to protect individuals who are susceptible to illness or sickness. The vaccination stimulus must be sufficient to prevent both infection and disease in order to be most effective in reducing disease.

7 Conclusion

Climate change is among the biggest threat to our planet and humanity in the twenty-first century. Beyond temperature rises, alterations in rainfall patterns, increase in the frequency of extreme events, or adjustments to ecosystems, climate change has far-reaching effects. Climate change has a significant impact on human health in many ways, including the emergence of respiratory and cardiovascular issues, heat stroke, dehydration and mental health problems. In addition, zoonotic diseases are becoming a serious threat as a result of the drastic changes in ecosystems and the unusually close contact that humans have with other species. As a result, research into how climate change affects zoonoses is expanding quickly. Millions of people, particularly those in developing nations, are in danger both directly and indirectly. Undoubtedly, we could better prepare for zoonotic disease management as well as mitigation and adaptation strategies for climate change if we understood how climate change affects the emergence and distribution of zoonosis. More efforts should be put in the management and control of disease by public and government partnership.

References

Acha, P. N., & Szyfres, B. (2001). *Zoonoses and communicable diseases common to man and animals. Volume 3: Parasitoses* (Vol. 580). Pan American Health Org.

Akritidis, N. (2011). Parasitic, fungal and prion zoonoses: An expanding universe of candidates for human disease. *Clinical Microbiology and Infection, 17*(3), 331–335.

Barreto, M. L., Teixeira, M. G., & Carmo, E. H. (2006). Infectious diseases epidemiology. *Journal of Epidemiology and Community Health, 60*(3), 192–195. https://doi.org/10.1136/jech.2003.011593

Bikbov, B., & Bikbov, A. (2021). Maximum incubation period for COVID-19 infection: Do we need to rethink the 14-day quarantine policy? *Travel Medicine and Infectious Disease, 40*, Article 101976. https://doi.org/10.1016/j.tmaid.2021.101976

Bueno-Marí, R., Almeida, A. P. G., & Navarro, J. C. (2015). Emerging zoonoses: Eco-epidemiology, involved mechanisms, and public health implications. *Frontiers in Public Health, 3*, 157.

Cascio, A., Bosilkovski, M., Rodriguez-Morales, A. J., & Pappas, G. (2011). The socio-ecology of zoonotic infections. *Clinical Microbiology and Infection, 17*(3), 336–342.

Centers for Disease Control and Prevention [CDC]. (2012). Introduction to epidemiology. Retrieved from https://www.cdc.gov/csels/dsepd/ss1978/lesson1/section10.html

Chevalier, V., Rocque, S. D. L., Baldet, T., Vial, L., & Roger, F. (2004). Epidemiological processes involved in the emergence of vector-borne diseases: West Nile fever, Rift valley fever, Japanese encephalitis and Crimean-Congo haemorrhagic fever. *Revue Scientifique Et Technique-Office International des Epizooties, 23*(2), 535–556.

Chomel, B. B., Boulouis, H. J., Maruyama, S., & Breitschwerdt, E. B. (2006). *Bartonella* spp. in pets and effect on human health. *Emerging Infectious Diseases, 12*(3), 389.

Chomel, B. B. (2009). Zoonoses. *Encyclopedia of Microbiology*, 820–829. https://doi.org/10.1016/B978-012373944-5.00213-3

Christian, U. A., Kayode, O. I., Chinenye, U. C., & Sule, U. B. (2021). *Environmental manipulation: A potential tool for mosquito vector control. The wonders of diptera-characteristics, diversity, and significance for the world's ecosystems* (pp. 153–155).

Christou, L. (2011). The global burden of bacterial and viral zoonotic infections. *Clinical Microbiology and Infection, 17*(3), 326–330.

Cutler, S. J., Fooks, A. R., & Van der Poel, W. H. (2010). Public health threat of new, reemerging, and neglected zoonoses in the industrialized world. *Emerging Infectious Diseases, 16*(1), 1.

Darvishi, M., Yaghoobi-Ershadi, M. R., Shahbazi, F., Akhavan, A. A., Jafari, R., Soleimani, H., & Arandian, M. H. (2015). Epidemiological study on sand flies in an endemic focus of cutaneous leishmaniasis, Bushehr city, southwestern Iran. *Frontiers in Public Health, 3*, 14.

Dhiman, R. C. (2014). Emerging vector-borne zoonoses: Eco-epidemiology and public health implications in India. *Frontiers in Public Health, 2*, 168.

Dubal, Z. B., Barbuddhe, S. B., & Singh, N. P. (2014). Important zoonotic diseases: Prevention and control. *Technical Bulletin, 39*, 1–38.

El-Sayed, A., & Awad, W. (2018). Brucellosis: Evolution and expected comeback. *International Journal of Veterinary Science and Medicine, 6*, S31–S35.

Finol, H. J., & Roschman-González, A. (2014). Ultrastructural study on tissue alterations caused by trypanosomatids in experimental murine infections. *Frontiers in Public Health, 2*, 75.

García-Alzate, R., Lozano-Arias, D., Reyes-Lugo, R. M., Morocoima, A., Herrera, L., & Mendoza-León, A. (2014). Triatoma maculata, the vector of *Trypanosoma cruzi*, in Venezuela. Phenotypic and genotypic variability as potential indicator of vector displacement into the domestic habitat. *Frontiers in Public Health, 2*, 170.

Ghazali, D. A., Guericolas, M., Thys, F., Sarasin, F., Arcos Gonzalez, P., & Casalino, E. (2018). Climate change impacts on disaster and emergency medicine focusing on mitigation disruptive effects: An international perspective. *International Journal of Environmental Research and Public Health, 15*(7), 1379.

Githeko, A. K., Lindsay, S. W., Confalonieri, U. E., & Patz, J. A. (2000). Climate change and vector-borne diseases: A regional analysis. *Bulletin of the World Health Organization, 78*(9), 1136–1147.

Gonzalez-Britez, N. E., Carrasco, H. J., Martínez Purroy, C. E., Feliciangeli, M. D., Maldonado, M., López, E., & Rojas de Arias, A. (2014). Genetic and morphometric variability of *Triatoma sordida* (Hemiptera: Reduviidae) from the eastern and western regions of Paraguay. *Frontiers in Public Health, 2*, 149.

Gould, E. A., & Higgs, S. (2009). Impact of climate change and other factors on emerging arbovirus diseases. *Transactions of the Royal Society of Tropical Medicine and Hygiene, 103*(2), 109–121.

Greer, A., Ng, V., & Fisman, D. (2008). Climate change and infectious diseases in North America: The road ahead. *CMAJ, 178*(6), 715–722.

Hambling, T., Weinstein, P., & Slaney, D. (2011). A review of frameworks for developing environmental health indicators for climate change and health. *International Journal of Environmental Research and Public Health, 8*(7), 2854–2875.

Hernández-Triana, L. M., Jeffries, C. L., Mansfield, K. L., Carnell, G., Fooks, A. R., & Johnson, N. (2014). Emergence of West Nile virus lineage 2 in Europe: A review on the introduction and spread of a mosquito-borne disease. *Frontiers in Public Health, 2*, 271.

Himsworth, C. G., Jardine, C. M., Parsons, K. L., Feng, A. Y., & Patrick, D. M. (2014). The characteristics of wild rat (Rattus spp.) populations from an inner-city neighborhood with a focus on factors critical to the understanding of rat-associated zoonoses. *PloS One, 9*(3), e91654.

Jones, T. C., Hunt, R. C., & King, N. W. (2006). *Veterinary pathology* (6th ed., pp. 655–656). Blackwell Publishing.

Katona, P., & Katona-Apte, J. (2008). The interaction between nutrition and infection. *Clinical Infectious Diseases, 46*(10), 1582–1588.

Lafferty, K. D. (2009). The ecology of climate change and infectious diseases. *Ecology, 90*(4), 888–900.

Madzingira, O. (2018). Animal welfare considerations in food-producing animals. *Animal Welfare*, 99.

Martin, V., Chevalier, V., Ceccato, P., Anyamba, A., De Simone, L., Lubroth, J., de La Rocque, S., & Domenech, J. (2008). The impact of climate change on the epidemiology and control of Rift Valley fever. *Revue scientifique et technique (International Office of Epizootics), 27*(2), 413–426.

Martin, E. A. (ed.) (2007). *Oxford concise medical dictionary*. Oxford University Press.

McBride, W. J. (2010). Chemoprophylaxis of tropical infectious diseases. *Pharmaceuticals, 3*(5), 1561–1575. https://doi.org/10.3390/ph3051561

McPhaden, M. J., Zebiak, S. E., & Glantz, M. H. (2006). ENSO as an integrating concept in earth science. *Science, 314*(5806), 1740–1745.

Medone, P., Ceccarelli, S., Parham, P. E., Figuera, A., & Rabinovich, J. E. (2015). The impact of climate change on the geographical distribution of two vectors of Chagas disease: Implications for the force of infection. *Philosophical Transactions of the Royal Society b: Biological Sciences, 370*(1665), 20130560.

Meslin, F. X., Stöhr, K., & Heymann, D. (2000). Public health implications of emerging zoonoses. *Revue Scientifique et Technique (International Office of Epizootics), 19*(1), 310–317.

Metzgar, D., Baynes, D., Myers, C. A., Kammerer, P., Unabia, M., Faix, D. J., & Blair, P. J. (2010). Initial identification and characterization of an emerging zoonotic influenza virus prior to pandemic spread. *Journal of Clinical Microbiology, 48*(11), 4228–4234.

Mills, J. N., Gage, K. L., & Khan, A. S. (2010). Potential influence of climate change on vector-borne and zoonotic diseases: A review and proposed research plan. *Environmental Health Perspectives, 118*(11), 1507–1514.

Myaing, T. T. (2011). Climate change and emerging zoonotic diseases. *KKU Veterinary Journal, 21*(2), 172–182.

Neira, M. V., Mahmood, F., Reisen, W. K., James, C. B., & Romoser, W. S. (2014). Comparative study of the pathological effects of western equine encephalomyelitis virus in four strains of *Culex tarsalis* Coquillett (Diptera: Culicidae). *Frontiers in Public Health, 2*, 184.

Obiegala, A., Heuser, E., Ryll, R., Imholt, C., Fürst, J., Prautsch, L. M., & Pfeffer, M. (2019). Norway and black rats in Europe: Potential reservoirs for zoonotic arthropod-borne pathogens? *Pest Management Science, 75*(6), 1556–1563.

Ogunseitan, O. (2011). Climate change. In *Green health: An A-to-Z guide* (Vol. 1, pp. 112–115). SAGE Publications, Inc. https://doi.org/10.4135/9781412974592.n34

Pappas, G. (2011). Of mice and men: Defining, categorizing and understanding the significance of zoonotic infections. *Clinical Microbiology and Infection, 17*(3), 321–325.

Pappas, G., Cascio, A., & Rodriguez-Morales, A. J. (2012). The immunology of zoonotic infections. *Clinical and Developmental Immunology*.

Parmar, N., Singh, K. H., Sharma, D., Singh, L., Kumar, P., Nanjundan, J., Khan, Y. J., Chauhan, D. K., & Thakur, A. K. (2017). Genetic engineering strategies for biotic and abiotic stress tolerance and quality enhancement in horticultural crops: A comprehensive review. *Biotech, 7*(4), 239. https://doi.org/10.1007/s13205-017-0870-y

Pollard, A. J., & Bijker, E. M. (2021). A guide to vaccinology: From basic principles to new developments. *Nature Reviews Immunology, 21*(2), 83–100.

Poon, L. L. M., Guan, Y., Nicholls, J. M., Yuen, K. Y., & Peiris, J. S. M. (2004). The aetiology, origins, and diagnosis of severe acute respiratory syndrome. *The Lancet Infectious Diseases, 4*(11), 663–671.

Prow, N. A., Hewlett, E. K., Faddy, H. M., Coiacetto, F., Wang, W., Cox, T., & Bielefeldt-Ohmann, H. (2014). The Australian public is still vulnerable to emerging virulent strains of West Nile virus. *Frontiers in Public Health, 2*, 146.

Qureshi, A. I. (2017). *Zika virus disease: From origin to outbreak*. Academic Press.

Rahman, M. T., Sobur, M. A., Islam, M. S., Ievy, S., Hossain, M. J., El Zowalaty, M. E., & Ashour, H. M. (2020). Zoonotic diseases: Etiology, impact, and control. *Microorganisms, 8*(9), 1405.

Ramasamy, R. (2014). Zoonotic malaria—Global overview and research and policy needs. *Frontiers in Public Health, 2*, 123.

Randolph, S. E. (2008). Tick-borne disease systems. *Rev. Sci. Tech. off. Int. Epiz., 27*(2), 1–15.

Ready, P. D. (2010). Leishmaniasis emergence in Europe. *Eurosurveillance, 15*(10), 19505.

Reiter, P. (2008). Climate change and mosquito-borne disease: Knowing the horse before hitching the cart. *Revue Scientifique et Technique (International Office of Epizootics), 27*(2), 383–398.

Resh, V. H., & Cardé, R. T. (eds.) (2009). *Encyclopedia of insects*. Academic Press.

Resnick, B. (2019). Melting permafrost in the Arctic is unlocking diseases and warping the landscape. Retrieved from https://www.vox.com/2017/9/6/16062174/permafrost-melting

Reta, D. H., Tessema, T. S., Ashenef, A. S., Desta, A. F., Labisso, W. L., Gizaw, S. T., & Reta, F. A. (2020). Molecular and immunological diagnostic techniques of medical viruses. *International Journal of Microbiology*.

Rezza, G., Nicoletti, L., Angelini, R., Romi, R., Finarelli, A. C., Panning, M., & CHIKV Study Group. (2007). Infection with chikungunya virus in Italy: An outbreak in a temperate region. *The Lancet, 370*(9602), 1840–1846.

Rothenburger, J. L., Himsworth, C. G., Nemeth, N. M., Pearl, D. L., Treuting, P. M., & Jardine, C. M. (2019). The devil is in the details—Host disease and co-infections are associated with zoonotic pathogen carriage in Norway rats (*Rattus norvegicus*). *Zoonoses and Public Health, 66*(6), 622–635.

Schaechter, M. (2009). *Encyclopedia of microbiology*. Academic Press.

Schellenberg, R. S., Tan, B. J., Irvine, J. D., Stockdale, D. R., Gajadhar, A. A., Serhir, B., Botha, J., Armstrong, C. A., Woods, S. A., Blondeau, J. M., & McNab, T. L. (2003). An outbreak of trichinellosis due to consumption of bear meat infected with *Trichinella nativa*, in 2 northern Saskatchewan communities. *The Journal of Infectious Diseases, 188*(6), 835–843. https://doi.org/10.1086/378094

Schlundt, J., Toyofuku, H., Jansen, J., & Herbst, S. A. (2004). Emerging food-borne zoonoses. *Revue Scientifique Et Technique-Office International des Epizooties, 23*(2), 513–534.

Schneider, D. S., & Ayres, J. S. (2008). Two ways to survive infection: What resistance and tolerance can teach us about treating infectious diseases: Nature reviews. *Immunology, 8*(11), 889–895. https://doi.org/10.1038/nri2432

Sejvar, J., Bancroft, E., Winthrop, K., Bettinger, J., Bajani, M., Bragg, S., & Rosenstein, N. (2003). Leptospirosis in "eco-challenge" athletes, Malaysian Borneo, 2000. *Emerging Infectious Diseases, 9*(6), 702.

Shiferaw, M. L., Doty, J. B., Maghlakelidze, G., Morgan, J., Khmaladze, E., Parkadze, O., & Reynolds, M. G. (2017). Frameworks for preventing, detecting, and controlling zoonotic diseases. *Emerging Infectious Diseases, 23*(Suppl 1), S71.

Singh, B., Sung, L. K., Matusop, A., Radhakrishnan, A., Shamsul, S. S., Cox-Singh, J., & Conway, D. J. (2004). A large focus of naturally acquired *Plasmodium knowlesi* infections in human beings. *The Lancet, 363*(9414), 1017–1024.

Tayde, R.S. (n.d.). Classification of Zoonoses. Retrieved from https://www.ndvsu.org/images/StudyMaterials/VPH/Zoonoses-Classification.pdf

Team, C. W., Pachauri, R. K., & Meyer, L. A. (2014). *Climate change 2014: Synthesis report. Contribution of working groups I, II and III to the fifth assessment report of the intergovernmental panel on climate change* (p. 151).

Teklehaimanot, H. D., Schwartz, J., Teklehaimanot, A., & Lipsitch, M. (2004). Alert threshold algorithms and malaria epidemic detection. *Emerging Infectious Diseases, 10*(7), 1220.

Thompson, A. A., Matamale, L., & Kharidza, S. D. (2012). Impact of climate change on children's health in Limpopo Province, South Africa. *International Journal of Environmental Research and Public Health, 9*(3), 831–854.

Tulchinsky, T. H., & Varavikova, E. A. (2014). Communicable diseases. *The New Public Health, 149*.

United Nations Framework Convention on Climate Change [UNFCCC]. (n.d.). Article 1 definitions. Retrieved from https://unfccc.int/resource/ccsites/zimbab/conven/text/art01.htm#:~:text=2.,observed%20over%20comparable%20time%20periods

Vora, N. (2008). Impact of anthropogenic environmental alterations on vector-borne diseases. *The Medscape Journal of Medicine, 10*(10), 238.

Voss, A., Loeffen, F., Bakker, J., Klaassen, C., & Wulf, M. (2005). Methicillin-resistant *Staphylococcus aureus* in pig farming. *Emerging Infectious Diseases, 11*(12), 1965.

Walsh, M. G., de Smalen, A. W., & Mor, S. M. (2018). Climatic influence on anthrax suitability in warming northern latitudes. *Scientific Reports, 8*(1), 1–9.

Wang, G., Minnis, R. B., Belant, J. L., & Wax, C. L. (2010). Dry weather induces outbreaks of human West Nile virus infections. *BMC Infectious Diseases, 10*(1), 1–7.

Weese, J. S. (2004). Barrier precautions, isolation protocols, and personal hygiene in veterinary hospitals. The veterinary clinics of North America. *Equine Practice, 20*(3), 543–559. https://doi.org/10.1016/j.cveq.2004.07.006

Weinstein, L. (1955). The chemoprophylaxis of infection. *Annals of Internal Medicine, 43*(2), 287–298.

Widders, A., Broom, A., & Broom, J. (2020). SARS-CoV-2: The viral shedding vs infectivity dilemma. *Infection, Disease & Health, 25*(3), 210–215. https://doi.org/10.1016/j.idh.2020.05.002

Wilder-Smith, A. (2006). The severe acute respiratory syndrome: Impact on travel and tourism. *Travel Medicine and Infectious Disease, 4*(2), 53–60.

Wolfe, N. D., Daszak, P., Kilpatrick, A. M., & Burke, D. S. (2005). Bushmeat hunting, deforestation, and prediction of zoonotic disease. *Emerging Infectious Diseases, 11*(12), 1822.

World Health Organization [WHO]. Considerations for quarantine of individuals in the context of containment for Coronavirus disease (COVID-19): Interim guidance, 19 March 2020. World Health Organization. Available online: https://apps.who.int/iris/bitstream/handle/10665/331497/WHO-2019-nCoV-IHR_Quarantine-2020.2-eng.pdf. Retrieved from 3 Feb 2021.

Chapter 4
Plague

1 Introduction

Yersinia pestis, a highly pathogenic bacterium causes an acute infectious zoonotic disease called plague (Wong et al., 2009). The occurrence of plague is significantly high in places with poor sanitation, heavy population density and a greater rodent population (Miarinjara et al., 2017). Even though person-to-person spread is possible if the pathogens are breathed, fleas called "*Xenopsylla cheopis*" that live on affected rodents are the main source of human outbreaks (Dean et al., 2018). Patients often experience an abrupt onset of fever, shivering, headaches, body pains, fatigue, diarrhea and nausea after an incubation time of 3–7 days (Riedel, 2005). Depending on the manner that *Y. pestis* infects a person, there are three different ways that the infection might be present: bubonic, septicaemic and pneumonic (Erova et al., 2013). One of most typical kind, caused by the bite of an infected flea, is bubonic (Stenseth et al., 2008). In majority of the cases of human plague, the disease first manifested as massive, pus-filled lymph node, or bubo (L. bubo = "groin," Gr. Boubon = "groin swelling"). Additionally, the bubonic plague carries military implications and is classified as a Category A bioterrorism weapon by the Centers for Disease Control and Prevention (Frith, 2012). Septicaemic plague, triggered by the presence of plague microbes in the bloodstream, results in rapid skin necrosis and haemorrhaging, accompanied by septic shock and mortality (Inglesby et al., 2000). Moreover, if the illness spreads to the lungs through the blood, it results in pneumonia, which is lethal, and thereafter causes pneumonic plague. This type can be spread directly from person-to-person by inhaling respiratory droplets that are contaminated (Meyer, 1961). Moreover, the mortality rate for bubonic plague is substantially lower (40–60%) than that for septicemic or pneumonic plague—both of which have mortality rates above 90% (Butler, 2012; Godfred-Cato et al., 2020).

Human civilization have experienced three significant plague pandemics. In the sixth century AD, the Justinian plague struck the Mediterranean region and is thought to have killed 50 million people. Again in fourteenth century, the Black Death claimed

M. H. Fulekar et al., *Pandemic Over the Centuries*,
https://doi.org/10.1007/978-3-031-94270-9_4

1/3rd of the European population. The most recent pandemic, which started in China in the late 1800s and spread across the world, is thought to have killed 12 million people by the year 1930 (Drancourt & Raoult, 2014). Since the rodents and people travelled across trade networks on land and sea during the pandemics, the infectious fleas were capable of spreading the plague over distant locations. They were also able to contaminate rice and wheat grains, clothes and other commodities. Following infection, an accumulation of bacteria obstructs the flea's proventriculus (foregut). The flea continues to feed while biting more frequently and viciously. In an effort to clear the blockage, the flea regurgitates a large amount of *Y. pestis* into the host's bloodstream along with the accumulated blood. The fleas proliferate on their host and leave quickly after the host dies, contaminating new hosts while establishing the conditions for an outbreak (Dewitte et al., 2020).

The plague bacterium still exists in wild animal reservoirs in certain regions of the world, and periodically escapes from such reservoirs to infect humans and other animals. The World Health Organization (WHO) receives reports of between 1,000 and 2,000 cases annually from all over the world, but the actual number is likely to be much greater because many infections are poorly detected (Centers for Disease Control and Prevention [CDC], 2021a). Although sporadic cases and epidemics can be found in any zone, however, Asia and Africa see the majority of episodes (CDC, 2021b). Furthermore, medical professionals and administrators must comprehend its historical background, clinical symptoms, epidemiology, and available treatments since it could be militarized for international bioterrorism (Das & Kataria, 2010; Glatter & Finkelman, 2021).

2 Three Pandemics of Plague: Historical Account

The Justinian Plague, which occurred between 541 and 544 AD, the European "Black Death, between 1347 and 1352, and the Third Pandemic of 1894 are all well-known historical outbreaks of the plague (Benedictow, 2021). Additionally, there are accounts of earlier outbreaks of plague, such as the Plague of Athens (430–427 BCE) and the Antonine Plague (165–180 CE), though these accounts are debated owing to an insufficiency of genetic evidence (Ansari et al., 2020).

2.1 The Justinian Plague

The Justinian Plague, bearing the name of Justinian I, the Roman Emperor of the Byzantine Empire at the time, started in Pelusium (Egypt) after travelling from Ethiopia. It lasted from 541 to 544 AD (Yavari et al., 2004). Bubonic plague, as determined by DNA testing of bones discovered in cemeteries, ravaged the civilization (Horgan, 2014). The black rat (*Rattus rattus*), which sailed on the grain ships and carts carried to Constantinople, served as the disease's primary carrier (McCormick,

Fig. 1 The Plague of Justinian. (*Source* https://www.sutori.com/en/story/justinian-plague-WddydjUEjVWc7PL7zEPSXu1Y)

2003). The grain, which was kept in big warehouses, served as the ideal habitat for rats and fleas, which are essential to the spread of the plague (Fig. 1).

Procopius of Caesarea, a Byzantium court chronicler, wrote about victims experiencing fevers; groyne, armpit and behind-the-ear swellings; hallucinations, and nightmares in his book "History of the Wars". According to Procopius, some patients went into comas while others had severe hallucinations. Many victims endured excruciating pain for days before passing away, while others passed away quite suddenly (Frith, 2012). He also noted the climatic changes that were occurring in southern Italy at the time, including uncommon occurrences of snow and ice during the summer, below-average temperatures, as well as less sunshine. Hence, commenced a lengthy cold spell that was marked by social unrest, war and the first known plague outbreak. The region's population was forced to move because of food shortages brought on by the colder-than-normal weather's impact on crop yields. Accompanying these reluctant migrants were plague infected fleas and rodents. The ideal conditions for a pandemic were produced by a combination of cold, exhausted and hungry people on the move, sickness and disease in the middle of battle, as well as an increasing rat population spreading a highly contagious disease. After becoming ill, people had two options, viz. seek medical attention or treat themselves at home. Lacking access to medical professionals, diseased people frequently used home cures, such as cold-water showers, powders "blessed" by monks, magical ornaments and other medications, particularly alkaloids. When all other forms of treatment at home failed, patients went to hospitals or were placed under quarantine (Horgan, 2014).

Additionally, the pandemic had a significant impact and radically changed western society. A huge disruption in food production led to an eight-year famine (Frith, 2012). The pandemic's social and economic upheaval signified the end of Roman rule and sparked the emergence of culturally different sociocultural groups that later led to the rise of mediaeval European states (Meier, 2016). The plague remained aggressive for about two centuries, until around 750 AD, killing over 50–100 million people and necessitating a long time to recuperate (Harper, 2017; Stone & Ozga, 2019; Wagner et al., 2014).

2.2 'Black Death' of Europe

The Black Death, also known as The Great Pestilence in various parts of Europe, was the second major pandemic (Ansari et al., 2020). Since it might turn the skin and lesions black, the infection received the moniker "Black Death." Buboes which were typically seen during the outbreak have been the term for the dark lesions that might appear anywhere on the body and were brought on by internal haemorrhages. Joint discomfort and a high fever were other symptoms (Cantor, 2001). Much likely from Central Asia via Genoese ships transporting flea-ridden rats, the disease is thought to have first arrived at Messina (Sicily) in October 1347. From there, it raged like a merciless wildfire across the majority of Europe (Glatter & Finkelman, 2021). In 1348, the plague began to spread across France and Spain before moving on to Germany, Austria and Switzerland. By 1350, it had ravaged northern England and Scandinavia after decimating London in 1349 (Bramanti et al., 2016; Riedel, 2005). In Europe, plague victims frequently passed away a week after becoming ill. In Naples and Genoa, two-third of the population perished from the illness in 1656–1657. Around 100,000 Britons perished from the plague in 1665–1666, and the same number died in Vienna in 1679 (Cohn & Alfani, 2007; Cohn & Cohn, 2002). In the end, the Black Death claimed at least 25 million lives in Europe between 1347 and 1352, accounting for at least one-third of the continent's population (Glatter & Finkelman, 2021). During the later years of the century, the plague was eliminated, however, it re-appeared as minor outbreak in the following 400 years across Europe.

A 100 year war between England and France, which had a devastating effect on Italy, Germany and Spain, began at the same time the disease first appeared. Given the enormous disruption brought about by moving troops and the forced displacement of the civilians that typified English tactics, population decline and economic collapse were further exacerbated in places that had already been severely affected by the plague (Jedwab et al., 2022). Social chaos was the plague's following consequence. One manifestation of this was the flagellant movement. Flagellants would move from town-to-town beating themselves in order to make amends for the sins they believed to be the main reason of the plague. The flagellant movement was seen by the political establishment as a catalyst for civil unrest, and Roman Emperor Charles V unsuccessfully urged the church to denounce it (NA, 2016). Third, the targetting of Jews as scapegoats was an immediate and well-researched effect of the disease. From 12th century onwards, ethnic cleansings, expulsions and acts of violence against Jews had become more frequent. However, during the Black Death, Europe experienced its worst-ever oppression. At least 363 cities in Europe had Jewish population on the advent of the Black Death. Half of these villages experienced the Black Death and were either slaughtered or forcefully expelled from their lands. Jews were held responsible for transmitting the disease, and communities that had long faced antisemitic prejudice were expropriated as a result of the upheaval and horror of the plague (Anderson et al., 2017) (Fig. 2).

Medieval practitioners were incapable to treat patients because they were unaware of microscopic creatures like bacteria, as well as being impeded in their efforts

Fig. 2 Triumph of Black Death. (*Source* https://schoolshistory.org.uk/topics/medicine-through-time/the-triumph-of-death-image-analysis-on-the-black-death/)

to avoid infections by the appalling state of sanitation in comparison to modern standards. But "plague doctors" used protective clothes that included a long cape, a mask and a bill-like component over the mouth and nose that was covered in aromatic substances—possibly an early iteration of the present hazmat suit—in part to block-off the foul smell of decomposing bodies (Riedel, 2005) (Fig. 3).

Fig. 3 Costume of Plague doctor. (*Courtesy* National Library of Medicine)

Public baths were closed because people believed that warm baths allowed plague miasma to enter human pores. Cats were subjected to execution as potential transmission agents, and victims' clothing and belongings believed to be tainted were burnt (Glatter & Finkelman, 2021). Physicians' first line of defence during the Black Death included lighting fires in the streets and homes, employing aromatic chemicals to combat stinking exhalations, creating barriers and blood-letting (Mullett, 1936). A further beneficial tactic would have been to quarantine areas, but because people rushed away in fear whenever a case of plague appeared, they inadvertently spread the illness with them and the rats took care of the remainder (Cartwright, 2023). Carts heaped up with bodies became a frequent sight throughout Europe as a result of the authorities' lack of knowledge regarding how to dispose the bodies and the overwhelming number of plague victims (Gottfried, 2010). It seems that the only option left was to remain in place, stay away from others and wait.

2.3 Third Pandemic of 1894

The third and final plague pandemic began in Yunnan Province in southwest China around 1855 and extended to Taiwan. It struck Canton in 1894, killing 70,000 people, and then showed up in Hong Kong the following year (Kohn, 2007). Between 1910 and 1920, ships transported it to Japan, India, Australia and North and South America. In India, the plague is believed to have killed 12 million people between 1898 and 1918 (Arnold & Peckham, 2015; Glatter & Finkelman, 2021). Major scientific discoveries were witnessed during this plague. The pathogen *Y. pestis* was first identified in 1894 by Alexandre Yersin in Hong Kong, and the disease's primary host, the brown rat, and its vector, the rat flea, were both identified in 1898 by Paul-Louis Simond in Karachi (Frith, 2012).

In the early stages of illness, the tongue typically swelled, turned bright-red at the tip and borders, and was coated in a furry, white–grey covering. The fur typically turned brownish or black on the second or third day after contracting the illness and formed a crust. In the third week of the illness, the tongue began to crack and fissure, quickly resembling the typhus or enteric fever. The mucosal membrane of the mouth and the pharynx was typically bright-red, and the lips quickly became dry and frequently fissured. The appetite also was suppressed. An area of the lower abdomen experienced pain and frequent, uncontrollable vomiting and intense thirst was noticed. Quite often the vomit was watery, other times it was bilious. Both, early in the disease and again in its later stages, diarrhoea was common. The stools frequently contained blood, mucus and epithelium. Additionally, during the second day, painful buboes or swellings made appearances in the armpits, neck or groyne (Pryor, 1975).

Unwavering attempts were made to identify the suspected cause, which was widely supposed to be unsanitary lifestyles. Chinese authorities developed hospitals called the "Hygeia," at the Kennedy Town Police Station and the Kennedy Town Glass Works. The first two hospitals were staffed by Europeans, while the third by

Chinese Tung Wah hospital employees (Pryor, 1975). Personnel from the garrison conducted a house-to-house check, and those where the plague had spread were cleaned and disinfected. Regulations were created mandating that all cases of plague be reported to the closest police station. Failure to comply with this requirement resulted in $25 fine at that time. Apparently, there were few instances where this regulation was rarely followed. In order to evade detection and subsequently sterilization, the inhabitants of contaminated households threw corpses at night. Cemetries were established, each with a designated individual in-charge of transporting and interring the plague deceased (Pryor, 1975).

Given the growing numbers, the scientists and researchers focused on the prevention of major diseases and the growing understanding of germ theory in the early 1900s, there was not much the worldwide institutions could do in terms of preventing a pandemic of the plague aside from establishing standardised protocols. Two conferences, first in Venice and the second in Paris, known as the International Sanitary Conferences, were conducted in 1897 and 1903 to help address the threat of future breakouts of the bubonic plague. The Office International D'hygiene Publique (OIHP) in Paris, one of the key forerunners of the League of Nations Health Organisation, oversaw the creation of an international disease convention from these conventions. The standard protocols for treating the bubonic plague throughout the early 1900s were developed from these standards (Echenberg, 2002). The three I's, viz. isolation, incineration and inoculation, generally summed up these old practises. The use of incineration was a disease control procedure that was most specifically utilised to treat the bubonic plague, despite the fact that isolation is a standard protocol for many existing disease outbreaks. Burning was frequently employed to combat the plague because it was thought to be the most efficient way to get rid of disease from areas where the infected lived. Unfortunately, it was also a troublesome approach since it caused a number of out-of-control flames that wreaked havoc on communities, the worst of which was the big fire in Honolulu, which obliterated the city's Chinatown community (Echenberg, 2002). Inoculation involved the use of plague vaccines, of which many were developed and some even proved to be incredibly efficient when used in India, where it was noted that they had an efficacy rate of over 50% (Hawgood, 2007).

3 Types of Plague

The plague is a deadly clinical infectious disease that, if not identified and treated promptly, can proceed fast to death. The patient's exposure to the plague bacterium determines the extent of the symptoms. Bubonic, pneumonic and septicemic plague are the most prevalent clinical manifestations of the disease. Following discusses these forms.

3.1 Bubonic Plague

Currently, bubonic form of the plague is the most prevalent, and it is easily recognised by patients and physicians due to its characteristic clinical presentation, particularly in areas where it is endemic (Andersson, 2017). When a flea carrying *Y. pestis* bites a host that has previously consumed an infected host, such as a rat, it spreads the plague (Prentice & Rahalison, 2007). *Y. pestis* builds a biofilm inside the flea that prevents entry to the flea's midgut. As a response, the flea regurgitates the obstructing biofilm into the host body while feeding, infecting the patient's skin with hundreds of microorganisms (Pechous et al., 2016). Bubonic plague is brought by skin contact and is distinguished by the onset of high fever, shivering, fatigue and headache all at same time (Dennis & Mead, 2010). The pathogens left behind from the flea bite move quickly to a local lymph node and multiply throughout the course of 2–6 day incubation period (Narayanan et al., 2018). A identifiable and excruciatingly painful bubo develops from the substantial bacterial multiplication and inflammation in a day after the onset of acute fever and chills. They are phagocytosed inside the lymph node but manage to avoid being destroyed, leading to necrosis of the lymph node's building structure. The buboes, which can be between 1 and 10 cm in diameter, are frequently so painful that they entirely impede mobility of the affected limb (Inglesby et al., 2000). Bubonic plague mortality can reach 60% if left untreated, however, it can be significantly decreased to less than 5% if caught early and treated with potent antimicrobial medication. In untreated, septicemia and subsequent pneumonic plague may develop (Narayanan et al., 2018).

3.2 Septicemic Plague

Flea bites into the capillaries or pathogens that skip the local lymph node and grow in the blood directly causing primary septicemic plague (Andersson, 2017; Pechous et al., 2016; Perry & Fetherston, 1997). Secondary septicemic plague develops when a concentrate of bacteria in the body spreads, such as haematological spread from a bubo in bubonic plague or from pneumonic plague (Pechous et al., 2016). Primary septicemic plague is the second most prevalent plague, characterised by the absence of palpable lymph nodes despite the presence of bacteremia (Yang, 2018). With high fever, rapidly increasing sepsis and multiorgan failure due to hypotension and shock, the disease can be lethal within few days (Inglesby et al., 2000). Once the infection has progressed to its final stages, vasculitis and disseminated intravascular coagulation result in gangrene, especially in the limbs, necessitating amputation. Patients turn black and quickly die giving "The Black Death" its moniker (Perry & Fetherston, 1997). Without proper care, septicemic plague inevitably results in death, with a 100% mortality rate. Even with treatment, the death rate might be as high as 30–50% (Narayanan et al., 2018) (Fig. 4).

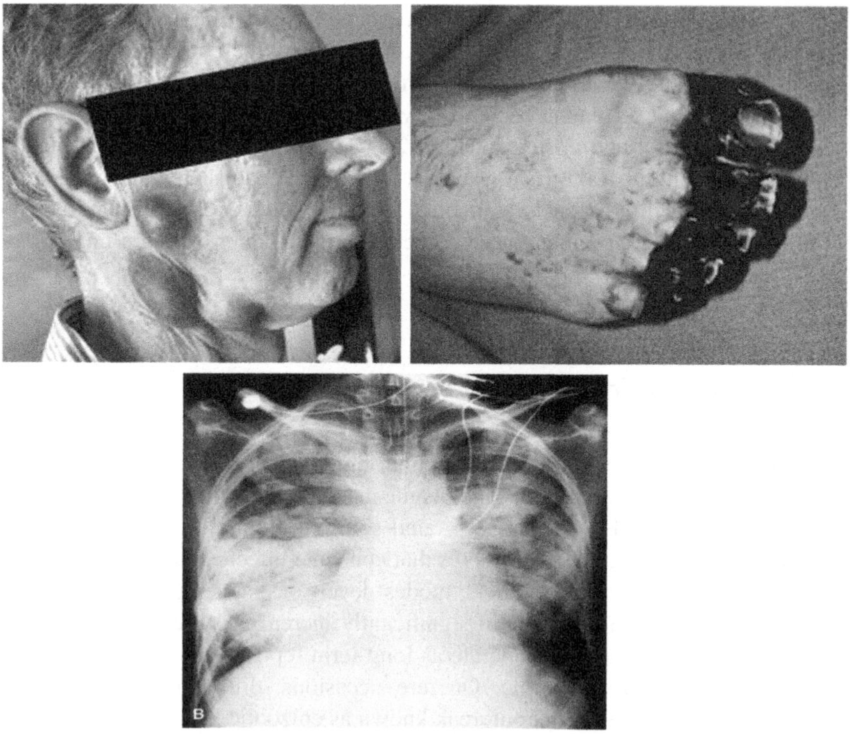

Fig. 4 Pneumonic plague (*Source* https://info.isradiology.org/tropical_deseases/tmcr/chapter24/radiological4.htm)

3.3 *Pneumonic Plague*

The third most prevalent and virulent form of plague is pneumonic plague that represents the most lethal presentation of the disease (Pechous et al., 2016). It is vital to note that human-to-human transmission, which accounts for the majority of reports of the primary pneumonic plague, results from the direct inhalation of *Y. pestis* aerosolized droplets, which are derived from ejected respiratory droplets of mammals (Doll et al., 1994; Wang et al., 2011). Secondary pneumonic plague is caused by the haematological transfer of pathogens from bubonic or septicemic plague to the lungs, and might be the first stage in the transmission of primary pneumonic plague in a community (Rabaan et al., 2019). The symptoms and signs of this disease, which manifests acutely within 1–4 days of inoculation and quickly advances, are non-specific in both humans and animals (Layton et al., 2011). The early infection shares many symptoms with influenza, including fever, headache, cold and fatigue (Andersson, 2017; Layton et al., 2011). In chest x-ray, severe lower respiratory tract symptoms such as dyspnea, tachypnea, coughing and hemoptysis can be seen within a day. Bilateral pulmonary infiltrates also evolve rapidly (Narayanan et al., 2018).

The initial purulent sputum may become acutely hemorrhagic or tainted with blood (Kwon et al., 2018). As the disease progresses, the cough becomes more copious and patients eventually cough up massive bright-red phlegm that contains "an infinite amount of plague bacteria in virtually pure culture" (Li et al., 2021). If antibiotics are not administered within 24 h in the case of pneumonic plague, fatality rates might hit 100% (Narayanan et al., 2018; Pechous et al., 2016). Due to the severity of infection, even with effective antibiotic therapy, up to 50% of patients could eventually die, and many need to be admitted in intensive care unit (Yang, 2018).

4 Ecology and Transmission

Yersinia pestis continues to exist in the environment through a cycle that includes rats and fleas (Eisen et al., 2021). The plague can impact a wide variety of species, including rock squirrels, wood rats, ground squirrels, prairie dogs, chimpanzees, mice, voles and rabbits. Wild carnivores that consume infected animals can contact the disease. According to researchers, modest levels of plague bacteria are present in some rodent populations without significantly increasing rodent mortality. The fleas of these afflicted animals operate as long-term repositories for the bacterium and is known as enzootic cycle. On rare occasions, different species become affected, resulting in an animal outbreak known as epizootic. Humans are typically more vulnerable during or immediately following a plague epizootic. According to research, the southwest of the United States has more epizootics in the cooler summers that follow wet winters. Epizootics are more likely to occur where a variety of rodent species coexist in dense populations over a range of habitats. In some parts of the developing world, urban centres with significant rat infestations might occasionally experience plague. Los Angeles experienced the last urban outbreak of the rat-associated plague in the USA between 1924 and 1925 (CDC, 2019).

The plague bacteria can infect humans as follows: Flea bites mostly infect the host with pathogens. During plague epizootics, several rodents perish, prompting fleas to search for alternative sources of blood. Flea bites can spread the plague to people and animals who roam in areas where rodents have recently perished from the disease. Infected fleas from dogs and cats may potentially enter the house. Exposure to flea bites can cause primary septicemic or bubonic plague. Contact with infected fluid or tissue might be a second reason of infection. By contacting body fluids or tissue of an animal that has plague, humans can be infected. For instance, a hunter who does not follow necessary procedures when skinning a rabbit or other infected animal, risks contacting the plague bacteria. The plagues that are most frequently brought on by this type of exposure are bubonic or septicemic. The infectious droplets or aerosols may act as another mode of transmission. People who have plague pneumonia may cough up airborne droplets of the plague bacteria. These bacteria-filled droplets can spread pneumonic plague if they are inhaled by others.

This usually necessitates close proximity to the pneumonic plague patient (CDC, 2019).

5 Prevention and Control Strategies

Following are the measures of plague long-term prevention and control: surveillance, rodent control and vector control (WHO, 2010).

5.1 Surveillance

Spread of the plague is not steady; instead, it is caused by a colony infection in wild rodents that eventually infects commensal rodents in their route (Biswas et al., 2020). Surveillance is the systematic collection of data for action that is constant and continuing. It is essential to determine local ecological factors or human activities that could raise the likelihood of human exposure to the plague and also to monitor patterns in the epidemiology of plague in a particular region. It is also necessary to incorporate pertinent and timely control measures, evaluate the impact of prevention strategies, and ensure early outbreak containment (Thacker & Berkelman, 1988). Therefore, as part of epidemic readiness and early warning systems, operational guidelines on plague surveillance, diagnosis, prevention and control should be established (WHO, 2010).

5.2 Rodent Control

The population of wild rodents is impossible to entirely control, and the fleas that feed on them are equally difficult to manage. Managing rodent populations in urban and rural areas is the foundation of prevention. Only during interepidemic periods can rodent control measures be taken. When rodents are killed during an epidemic, lot of fleas escape from the dead mice and bite people, spreading the disease. Therefore, it is never advisable to implement rodent control measures during outbreaks.

Following techniques are used to control rodents.

5.2.1 Environmental Sanitation

Rat-proof food storage, correct drainage system, proper garbage disposal, and rat-proofing of building structures are all part of the environmental control of household rodents.

5.2.2 Physical Methods

By erecting fences surrounding crops, a trap barrier system (TBS) employs physical control of field rats. Fencing around trap crops planted prior to the main crop improves this strategy, which is referred to as eco-friendly. The rodents are drawn to the trap crop from the surrounding areas and are captured in large numbers.

5.2.3 Chemical Methods

Chronic or multiple-dose poisons take longer to effect and may need to be administered with several meals over few days. The anticoagulants are the poisons that are currently and often utilised. Warfarin is the most well-known, although there are others that also work.

5.2.4 Biological Methods

Biocontrol potential exists for pathogenic bacteria, viruses, protozoans as Microgen, and helminths, worms, and arthropodes as microparasites.

5.3 Vector Control

Following circumstances call for taking flea control measures, viz. when a locality reports of rat population decline or an increase in flea population, active surveillance in areas with known natural plague foci shows a Specific Flea Index to be greater than 1.0.

Upon receiving detailed information about a flea infestation from state or federal government officials, following actions are advised for preventing the spread of the plague.

5.3.1 Personal Prophylactic Measures

It incorporates the utilization of repellents like benzylbenzoate, diethyltolumide (DEET), dimethylphtholate (DMP) on body or on clothing to prevent flea bites. Use of high-necked footwears or socks upto the knees and sleeping on beds at least 0.5 m above ground help in plague prevention.

5.3.2 Insufflation

The method of treating rat runs and rodent burrows with 5% Malathion dust powder or 10% DDT.

5.3.3 Residual Insecticide Spray

The insecticides Malathion, deltamethrin, cyfluthrin and lambdacyhalothrin are used for residual vector control.

5.3.4 Health Education and Community Participation

People who live in endemic areas should receive health education about the relevance of rat falls, handling rat remains, mechanisms of infection, symptoms of the plague, and the significance of immediately reporting any suspected case to the nearest medical facility. Additionally, they must be educated on prevalent pandemic beliefs and misunderstandings. The public needs to be made aware about plague management and prevention by organizing discussions and involving the media, public relations firms, educational institutions, community-based organisations, and nongovernmental organisations (NGOs) to promote health education.

6 Diagnosis and Treatment

Any patient who exhibits clinical symptoms of the disease and has lived in or travelled to a plague-affected area should be evaluated for the disease. The most frequent primary symptom is bubonic plague, with a bubo typically appearing in the groyne, axilla, or cervical nodes. The bubonic plague takes 2–8 days to incubate. If the bubonic plague is left untreated, *Y. pestis* can enter the bloodstream, spread quickly, and cause sepsis; if it seeds the lungs, it can also cause pneumonia. Pneumonia and sepsis can also be first signs of plague. While the incubation period for pneumonic plague is often approx. 1–3 days, the incubation period for septicemic plague is poorly established but presumably occurs within days after exposure. Pneumonic plague symptoms include high fever, chills, coughing blood, dyspnea and dyspnea. Patients with pneumonic or septicemic plague risk early death if proper antibiotic therapy is not administered. Even while a bubo is the most common plague sign, some patients may also have other symptoms. For instance, the gastrointestinal symptoms of septicemic plague, which include nausea, vomiting, diarrhoea and abdominal pain, can be very noticeable (CDC, 2006). Pharyngeal, meningeal and cutaneous plagues are also uncommon types. Blood cultures, lymph node aspirates, if practical and/or sputum, if necessary, are acceptable diagnostic samples. Local and state health departments should be informed immediately if plague is detected. To lessen the potential of plague transmission from person-to-person, the patient should also be separated and put on droplet precautions if they exhibit symptoms of pneumonic plague (CDC, 2022).

6.1 Clinical Diagnosis

A suspected, presumptive, or confirmed case of plague is what the WHO defines in its 2006 recommendation. A visit to or habitation in a region where the disease is endemic within 10 days of the commencement of symptoms, contact with plague patients or infected animals, and/or a clear history of flea bites should always be present alongside clinical symptoms and signs in a suspected case. The aforementioned clinical and epidemiological signs should be watched for in a presumed case. If the patient also fulfils the requirements, viz. Giemsa or Wayson staining of Gram-negative coccobacilli in samples from bubo aspirate, blood or sputum, with a bipolar appearance evocative of a safety pin; detection of F1 antigen from sputum, bubo aspirate or blood; detection of serum anti-F1 antibody without a history of prior plague infection or immunisation; and PCR detection of Y Having two positive results from the aforementioned laboratory tests should be regarded evidence of infection in a new or reemerging plague focus if the patient is not from an area where the plague is endemic. One positive test can identify the patient as a presumptive case if they come from an area where the plague is endemic. The patient must meet the following requirements for a confirmed case in addition to those for a suspected case: *Y. pestis* isolated from bubo aspirate, blood or sputum; morphological, biochemical, phage lysis, F1 antigen detection, and PCR tests used to identify *Y. pestis*; and a fourfold rise in anti-F1 antibody titer in associated serum samples. Additionally, in regions where plague is common and when laboratory support is lacking for the execution of other confirmation tests, fast procedures, such as an immunochromatographic strip test, may be employed to establish the existence of F1 antigen from clinical specimens.

6.2 Laboratory Testing (CDC, 2022)

If plague is suspected, it is best to collect samples prior to treatment but not to postpone it. Depending on the clinical presentation, specimens should be collected from suitable areas for identifying the bacteria, and may include:

- *Lymph node aspirate*: A bubo that is impacted often contains a variety of organisms that can be examined under a microscope and through culture.
- *Blood cultures*: Regular blood cultures are accurate to find the plague. When a disease is advanced, bacteremia levels can be so high that sometimes, organisms can be recognised on blood smears.

- *Sputum*: Sputum from severely ill pneumonic plague patients can be cultured, but blood is culture-positive at this stage.
- Patients suspected of having pneumonic plague may have bronchial/tracheal wash samples obtained. Throat samples are not the best for isolating *Y. pestis* since they frequently contain numerous other bacteria that can obscure its presence.

Besides, when impossible to culture live organisms, direct detection techniques like direct fluorescent antibody (DFA) or PCR may be used to detect *Y. pestis* in lymphoid, spleen, lung, liver, or bone marrow samples. Gram, Wright, Giemsa, or Wayson's stained smears of peripheral blood, lymph node specimens, or sputum can all be used to microscopically identify *Y. pestis*. Serologic testing can be used to confirm the diagnosis if cultures show no evidence of plague but the infection is still thought to exist. A convalescent sample should be collected 4–6 weeks or more after the start of the sickness after the first serum sample, which should be taken earliest in the illness.

6.3 Treatment (CDC, 2022)

Early detection and prompt administration of potent antibiotics are crucial for the successful treatment of plague. In most cases, it will be lethal for the patients if the delivery of appropriate antibiotics and antishock medicines is delayed by more than 24 h. In the USA, gentamicin and fluoroquinolones are first-line medications. Doxycycline, ciprofloxacin, and chloramphenicol can also be used for both adults and children as an alternative. The course of treatment lasts 10–14 days, but may be prolonged for individuals who continue to have fever or exhibit other unsettling symptoms. Depending on the severity of the illness and other clinical considerations, patients may get intravenous or oral antibiotic treatments. The course of treatment will vary with age, medical history, underlying diseases, allergies, and other factors.

In addition to administering antibiotics, supportive therapy for severe symptoms like shock should not be disregarded when treating plague. Other treatments such immunotherapy, phage therapy, bacteriocin therapy, and the administration of virulence factor inhibitors have been documented, although they are not employed in clinics.

The majority of *Y. pestis* isolates in the world are streptomycin-sensitive, however, a multidrug-resistant (MDR) strain was discovered in Madagascar. Fortunately, in this area and other parts of the world, this MDR strain has never reemerged naturally. This MDR strain, which is resistant to streptomycin, chloramphenicol, ampicillin, spectinomycin, kanamycin, tetracycline, sulfonamides, and minocycline, has raised concerns that bioterrorists may find it appealing and use it in future. When choosing an antibiotic that will be effective in the event that infection with this strain does arise, care must be taken to avoid administering the medications listed above.

6.4 Vaccines

Currently, no plague vaccines have been licenced by the Food and Drug Administration (FDA) for mass vaccination in humans. To protect the population from these biothreats, a safe and effective subunit vaccine has been a top focus.

A licenced anthrax vaccine that contains filtered crude culture supernatant of *Bacillus anthracis* strain V770-NP1-R adsorbed to alum adjuvant is available, but it is only given to military personnel and high-risk health professionals due to the vaccine's significant reactogenicity in those who have already received it. Notably, a suspension of the formaldehyde-killed, encapsulated form of *Y. pestis* is the plague vaccine approved for use in the USA (Tao & Rao, 2021).

7 Plague as Biowarfare Weapon

A biological attack, often known as bioterrorism, is the deliberate introduction of viruses, bacteria, or other germs capable of infecting or killing people, cattle or crops. When intentionally spread in an area, these agents and the toxins they produce, can cause large-scale death, morbidity, and can render a large number of people incapable in short time while also having detrimental impact on human health (Thavaselvam & Vijayaraghavan, 2010).

The CDC (2018) classified plague as "Category A" agent, which shows that it received top priority owing to its potential to endanger national security. *Y. pestis* is a readily available bacterium that can be isolated and grown in large quantities in a laboratory, making it a potential candidate for a bioweapon (Occupational Safety and Health Administration [OSHA], n.d.). A biological strike with such an agent could resemble a natural occurrence and make it more difficult to analyse the situation and respond to it in terms of public health. High-threat diseases laboratory may be targeted during war and conflict, which could have detrimental effects on the public health. Consequently, biological weapons are a subset of a broader category of weapons that is often referred to as unconventional weapons or of mass destruction and includes chemical, nuclear and radiological weapons. The employment of biological weapons raises severe concerns, and there is an increasing likelihood that such weapons may be used by terrorist (WHO, n.d.).

References

Anderson, R. W., Johnson, N. D., & Koyama, M. (2017). Jewish persecutions and weather shocks: 1100–1800. *The Economic Journal, 127*(602), 924–958.

Andersson, J. (2017). *Combating plague: Fighting a war on two fronts through vaccine design and new therapeutic intervention* [Doctoral dissertation].

Ansari, I., Grier, G., & Byers, M. (2020). Deliberate release: Plague—a review. *Journal of Biosafety and Biosecurity, 2*(1), 10–22.

Arnold, D., & Peckham, R. (2015). Epidemics and the modern world. In R. Peckham (Ed.), *Epidemics in modern Asia* (pp. 33–54). Cambridge University Press.

Benedictow, O. J. (2021). *The complete history of the Black Death*. Boydell & Brewer.

Biswas, S., Mittal, V., Lal, S., & Kumar, S. (2020). Plague-needs continuous surveillance system. *Journal of Communicable Diseases, 52*(4), 49–56. E-ISSN: 2581-351X & P-ISSN: 0019-5138.

Bramanti, B., Stenseth, N. C., Walløe, L., & Lei, X. (2016). Plague: A disease which changed the path of human civilization. In *Yersinia pestis: Retrospective and perspective* (pp. 1–26). Springe.

Butler, T. (2012). *Plague and other Yersinia infections*. Springer Science & Business Media.

Cantor, N. F. (2001). *In the wake of the Plague: The Black Death and the world it made*. Simon and Schuster.

Cartwright, M. (2023). Black Death. *Ancient History Encyclopedia, 28*.

Centers for Disease Control and Prevention [CDC]. (2006). Fatal human plague—Arizona, 2007. *Morbidity and Mortality Weekly Report, 57*(14), 393–396.

Centers for Disease Control and Prevention [CDC]. (2018). *Bioterrorism agents/diseases*. Retrieved from https://emergency.cdc.gov/agent/agentlist-category.asp

Centers for Disease Control and Prevention [CDC]. (2019). *Ecology and transmission*. Retrieved from https://www.cdc.gov/plague/transmission/index.html

Centers for Disease Control and Prevention [CDC]. (2021a). *Frequently asked questions*. Retrieved from https://www.cdc.gov/plague/faq/index.html

Centers for Disease Control and Prevention [CDC]. (2021b). *Plague*. Retrieved from https://www.cdc.gov/plague/index.html

Centers for Disease Control and Prevention [CDC]. (2022). *Resources for clinicians*. Retrieved from https://www.cdc.gov/plague/healthcare/clinicians.html

Cohn, S. K., Jr., & Alfani, G. (2007). Households and plague in early modern Italy. *Journal of Interdisciplinary History, 38*(2), 177–205.

Cohn, S. K., & Cohn, S. K. (2002). *The Black Death transformed: Disease and culture in early renaissance Europe* (p. 75). Arnold.

Das, S., & Kataria, V. K. (2010). Bioterrorism: A public health perspective. *Medical Journal, Armed Forces India, 66*(3), 255–260. https://doi.org/10.1016/S0377-1237(10)80051-6

Dean, K. R., Krauer, F., Walløe, L., Lingjærde, O. C., Bramanti, B., Stenseth, N. C., & Schmid, B. V. (2018). Human ectoparasites and the spread of plague in Europe during the second pandemic. *Proceedings of the National Academy of Sciences, 115*(6), 1304–1309.

Dennis, D. T., & Mead, P. S. (2010). *Yersinia species, including Plague. Principles and practice of infectious diseases* (7th ed., 2943–2953). Elsevier.

Dewitte, A., Bouvenot, T., Pierre, F., Ricard, I., Pradel, E., Barois, N., & Sebbane, F. (2020). A refined model of how *Yersinia pestis* produces a transmissible infection in its flea vector. *PLoS Pathogens, 16*(4), Article e1008440.

Doll, J. M., Zeitz, P. S., Ettestad, P., Bucholtz, A. L., Davis, T., & Gage, K. (1994). Cat-transmitted fatal pneumonic plague in a person who traveled from Colorado to Arizona. *The American Journal of Tropical Medicine and Hygiene, 51*(1), 109–114. https://doi.org/10.4269/ajtmh.1994.51.109

Drancourt, M., & Raoult, D. (2014). *Yersinia pestis* and the three plague pandemics. *The Lancet Infectious Diseases, 14*(10), 918–919.

Echenberg, M. (2002). Pestis redux: The initial years of the third bubonic plague pandemic, 1894–1901. *Journal of World History*, 429–449.

Eisen, R. J., Atiku, L. A., Enscore, R. E., Mpanga, J. T., Acayo, S., Mead, P. S., & Gage, K. L. (2021). Epidemiology, ecology and prevention of plague in the West Nile Region of Uganda: The value of long-term field studies. *The American Journal of Tropical Medicine and Hygiene, 105*(1), 18.

Erova, T. E., Rosenzweig, J. A., Sha, J., Suarez, G., Sierra, J. C., Kirtley, M. L., & Chopra, A. K. (2013). Evaluation of protective potential of *Yersinia pestis* outer membrane protein antigens as possible candidates for a new-generation recombinant plague vaccine. *Clinical and Vaccine Immunology, 20*(2), 227–238.

Frith, J. (2012). The history of plague—Part 1: The three great pandemics. *Journal of Military and Veterans Health, 20*(2), 11–16.

Glatter, K. A., & Finkelman, P. (2021). History of the plague: An ancient pandemic for the age of COVID-19. *The American Journal of Medicine, 134*(2), 176–181.

Godfred-Cato, S., Cooley, K. M., Fleck-Derderian, S., Becksted, H. A., Russell, Z., Meaney-Delman, D., & Nelson, C. A. (2020). Treatment of human plague: A systematic review of published aggregate data on antimicrobial efficacy, 1939–2019. *Clinical Infectious Diseases, 70*(Supplement_1), S11–S19.

Gottfried, R. S. (2010). *Black Death*. Simon and Schuster.

Harper, K. (2017). The fate of Rome. In *The fate of Rome*. Princeton University Press.

Hawgood, B. J. (2007). Waldemar Mordecai Haffkine, CIE (1860–1930): Prophylactic vaccination against cholera and bubonic plague in British India. *Journal of Medical Biography, 15*(1), 9–19.

Horgan, J. (2014). Justinian's Plague (541–542 CE). *World History Encyclopedia, 26*, 12.

Inglesby, T. V., Dennis, D. T., Henderson, D. A., Bartlett, J. G., Ascher, M. S., Eitzen, E., & Working Group on Civilian Biodefense. (2000). Plague as a biological weapon: Medical and public health management. *Jama, 283*(17), 2281–2290.

Jedwab, R., Johnson, N. D., & Koyama, M. (2022). The economic impact of the Black Death. *Journal of Economic Literature, 60*(1), 132–178.

Kohn, G. C. (2007). *Encyclopedia of plague and pestilence: From ancient times to the present*. Infobase Publishing.

Kwon, E. H., Reisler, R. B., Cardile, A. P., Cieslak, T. J., D'Onofrio, M. J., Hewlett, A. L., & Korte-peter, M. G. (2018). Distinguishing respiratory features of category A/B potential bioterrorism agents from community-acquired pneumonia. *Health Security, 16*(4), 224–238.

Layton, R. C., Mega, W., McDonald, J. D., Brasel, T. L., Barr, E. B., Gigliotti, A. P., & Koster, F. (2011). Levofloxacin cures experimental pneumonic plague in African green monkeys. *PLoS Neglected Tropical Diseases, 5*(2), Article e959.

Li, M., Guo, P., Chen, C., Feng, H., Zhang, W., Gu, C., & Tao, P. (2021). Bacteriophage T4 vaccine platform for next-generation influenza vaccine development. *Frontiers in Immunology, 12*, 1–12.

McCormick, M. (2003). Rats, communications, and plague: Toward an ecological history. *Journal of Interdisciplinary History, 34*(1), 1–25.

Meier, M. (2016). The 'Justinianic Plague': The economic consequences of the pandemic in the eastern Roman empire and its cultural and religious effects. *Early Medieval Europe, 24*(3), 267–292.

Meyer, K. F. (1961). Pneumonic plague. *Bacteriological Reviews, 25*(3), 249–261.

Miarinjara, A., Vergain, J., Kavaruganda, J. M., Rajerison, M., & Boyer, S. (2017). Plague risk in vulnerable community: Assessment of *Xenopsylla cheopis* susceptibility to insecticides in Malagasy prisons. *Infectious Diseases of Poverty, 6*(1), 1–7.

Mullett, C. F. (1936). The English Plague scare of 1720–23. *Osiris, 2*, 484–516.

NA, N. (2016). *The Black Death: The great mortality of 1348–1350: A brief history with documents*. Springer.

Narayanan, N., Lacy, C. R., Cruz, J. E., Nahass, M., Karp, J., Barone, J. A., & Hermes-DeSantis, E. R. (2018). Disaster preparedness: Biological threats and treatment options. *Pharmacotherapy: The Journal of Human Pharmacology and Drug Therapy, 38*(2), 217–234.

Pechous, R. D., Sivaraman, V., Stasulli, N. M., & Goldman, W. E. (2016). Pneumonic plague: The darker side of *Yersinia pestis*. *Trends in Microbiology, 24*(3), 190–197.

Perry, R. D., & Fetherston, J. D. (1997). *Yersinia pestis*—Etiologic agent of plague. *Clinical Microbiology Reviews, 10*(1), 35–66.

Prentice, M. B., & Rahalison, L. (2007). Plague. *The Lancet, 369*(9568), 1196–1207.

Pryor, E. G. (1975). The great plague of Hong Kong. *Journal of the Hong Kong Branch of the Royal Asiatic Society*, 61–70.

Rabaan, A. A., Al-Ahmed, S. H., Alsuliman, S. A., Aldrazi, F. A., Alfouzan, W. A., & Haque, S. (2019). The rise of pneumonic plague in Madagascar: Current plague outbreak breaks usual seasonal mould. *Journal of Medical Microbiology, 68*(3), 292–302.

Riedel, S. (2005). Plague: From natural disease to bioterrorism. *Baylor University Medical Center Proceedings, 18*(2), 116–124.

Occupational Safety and Health Administration [OSHA]. (n.d.). *Plague as a Bioweapon*. Retrieved from https://www.osha.gov/plague/bioweapon

Stenseth, N. C., Atshabar, B. B., Begon, M., Belmain, S. R., Bertherat, E., Carniel, E., & Rahalison, L. (2008). Plague: Past, present, and future. *PLoS Medicine, 5*(1), Article e3.

Stone, A. C., & Ozga, A. T. (2019). Ancient DNA in the study of ancient disease. In *Ortner's identification of pathological conditions in human skeletal remains* (pp. 183–210). Academic Press.

Tao, P., Rao, V. B. (2021). *Bacteriophage vaccines*.

Thacker, S. B., & Berkelman, R. L. (1988). Public health surveillance in the United States. *Epidemiologic Reviews, 10*, 164–190. https://doi.org/10.1093/oxfordjournals.epirev.a036021

Thavaselvam, D., & Vijayaraghavan, R. (2010). Biological warfare agents. *Journal of Pharmacy & Bioallied Sciences, 2*(3), 179–188. https://doi.org/10.4103/0975-7406.68499

Wagner, D. M., Klunk, J., Harbeck, M., Devault, A., Waglechner, N., Sahl, J. W., & Poinar, H. (2014). *Yersinia pestis* and the plague of Justinian 541–543 AD: A genomic analysis. *The Lancet Infectious Diseases, 14*(4), 319–326.

Wang, H., Cui, Y., Wang, Z., Wang, X., Guo, Z., Yan, Y., Li, C., Cui, B., Xiao, X., Yang, Y., Qi, Z., Wang, G., Wei, B., Yu, S., He, D., Chen, H., Chen, G., Song, Y., & Yang, R. (2011). A dog-associated primary pneumonic plague in Qinghai Province, China. *Clinical Infectious Diseases: An Official Publication of the Infectious Diseases Society of America, 52*(2), 185–190. https://doi.org/10.1093/cid/ciq107

Wong, D., Wild, M. A., Walburger, M. A., Higgins, C. L., Callahan, M., Czarnecki, L. A., & Mead, P. S. (2009). Primary pneumonic plague contracted from a mountain lion carcass. *Clinical Infectious Diseases, 49*(3), e33–e38.

World Health Organization [WHO]. (2006). International meeting on preventing and controlling plague: The old calamity still has a future. *Weekly Epidemiological Record = Relevé épidémiologique hebdomadaire, 81*(28), 278–284.

World Health Organization [WHO]. (2010). *Operational guidelines on plague surveillance, diagnosis, prevention and control*. Retrieved from https://www.who.int/publications/i/item/operational-guidelines-on-plague-surveillance-diagnosis-prevention-and-control

World Health Organization [WHO]. (n.d.). *Biological weapons*. Retrieved from https://www.who.int/health-topics/biological-weapons/safeguarding-biosafety-and-biosecurity-in-laboratories#tab=tab_1

Yang, R. (2018). Plague: Recognition, treatment, and prevention. *Journal of Clinical Microbiology, 56*(1), e01519-e1617.

Yavari, N., Potter, L. G., & Oppenheim, J. M. R. (2004). Views from the edge. *Essays in honor of Richard W. Bulliet* (pp. 255–279). Columbia University Press.

Chapter 5
Spanish Flu

1 Introduction

The influenza virus, also alluded as "flu", belongs to the Orthomyxoviridae family and has multiple negative-strand RNA sequences (Nayak et al., 2004). Depending on the antigenicity of the surface nucleoproteins and matrix proteins, these viruses are split into four genera A, B, C and D (Centers for Disease Control & Prevention [CDC], 2021a). Hemagglutinin (HA) and neuraminidase (NA), the two proteins present on the surface of type A, categorise them into subtypes. There are eleven NAs and eighteen HAs that have been identified. In order to categorise influenza A viruses, H and N subtypes used are H1N1 and H3N2. There are variants and lineages of influenza B (B/Yamagata and B/Victoria) (Boktor & Hafner, 2022). Presently, influenza C does not constitute a threat to the public health due to its rarity and only propensity to cause minor infections. On the other hand, influenza D does not seems to spread or have an effect on humans and is thought to primarily infect wildlife (World Health Organization [WHO], 2018a). Human influenza A and B viruses spread around the world almost every winter, triggering seasonal epidemics. Nevertheless, influenza A viruses alone are well-known for triggering pandemics, with the "Spanish flu" being one of the most destructive (Zambon, 2001). Prominent influenza A repositories include aquatic birds and pigs, which pose challenges to control efforts and create possibilities for virus mutation and reassortment (Lofgren et al., 2007).

Antigenic drift and shift in the surface glycoproteins of these viruses result in their evolution (Kim et al., 2018). Antigenic drift in the viral genome causes seasonal influenza. Minor alterations in the genomes of these viruses termed as "drift" could result in modifications of the surface proteins (CDC, 2021b). In contrast, a pandemic influenza outbreak is brought by an antigenic shift, proves sudden changes in flu virus can emerge from genetic reassortment. This begins when two distinct influenza viruses simultaneously co-infect the same host, producing a new viral strain (CDC, 2021b; Zambon, 1999). Antigenic drift enables influenza type A viruses to alter

M. H. Fulekar et al., *Pandemic Over the Centuries*,
https://doi.org/10.1007/978-3-031-94270-9_5

regularly, whereas antigenic shift occurs rarely. Four pandemics have occurred in the past century as a consequence of the introduction of unique flu strains. These pandemics were H1N1 swine flu (2009), H2N2 Asian flu (1957), H3N2 Hong Kong flu (1968) and H1N1 Spanish flu (1918) (CDC, 2021b).

Influenza symptoms are onset of fever, muscle aches, migraine, fatigue, chest infection, sores and sinus infections (Lam et al., 2016; Monto et al., 2000). Furthermore, other common symptoms include gastrointestinal discomforts such as nausea, vomiting and diarrhoea (Minodier et al., 2015). Incubation period of these viruses varies from 1 to 4 days (Lessler et al., 2009). Whenever the virus is contagious, shedding often takes place 1 day ahead from the appearance of the symptoms and continues for 5–7 days thereafter (Killingley et al., 2016; Lau et al., 2010). WHO's projection for 2022 was that seasonal influenza may cause 290,000–650,000 fatalities annually from respiratory illnesses. This chapter discusses a historical overview of the "Spanish flu," together with information on symptoms, epidemiology and available treatments.

2 The Great Influenza Pandemic of 1918

First World War (WWI) end brought the great Influenza pandemic which was later called Spanish flu. The civilization that had been predisposed to vulnerability and agility was devastated by a virus unlike any other in recorded history. The 1918 pandemic, also called as "La Grippe" in French, has been dubbed as the "worst health catastrophe in existence." H1N1 virus was the pathogen responsible, which originated in specific bird species (Torrens & Castellano, 2020; Waring, 1964). The pandemic is usually believed to have three distinguished waves, viz. spring (1918), fall (1918) and winter (1918–19) (Humphries, 2014). While the first and third wave were relatively minor, the second caused devastating damages and millions of fatalities worldwide. During 1918 and 1919, the outbreak was estimated to affect 500 million worldwide (about 33% of the population), resulting in 50–100 million fatalities with a case-fatality rate above 2.5% (Johnson & Mueller, 2002; Morens & Fauci, 2007; Rosenau & Last, 1980; Taubenberger & Morens, 2006).

2.1 Origin and Transmission

Spanish flu genesis has been the topic of ambiguity and discussion, with most hypotheses focusing on the USA or China (Barry, 2004; Potter, 2001). The very first concept emphasises on the Chinese region of Canton. According to this belief, this virus emerged in China and expanded to the USA because of large immigration of Chinese to North America (Reid & Taubenberger, 2003). Before spreading across the world, the virus first erupted in the USA (Iezzoni, 1999). The second possibility was that the virus originated solely in the USA. The first wave of spring

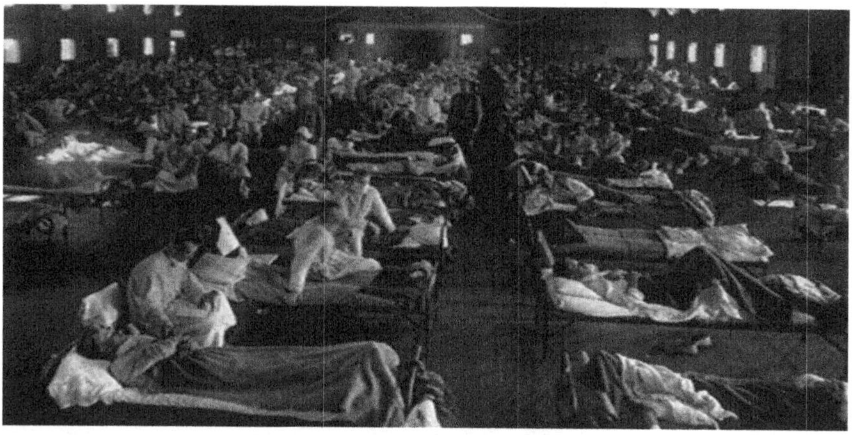

Fig. 1 Influenza victims in an emergency hospital near Camp Funston (now Fort Riley) in Kansas (USA) in 1918. *Photograph*: AP Photo/National Museum of Health. (*Source* The Guardian https://www.theguardian.com/world/2018/sep/09/spanish-flu-pandemic-centenary-first-world-war)

was thought to commence in the USA during March 1918 (Barry, 2004; Crosby, 1989; Jordan, 1927). However, it is challenging to precise geographic site of origin because influenza appeared in North America, Europe and Asia almost simultaneously in March and April 1918 (Jordan, 1927), probably in late summer of 1918, a mutation or reassortment took place that greatly increased virulence (Jordan, 1927) (Fig. 1).

Spanish flu spread more quickly than earlier pandemics in military setting where it emerged. Past pandemics generally spread by commerce and communications networks. In the meantime, congestion, poor sanitation and lack of medical facilities made trench warfare in Europe the real environment for disease spread (Humphries, 2014). At a military outpost in the agricultural state of Kansas, the spring wave began. Over the following 6 months, it expanded from American troops aboard public transport and military ships across the country and abroad. It soon spread in Europe's crowded military bases before moving to east and South Asia, affecting all factions of the war regardless of ethnicity. This initial outbreak was not deadly, but was mild. Those affected were fortunate as it seemed to have provided some immunising resistance for the pathogen's subsequent aggressive strains (Vaughan et al., 1922). However, overwhelming infections and fatalities related to the pandemic were caused by the fall wave, which first appeared in Southern England, more notably in Plymouth and Devonpor (Humphries, 2014). This wave, which lasted for nearly 6 weeks, travelled from North America to Central and South America, from Freetown to West and South Africa in September 1918 and to the Horn of Africa in November. The disease emerged in Russia and travelled over northern Asia before arriving in India around September and resurfacing in China in October. Spain was also affected by the pandemic. Data was freely available, and Spain's position on battle was neutral. Since the virus was extensively addressed in Spanish media, the

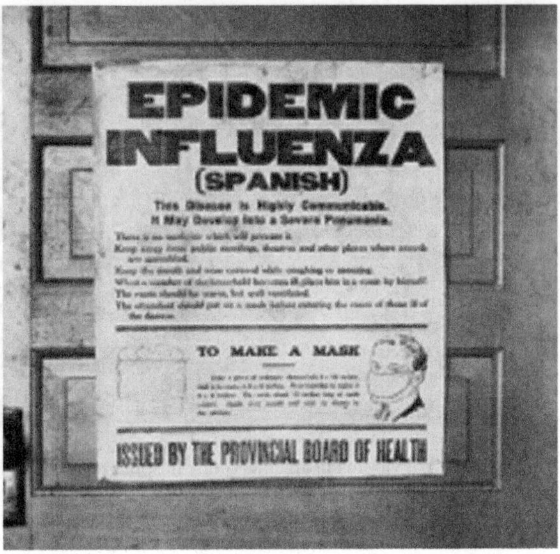

Fig. 2 Classic poster of the 1918 pandemic in US, calling its name "Spanish Influenza". (*Source* Gateway house https://www.gatewayhouse.in/1918-flu-india/)

phrase "Spanish flu" began commonly used to refer to the disease and hence the pandemic got a moniker. The pandemic mortality toll increased during the winter wave, which then further decreased in the summer of 1919. Australia experienced a big impact from this recent wave. In an effort to avert the outbreak, Australia decided to isolate itself from rest of the world. Australia's fatality rate was far higher than other countries affected by the virus's initial two waves in 1919 (Martini et al., 2019) (Fig. 2).

2.2 *Diagnosis and Symptomology*

The clinical features for influenza and pneumonia were ambiguous in 1918, and clinicians depended on the disease's signs and symptoms. Additionally, influenza was not a nationally reportable disease. Army personnel in Kansas (USA), recorded the initial symptoms, which included fever, coughing and headaches; other signs and symptoms included coma, pneumonia, encephalitis, blood in the urine and nasal bleeds (Tsoucalas et al., 2015). Also, there was cutaneous bleeding from the petechiae and ears (Knobler et al., 2005). According to Patterson and Pyle (1991) the cytokine storm (abnormal cytokine release in response to infection and other stimuli) was considered the reason why the symptoms were so severe. In 1918, doctors generally assumed that influenza was triggered by the *Bacillus influenzae* bacterium. Several questions were expressed when doctors could not detect any bacteria during autopsies (Van Epps, 2006). Additionally, physicians struggled to recognize influenza infections, frequently misdiagnosing for the cold or flu, cholera or bubonic plague (United States Department of Health and Human Services [USDHHS], n.d.). However, doctors were well-aware that infectious respiratory particles from the nose and mouth are indeed the main channel of infection. In addition, the majority of those who died during the pandemic suffered from secondary bacterial pneumonia brought by *Staphylococcus aureus*, *Streptococcus pneumoniae* and other bacteria (LeCOUNT, 1919; Winternitz et al., 1920).

3 Spanish Flu: Environmental Perspectives

The environment had certain effect on the spread, severity, prevention and control of the Spanish flu. When the pandemic hit the USA, medical experts converted classrooms, public spaces, and massive properties into outdoor emergency centres. Consequently, open-air hospitals became regular practise by 1918 pandemic. Due to sunlight and fresh air, less infections and causalities were observed at open-air facilities. Doctors found that serious patients recovered more promptly outdoors compared to when they were given regular food, warmth and plenty of fresh air and sunlight indoors. Additionally, researchers found that hospitals with outdoor facilities had the same effect as establishments with higher ceilings and proper ventilation indoors and natural ventilation was highly effective in preventing the transmission of the virus (Hobday & Cason, 2009; Hobday & Dancer, 2013).

Though records frequently mention abnormally heavy rain and freezing temperatures in the years before the pandemic began and especially during its lethal wave in 1918, the environmental conditions and specifically the weather patterns during which the pandemic originated has received less attention in the published research. High precipitation, troops soldiers on the battlefields, unsanitary surroundings, and the usage of chlorine gas as a chemical weapon during the fall and winter of 1918–1919 is believed to have contributed towards the deadliest wave of the outbreak. The optimum conditions for the virus to thrive and proliferate were provided by torrential rain and falling temperatures (Brown et al., 2009; Tang, 2009). As per latest reports, prolonged exposure to colder temperatures may also have contributed to the proliferation of pneumococcal co-infections, which became more usual than earlier thought during WWI (Iwasaki et al., 2017; Meissner, 2016; Worobey et al., 2014). Recent epidemiological analyses have shown that a significant contributor to the pandemic's high fatality rate was pneumococcal co-infections, which infected upto one-fifth of those who got the flu and culminated in death rate of 34%.

Conversely, another exacerbating element may have been the effect of weather anomalies on bird migratory patterns. The abnormality, in particular, may have disrupted the movement of the mallard duck carrying H1N1 avian influenza virus. This alteration might have led to an increase in the emergence of this life-form across Europe during the autumn of 1918. Their closer proximity to both military and civilians and to farm animals also might have accelerated the spread (Oxford et al., 2005; Saunders-Hastings & Krewski, 2016; Worobey et al., 2014).

Consequently, the influence of environmental alteration and human behavior on pandemic severity, mortality and transmission cannot be underestimated. Given the recurrent cyclical recurrence of H1N1 and severe weather anomalies brought by global climate change, future outbreaks are inevitable. Hence, the chronology of the Spanish influenza pandemic is a caution about the continuing hazards of animal trade, humanitarian disaster, conflict zones and unsanitary environments as disease incubators a century later.

4 Prevention and Control of Spanish Flu

Until 1918, there were no effective influenza vaccines or medicines to cure secondary infectious diseases. Pneumonia was not anticipated for many years. Few doctors and scientists were available due to WWI. Also, there were not enough beds available, so nearby communities and schools served as emergency facilities. Given that there was no established practice for the disease cure, the only options for treatment were symptomatic therapy and home remedies (USDHHS, n.d.). Since the beginning of the first pandemic wave, the migration of women working in non-domestic settings, the exodus of refugees, and the withdrawal of soldiers from the war have all contributed in the virus' rapid spread. As a result, it was essential to implement precautionary public health strategies in an effort to stop the spread of the virus (Reid et al., 2004). The very first formal precautionary measures were taken during August 1918 that included community surveillance in areas like childcare centers, primary schools, and regiments with compulsory reporting of suspected cases (Martini et al., 2019).

When the virus began to spread, people would test a number of ineffective over-the-counter drugs and home remedies. These included topical creams or a mixture of water, salt with coal oil (USDHHS, n.d.). Doctors also relied on haphazard mix of medications with widely varying degrees of efficacy. The medicines prescribed were aspirin, quinine, arsenics, digitalis, strychnine, epsom salts, castor oil and iodine. Additionally, traditional medical practises like bloodletting, gargling with salt water, Ayurveda, and kampo were employed (Spanyol & Valencia, 2018). Alcohol demand increased as certain doctors recommended alcohol to prevent infection (USDHHS, n.d).

To contain the pandemic, non-pharmaceutical interventions (NPIs) were employed, including quarantines, school closures, ban on social assemblies, and infection preventive measures like facemasks, and safe sneezing and coughing practices (Markel et al., 2006). Success at different levels were seen with these approaches. For instance, many people argued against the use of gauze face-masks since they could only limit bacterial infection and were useless at stopping viral penetration. Governmental warnings regarding the dangers of influenza were routinely ignored or misinterpreted (USDHHS, n.d). NPI strategies, meanwhile, were employed in the USA during the second and third waves. Recent investigations found that they were strongly connected to mortality decrease and deferred pandemic spikes (Markel et al., 2006). The proclamation of marine quarantines such as on Iceland, Australia and American Samoa protected several people. Socially isolating practices were introduced, including closing of schools, movie theatres and places of worship, limiting access to public transportation, and forbidding large public gatherings. In others like Japan, using face masks became common, although there usefulness was doubtful (McLeod et al., 2008).

Along with public gathering ban and limiting the number of commuters on public transports, street cleaning and disinfecting public places like churches, theatres and workshops were seen as crucial practice for preventing the spread of virus. They were not, however, particularly successful. Local health authorities initiated proficiency in waste disposal, management of public toilets, and the safety checks of dairy and other food items among other essential health interventions. Spitting in public was prohibited, which led to the diffusion of pocket spittoons, and press releases in newspapers and leaflets promoted the prevention and treatment measures (Martini et al., 2019).

5 Influenza Diagnosis, Prevention and Control Strategies: Present Status

Human infection with zoonotic influenza requires laboratory testing. WHO constantly updates technical advisory for the diagnosis of zoonotic influenza in public employing molecular techniques, viz. reverse transcription polymerase chain reaction (RT-PCR) and others through its Global Influenza Surveillance and Response System (GISRS). Virus culture, serological tests, quick antigen test, (RT-PCR), immunofluorescence assays, and rapid molecular assays are among the diagnostics available for influenza. Depending on the type of test being performed, nasopharyngeal or nasal swabs, and nasal wash or aspirate, are the recommended respiratory specimens for influenza detection. During the first 3–4 days of infection, samples should be collected. Results from rapid influenza diagnostic tests (RIDTs) can be obtained within 15 min; those from virus cultures take 3–10 days. The majority of in-office quick diagnostic tests for influenza are between 50 and 70% sensitive and more than 90% specific for identifying influenza. Owing to this, inaccurate negative results are frequent than false positive results, particularly when the regional influenza outbreak is at its peak. Rapid molecular assays have higher sensitivity and accurate and can deliver results within 20 min. More and more molecular analyses are readily accessible, and they can deliver results in 60–80 min or less with extremely high accuracy and specificity. Influenza viral shed in the lower airtracts may be more persistent than in the upper respiratory tract. Lower airway specimens (endotracheal aspirate or bronchioalveolar lavage fluid) from hospitalised patients with probable influenza should be acquired and analyzed for influenza viruses using RT-PCR if the patient is critically ill as well as on ventilator plus tested negative on an upper respiratory tract sample (CDC, 2020).

There are many difficulties in treating severe influenza. Antiviral medicine must be started as soon as practicable following the development of the infection for patients with influenza as this is important for their treatment (Uyeki et al., 2019).

NA inhibitors (NAIs) (such as oral oseltamivir, inhaled zanamivir and intravenous peramivir), cap-dependent endonuclease inhibitors (like baloxavir marboxil) and adamantanes are among the antiviral drugs now available and have received FDA approval. Baloxavir and NAIs are both active against influenza A and B viruses. Due to significant resistance among the seasonal influenza A virus strains that are now proliferating, adamantanes only have efficacy against influenza A viruses and are not advised for the treatment of influenza.

In addition to antiviral medication, the management of public health also includes personal preventive measures, e.g. systematic hand washing and appropriate hand drying; utilising tissues and their proper disposal, protecting mouth and nose while coughing or sneezing, and practising good pulmonary etiquette; early isolation of those experiencing illness, fever and other influenza-related symptoms; avoiding direct contact with the patient; and refraining from touching eyes, nose or mouth must be strictly observed.

Health care professionals should take airborne safeguards when performing aerosol-generating operations. During outbreaks, the proper personal protective equipment (PPE) should be made accessible and used together with standard contact and droplet procedures. If necessary, visitors where avian influenza outbreaks are known to have occurred should stay away from poultry farms, interacting with animals in live poultry markets, going to places where poultry gets butchered, and coming in contact with any interfaces that seem to have been tainted by the faecal matter of poultry or other animals. Following good food safety and sanitation procedures, like washing hands with soap and water, is recommended. Returning visitors from affected areas should contact local health departments if they experience breathing problems that could be related to zoonotic influenza virus infection (WHO, 2018b).

6 World Health Organization: Global Influenza Programme (GIP) and the Global Influenza Surveillance and Response System (GISRS)

The Interim Committee of the WHO launched an international coalition for influenza surveillance, research and management in 1947, following comprehending the disease's severity. This led to the creation of Global Influenza Programme (GIP) and the Global Influenza Surveillance and Response System (GISRS). The GIP provides Member States with strategic direction, technical advice, and coordinated actions in order to increase Member States' health systems' preparedness for seasonal, zoonotic and pandemic flu threat to people. Given Member States' dedication to a worldwide public health strategy, the GISRS is a framework that has been effectively working and integrating pathogens, data and benefits for more than 50 years.

The Pandemic Influenza Preparedness (PIP), that was accepted by WHO Member Countries in 2011, brings together Member States, enterprise, society, and others interested to encourage the prompt and instantaneous information exchange on influenza viruses with the potential to trigger a human pandemic and sharing of benefits arising from such disclosure. Substantial advancements in pandemic preparedness have also been made owing to the PIP Framework.

Although influenza preparedness has come a long way, there has been no realistic solution that brings together all the people and the initiatives to unite towards a specific objective. Consequently, WHO released the Global Influenza Strategy 2019–2030 (GIS) in March 2019 to serve as the overarching framework for WHO, Member States and relevant stakeholders. WHO encourages a holistic approach to influenza by mobilizing such global resources by creating seasonal influenza capacities as the pillars of pandemic influenza preparedness and response. The GIS sets two high-level outcomes for 2030, with the goal of achieving the maximum influenza prevention, control and preparation to safeguard the health of people: (1) better global tools, such that a targeted, consensus-driven strategy results in increased investigation, advancement and accessibility of new and upgraded techniques for the prevention, identification, regulation and treatment of influenza; and (2) stronger country capabilities, such that every country is capable of preventing and treating the infection efficiently.

7 Conclusion

The 1918 pandemic influenza was a global health disaster, resulting in one of the biggest infectious ailment fatality ever recorded. Flu and other health emergencies have shown that building new systems during an epidemic or pandemic is impossible. The only way to prepare for a pandemic is to develop standard procedures that are constantly evaluated and updated, and seasonal influenza is the perfect disease to test those mechanisms since it impacts every country each year.

References

Barry, J. M. (2004). The site of origin of the 1918 influenza pandemic and its public health implications. *Journal of Translational Medicine, 2*(1), 1–4.

Boktor, S. W., & Hafner, J. W. (2022). Influenza. StatPearls [internet].

Brown, J. D., Goekjian, G., Poulson, R., Valeika, S., & Stallknecht, D. E. (2009). Avian influenza virus in water: Infectivity is dependent on pH, salinity and temperature. *Veterinary Microbiology, 136*(1–2), 20–26.

Centers for Disease Control and Prevention [CDC]. (2020). *Influenza signs and symptoms and the role of laboratory diagnostics*. Retrieved from https://www.cdc.gov/flu/professionals/diagnosis/labrolesprocedures.htm

Centers for Disease Control and Prevention [CDC]. (2021a). *Types of influenza viruses*. Retrieved from https://www.cdc.gov/flu/about/viruses/types.htm

Centers for Disease Control and Prevention [CDC]. (2021b). *How flu viruses can change: "Drift" and "Shift"*. Retrieved from https://www.cdc.gov/flu/about/viruses/change.htm

Crosby, A. (1989). *America's forgotten pandemic*. Cambridge, UK.
Hobday, R. A., & Cason, J. W. (2009). The open-air treatment of pandemic influenza. *American Journal of Public Health, 99*(Suppl 2), S236–S242. https://doi.org/10.2105/AJPH.2008.134627
Hobday, R. A., & Dancer, S. J. (2013). Roles of sunlight and natural ventilation for controlling infection: Historical and current perspectives. *The Journal of Hospital Infection, 84*(4), 271–282. https://doi.org/10.1016/j.jhin.2013.04.011
Humphries, M. O. (2014). Paths of infection: The First World War and the origins of the 1918 influenza pandemic. *War in History, 21*(1), 55–81.
Iezzoni, L. (1999). *Influenza 1918: The worst epidemic in American history*. TV Books Incorporated.
Iwasaki, A., Foxman, E. F., & Molony, R. D. (2017). Early local immune defences in the respiratory tract. *Nature Reviews Immunology, 17*(1), 7–20.
Johnson, N. P., & Mueller, J. (2002). Updating the accounts: Global mortality of the 1918–1920 "Spanish" influenza pandemic. *Bulletin of the History of Medicine*, 105–115.
Jordan, E. O. (1927). The influenza epidemic of 1918: III. *Journal of the American Medical Association, 89*(21), 1779–1783.
Killingley, B., Greatorex, J., Digard, P., Wise, H., Garcia, F., Varsani, H., & Nguyen-Van-Tam, J. S. (2016). The environmental deposition of influenza virus from patients infected with influenza A (H1N1) pdm09: Implications for infection prevention and control. *Journal of Infection and Public Health, 9*(3), 278–288.
Kim, H., Webster, R. G., & Webby, R. J. (2018). Influenza virus: Dealing with a drifting and shifting pathogen. *Viral Immunology, 31*(2), 174–183.
Knobler, S. L., Mack, A., Mahmoud, A., & Lemon, S. M. (2005). *The threat of pandemic influenza: Are we ready?* Workshop Summary.
Lam, P. P., Coleman, B. L., Green, K., Powis, J., Richardson, D., Katz, K., & McGeer, A. (2016). Predictors of influenza among older adults in the emergency department. *BMC Infectious Diseases, 16*(1), 1–9.
Lau, L. L., Cowling, B. J., Fang, V. J., Chan, K. H., Lau, E. H., Lipsitch, M., & Leung, G. M. (2010). Viral shedding and clinical illness in naturally acquired influenza virus infections. *The Journal of Infectious Diseases, 201*(10), 1509–1516.
LeCOUNT, E. R. (1919). The pathologic anatomy of influenzal bronchopneumonia. *Journal of the American Medical Association, 72*(9), 650–652.
Lessler, J., Reich, N. G., Brookmeyer, R., Perl, T. M., Nelson, K. E., & Cummings, D. A. (2009). Incubation periods of acute respiratory viral infections: A systematic review. *The Lancet Infectious Diseases, 9*(5), 291–300.
Lofgren, E., Fefferman, N. H., Naumov, Y. N., Gorski, J., & Naumova, E. N. (2007). Influenza seasonality: Underlying causes and modeling theories. *Journal of Virology, 81*(11), 5429–5436.
Markel, H., Stern, A. M., Navarro, J. A., Michalsen, J. R., Monto, A. S., & DiGiovanni, C., Jr. (2006). Nonpharmaceutical influenza mitigation strategies, US communities, 1918–1920 pandemic. *Emerging Infectious Diseases, 12*(12), 1961.
Martini, M., Gazzaniga, V., Bragazzi, N. L., & Barberis, I. (2019). The Spanish influenza pandemic: A lesson from history 100 years after 1918. *Journal of Preventive Medicine and Hygiene, 60*(1), E64–E67. https://doi.org/10.15167/2421-4248/jpmh2019.60.1.1205
McLeod, M. A., Baker, M., Wilson, N., Kelly, H., Kiedrzynski, T., & Kool, J. L. (2008). Protective effect of maritime quarantine in South Pacific jurisdictions, 1918–19 influenza pandemic. *Emerging Infectious Diseases, 14*(3), 468–470. https://doi.org/10.3201/eid1403.070927
Meissner, H. C. (2016). Viral bronchiolitis in children. *New England Journal of Medicine, 374*(1), 62–72.
Minodier, L., Charrel, R. N., Ceccaldi, P. E., Van Der Werf, S., Blanchon, T., Hanslik, T., & Falchi, A. (2015). Prevalence of gastrointestinal symptoms in patients with influenza, clinical significance, and pathophysiology of human influenza viruses in faecal samples: What do we know? *Virology Journal, 12*(1), 1–9.
Monto, A. S., Gravenstein, S., Elliott, M., Colopy, M., & Schweinle, J. (2000). Clinical signs and symptoms predicting influenza infection. *Archives of Internal Medicine, 160*(21), 3243–3247.

Morens, D. M., & Fauci, A. S. (2007). The 1918 influenza pandemic: Insights for the 21st century. *The Journal of Infectious Diseases, 195*(7), 1018–1028.

Nayak, D. P., Hui, E. K. W., & Barman, S. (2004). Assembly and budding of influenza virus. *Virus Research, 106*(2), 147–165.

Oxford, J. S., Lambkin, R., Sefton, A., Daniels, R., Elliot, A., Brown, R., & Gill, D. (2005). A hypothesis: The conjunction of soldiers, gas, pigs, ducks, geese and horses in Northern France during the Great War provided the conditions for the emergence of the "Spanish" influenza pandemic of 1918–1919. *Vaccine, 23*(7), 940–945.

Patterson, K. D., & Pyle, G. F. (1991). The geography and mortality of the 1918 influenza pandemic. *Bulletin of the History of Medicine, 65*(1), 4–21.

Potter, C. W. (2001). A history of influenza. *Journal of Applied Microbiology, 91*(4), 572–579.

Reid, A. H., & Taubenberger, J. K. (2003). The origin of the 1918 pandemic influenza virus: A continuing enigma. *The Journal of General Virology, 84*(Pt 9), 2285–2292. https://doi.org/10.1099/vir.0.19302-0

Reid, A. H., Taubenberger, J. K., & Fanning, T. G. (2004). Evidence of an absence: The genetic origins of the 1918 pandemic influenza virus. *Nature Reviews Microbiology, 2*(11), 909–914.

Rosenau, M. J., & Last, J. M. (1980). *Maxcy-Rosenau preventative medicine and public health.* Appleton-Century-Crofts.

Saunders-Hastings, P. R., & Krewski, D. (2016). Reviewing the history of pandemic influenza: Understanding patterns of emergence and transmission. *Pathogens, 5*(4), 66.

Spanyol, D., & Valencia, W. (2018). *Info Beasiswa.*

Tang, J. W. (2009). The effect of environmental parameters on the survival of airborne infectious agents. *Journal of the Royal Society Interface, 6*(suppl_6), S737–S746.

Taubenberger, J. K., & Morens, D. M. (2006). 1918 Influenza: The mother of all pandemics. *Revista Biomedica, 17*(1), 69–79.

Torrens, F., & Castellano, G. (2020). Swine, avian, and human flu, and origins of a (H1N1) flu pandemic strain. In *Biochemistry, biophysics, and molecular chemistry* (pp. 53–68). Apple Academic Press.

Tsoucalas, G., Karachaliou, F., Kalogirou, V., Gatos, G., Mavrogiannaki, E., Antoniou, A., & Gatos, K. (2015). The first announcement about the 1918 "Spanish flu" pandemic in Greece through the writings of the pioneer newspaper "Thessalia" almost a century ago. *Le Infezioni in Medicina, 23*(1), 79–82.

United States Department of Health and Human Services [USDHHS]. (n.d.). *The great pandemic: The United States in 1918–1919.* Retrieved from http://www.flu.gov/pandemic/history/1918/index.html

Uyeki, T. M., Bernstein, H. H., Bradley, J. S., Englund, J. A., File, T. M., Jr., Fry, A. M., & Pavia, A. T. (2019). Clinical practice guidelines by the Infectious Diseases Society of America: 2018 update on diagnosis, treatment, chemoprophylaxis, and institutional outbreak management of seasonal influenza. *Clinical Infectious Diseases, 68*(6), e1–e47.

Van Epps, H. L. (2006). Influenza: Exposing the true killer. *The Journal of Experimental Medicine, 203*(4), 803.

Vaughan, V. C., Vaughan, H. F., & Palmer, G. T. (1922). *Epidemiology and public health: Respiratory infections* (Vol. 1). CV Mosby Company.

Waring, J. I. (1964). *A history of medicine in South Carolina, 1670–1825.* South Carolina Medical Association.

Winternitz, M. C., McNamara, F. P., & Wason, I. M. (1920). *The pathology of influenza* (No. 4). Yale University Press.

World Health Organization [WHO]. (2018a). Influenza (seasonal). Retrieved from https://www.who.int/news-room/fact-sheets/detail/influenza-(seasonal)

World Health Organization [WHO]. (2018b). *Influenza (Avian and other zoonotic).* Retrieved from https://www.who.int/news-room/fact-sheets/detail/influenza-(avian-and-other-zoonotic)?gclid=Cj0KCQjw48OaBhDWARIsAMd966BukeWP6B6H3AOAezqiEyb22LLNXrMHDZolr3AYKnpo4eY_AEHmnucaAltPEALw_wcB

World Health Organization [WHO]. (2022). *Estimate of influenza deaths due to respiratory disease*. Retrieved from https://www.who.int/teams/global-influenza-programme/surveillance-and-monitoring/burden-of-disease#:~:text=WHO%20estimates%20that%20seasonal%20influenza,which%20can%20be%20influenza%2Drelated

Worobey, M., Han, G. Z., & Rambaut, A. (2014). Genesis and pathogenesis of the 1918 pandemic H1N1 influenza A virus. *Proceedings of the National Academy of Sciences, 111*(22), 8107–8112.

Zambon, M. C. (2001). The pathogenesis of influenza in humans. *Reviews in Medical Virology, 11*(4), 227–241.

Zambon, M. C. (1999). Epidemiology and pathogenesis of influenza. *Journal of Antimicrobial Chemotherapy, 44*(suppl_2), 3–9.

Chapter 6
Cholera

1 Introduction

Cholera has ravaged human civilization for at least a millennium. Researchers suggest that the disease persisted in India since the historical era (Pollitzer, 1959). Ingesting food or water that has been infected with the *Vibrio cholerae* bacterium can lead to this severe disease which causes diarrhoea. The World Health Organization [WHO] in 2022 reported that approximately 2 billion people across the globe are bound to ingest water that has been tainted with faecal matter. The Ganges delta in India is accounted as cholera's prime reserve before it began to spread globally during 1800s. Millions of lives were affected due to six consecutive pandemics between 1817 and 1923 across the continents. Breakthrough scientific revelations related to the infection were accorded during the century. British physician John Snow is credited with identification of the source of this infection. In his studies conducted in Soho (London) in 1854 he found that the contaminated water was behind the pandemics. Cholera's causative agent was initially identified by the Italian physician Filippo Pacini Florence (Italy) in 1854. It was later identified by Robert Koch in India in 1883. Further, in 1885, Spanish bacteriologist Jaume Ferran I Clua formulated the first cholera live whole-cell injectable vaccine for the infection (Koch, 1884; Mukerjee, 1963; Pollitzer & Burrows, 1955; Tulchinsky, 2018). The seventh pandemic, that is currently active, started in South Asia during 1961 and travelled to Africa around 1971 then to America in 1991 (WHO, 2022). It is to be noted that the previous six pandemics were brought by the classic *O1* biotype. However, the ongoing cholera pandemic is accorded to the *El Tor O1* biotype, that has overtaken the classic form since 1993 (Barua, 1992; Morales et al., 2008). As per studies, cholera causes around 21,000 to 143,000 fatalities worldwide annually and impacts approximately 1.3 to 4.0 million people (Ali et al., 2015). Conforming to the reports by Mondiale de la Santé and WHO (2021), total 24 countries recorded 323,369 cases and 857 deaths in 2020. All the continents now have endemic foci harbouring cholera, proving that the disease outlasted the transition from ancient to the modern

M. H. Fulekar et al., *Pandemic Over the Centuries*,
https://doi.org/10.1007/978-3-031-94270-9_6

world. Cholera pandemic prevention and control strategies were same as influenza outbreaks. Enemas, castor oil, calomel (mercurous chloride; a purgative), stomach washing, venesection (bloodletting), opium, brandy and plugging to stop fluid from leaving, were all part of traditional treatment. Oral rehydration solution (ORS) and evidence-based rehydration therapy also developed as a result of human response to cholera. Unfortunately, the prevalence of water-borne outbreaks have escalated as billions of individuals even today do not have access to clean drinking water, adequate sanitation and hygiene. As seen in the recent outbreak in Yemen (2016–2021), these epidemics are becoming more intense, prolonged and frequent, placing a great deal of strain on medical systems and communities, and raising questions about their prevention and control (Ng et al., 2020). This chapter analyse the epidemiology and historical context of the pandemic and describe the prevention and control techniques.

2 Seven Pandemics of Cholera

2.1 First Cholera Pandemic

Often known as Asiatic cholera, this pandemic struck between 1817 and 1824. Infested rice helped in the transmission of the disease from its Ganges Delta reservoir, resulting in an outbreak in Jessore (India) during 1817 (Mulhern, 2020). After emerging in India, the disease spread to South East Asia, the Chinese Empire, and by 1822 ravaged Japan. At the same time, it expanded westward, impacting eastern Africa, Mesopotamia and Persia, and eventually reached Russia and the West (Hays, 2005). According to epidemiologists and medical historians, the Kumbh Mela, a Hindu pilgrimage on the upper Ganges River, may be the cause for its global spread (Holman, 2015; McGrew, 1960). During the original outbreak and after subsequent outbreaks decades later, this cholera pandemic contributed to an increase in anti-Asian attitude in the West, particularly towards Indian people and their culture. Later, the infection was linked to Asia and South Asia was considered responsible for cholera (JSTOR Daily, 2020). Whilst the exact number of casualties by the pandemic remained undetermined, experts believe millions of people died.

2.2 Second Cholera Pandemic

This (started in 1829 and ended in 1837) was the first to spread to Europe and the America. Prior to reaching Finland and Poland in 1830, the disease travelled to Moscow and St. Petersburg. Artisans commuting through trade routes spread it to the port of Hamburg in north Germany, and eventually to England around 1831. During 1832, it reached the Western World. The infection gradually travelled from Canada to the USA, inflicting severe impact on day-to-day activities in most of the eastern coastal areas, notably New Orleans, Louisiana, experiencing the worst (Snowden, 1995; Wendt, 1885). It was widely believed that cholera evolved during the second pandemic amongst morally inferior and in settings marked by deprivation, filth, suffering and immorality (Lippi & Gotuzzo, 2014).

2.3 *Third Cholera Pandemic*

This pandemic (1846–1860) is widely perceived as the harshest. As per some literature, it originated in India and eventually travelled to Persia (Iran), Europe, the USA and to the remaining world. Africa was wrecked by the infection, which moved from the continent's eastern coast towards Ethiopia and Uganda. Over 23,000 deaths from cholera in just Great Britain (1854) had been darkest year. In that year, John Snow, a British doctor practising in an impoverished part of London, determined that the disease was spread through contaminated water. He noticed cholera cases in London's Soho neighbourhood following the Broad Street cholera outbreak and had found a number of cases in neighbourhood of a water pump. He persuaded authorities to take the pump handle out as a test of his theory, and the incidence of cholera in the region soon decreased. His discovery contributed to the eventual suppression of the outbreak (Snow, 1855). Cholera killed roughly 3,500 individuals in Chicago in 1854, accounting for 5.5% of the city's total population (the global death toll is still unknown) (Rosenberg, 2009). In Spain, during the cholera outbreak of 1854–1855, nearly 236,000 people perished (Kohn, 2007). Additionally, in 1855, the disease spread to Puerto Rico and claimed 25,820 lives (Lugo, 2008) (Fig. 1).

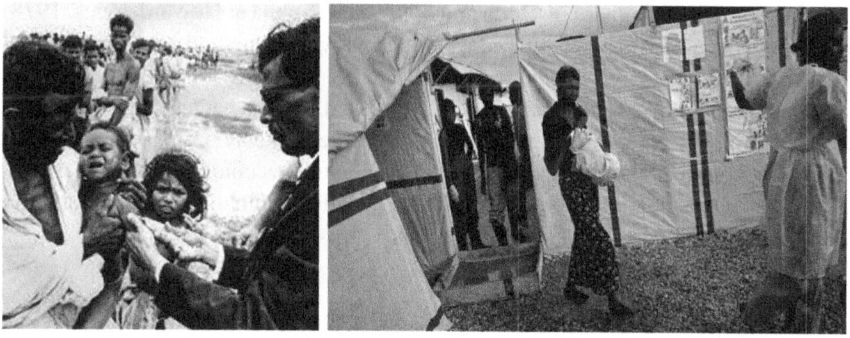

Fig. 1 Havoc of cholera

2.4 Fourth Cholera Pandemic

Fourth pandemic (1863–1875) which began in Bengal's Ganges Delta and spread to Mecca with Muslim travellers. Within its first year, the disease killed 30,000 out of 90,000 hajis (travellers to Mecca). The Middle East's cholera spread to the Russian Empire, Europe, Africa and North America by the travellers visiting port cities and those travelling through inland waterways. The outbreak, which originated in Northern Africa and migrated to Sub-Saharan Africa in 1865, reached Zanzibar in 1869–1870, affecting 70,000 people (Byrne, 2008). An estimated 165,000 people perished across the Austrian Empire by the disease, which progressed with the Austro-Prussian War (1866), especially 30,000 in Hungary, 20,000 in the Netherlands and 30,000 in Belgium (Smallman-Raynor & Cliff, 2004).

2.5 The Pandemic's Fifth Wave

The pandemic (1881–1896) continued to wreak havoc in both east and west. The *V. cholerae* bacterium, which causes cholera, was eventually discovered by Filippo Pacini in 1884 who analysed the situation to be caused by a substantial loss of fluids and electrolytes resulting in the pathogen's symptoms on the intestinal epithelium. For serious patients, he suggested intravenous administration of sodium chloride dissolved in water, that eventually proved to be highly beneficial. As cholera is a "quantity disorder" rather than "quality disorder", according to mathematical model introduced by Pacini, "all its mechanism is nothing but an issue of losses and reparations" (Bianchi, 1889). Even though Pacini found the vibrion, the scientific community disregarded his cholera research and opposed it (Subba & Howard-Jones, 1978). Robert Koch, a German scientist, developed innovative methods for growing and identifying microbes. Koch went to Calcutta (India) during 1883, when the infection was still rampant (Yoshifumi, 2010). Koch reported that the bacterium was successfully isolated in pure culture and was "a little curved, like a comma" on January 7, 1884 (Koch, 1884). Ferran developed the first cholera vaccine during this period in 1885, which was employed extensively in vaccination campaigns throughout Spain (Pollitzer & Burrows, 1955).

2.6 Sixth Cholera Pandemic

It (1899–1923) impacted less to the western Europe owing to the advancement in hygiene and healthcare. In Bengal (1899), the sixth pandemic havoc first started. Then, along with two well-established bacterial routes, it continued to advance inexorably westward. The disease travelled north along the trade routes, reaching the Punjab (India), Afghanistan and Persia. It travelled across the Caspian Sea and to the Volga, Moscow and the west. The cholera also spread in southern direction at the same time, travelling by land to Bombay and Madras and by sea to Ceylon and Jeddah. Then (1902), it travelled to Mecca with the pilgrims and split up with them as they crossed the Red Sea to Egypt, the Middle East and North Africa (Snowden, 1995). Cholera associated with foreigners or marginalised people in cultures during the pandemic because it was spread by immigrants and travellers. Anglo-Indians in British India criticized Hindu devotees for the transmission of cholera and citizens of Italy accused Jews and Romans, and many Americans believed Filipino immigrants brought the infection to the USA (Hays, 2005). Two pilgrims coming from Mecca's mosque had vibrio in their stool samples, according to Felix Gotschlich, a German doctor at the El Tor station in 1905. It is believed that the sixth and current pandemic was brought by the El Tor form (Nemmara & Thompson, 2018).

2.7 Seventh Pandemic of Cholera

The seventh pandemic marked its first appearance in 1966 that ravaged most of Asia. The prevalence of cholera then marginally declined until 1971, when a rise was seen in Africa and Europe, where the disease was eradicated for more than a century (Reeves & Lan, 1998). Cholera frequency remained stagnant and was confined to Asia and Africa during 1980s. Conversely, there have been two big cholera pandemics in 1990s: first was a resurgence of infection in Africa, and second was a sequence of epidemics that began in Peru and culminated in the first cholera pandemic across Latin America until 1895 (Seas et al., 2000). The current outbreak was thought to be caused by *V. cholerae* O1 of the El Tor (Zhang et al., 2014). Recent cholera outbreak in Yemen was the deadliest in modern times, with 2,538,677 confirmed cases and 3997 deaths (Reliefweb, 2021). Despite the fact that cholera has spread to many countries and is now endemic, the WHO still categorise current episode as a pandemic.

3 Cholera: Epidemiology

3.1 Global Distribution

Cholera is more prevalent in communities with poor access to reliable drinking water and sanitary facilities. Over 50 countries, particularly in Asia and Africa, have a high cholera incidence rate (Ali et al., 2015). Serious *V. cholerae* episodes have also been recorded in South and Central America, the Middle East, Africa, Asia and the Caribbean (Harris et al., 2012). Cases in resource-rich environments are typically brought by travellers from endemic or epidemic regions (Newton et al., 2011).

3.2 Vibrio Cholerae *Strain and Transmission of Cholera*

Vibrio cholerae is a Gram-negative facultative comma-shaped bacterium found in marshes, streams and other aquatic bodies (Reen et al., 2006). Based on the makeup of the bacterium's major surface antigen (O), which is produced from lipopolysaccharide, there are about 206 serogroups. The O1 and O139 serogroups have been associated with cholera. Depending on biochemical characteristics and sensitivity to bacteriophages, the O1 is further divided into El Tor and classical biotypes (Sharma et al., 2007). By splitting both biotypes even more, two additional serotypes, Inaba and Ogawa, can be developed (Wachsmuth et al., 1994). The O139 originally appeared in 1992 due to a multi-gene alteration in a parent O1 El Tor strain's O antigen-coding area (Waldor et al., 1994). Further, the El Tor strain continued to be the most predominant globally notwithstanding the outbreaks caused by O139 serogroup in the nineties (Longini et al., 2002; Sack et al., 2003; Sharma et al., 2007).

Among the crucial virulence determinants found in such pathogenic organisms is the toxin co-regulated pilus (TCP), a self-binding pilus which holds bacterial cells together to withstand shear stresses in the host intestinal tract (Kirn et al., 2000; Taylor et al., 1987). This toxin is secreted by the bacteria once it establishes in small intestine after being absorbed via food. The toxin is the causative factor of a profuse diarrhoeal disease that could also cause fatality within a day (Mekalanos et al., 1997). Large numbers of *V. cholerae* cells are introduced into the environment throughout infestation via bowel movements, enabling them to grow as free-living cells, whilst forming biofilms on the chitin wall of crustaceans, and by inhabiting into the digestive tract of birds and fish until they are again ingested by living beings, completing their life span (Haan & Hirst, 2004; Silva & Benitez, 2016; Yahr, 2006).

3.3 *Risk Factors*

Cholera is associated with inequality and lack of availability of nutritious, potable water and sanitary conditions (Richterman et al., 2018). Poor hygiene practises are still largely cause for the disease's rising occurrence, especially in rural areas where open defecation is still frequent (Fredrick et al., 2015). Large cholera epidemics frequently target communities in areas affected by natural disasters or war (Mohareb & Ivers, 2019). These relationships highlight the fundamental method of transmission through contaminated food and water. For instance, using soap reduces the risk of infection but drinking untreated or un-boiled water is a frequently recognised risk factor for cholera (O'Connor et al., 2011). Risk factors for some outbreaks of cholera, mainly those caused by food, may include certain foods like rice products, or certain vegetables or fruits (Rabbani & Greenough III, 1999). The preponderance of cholera cases in regions where it happens sometimes are linked to shellfish intake. The biological interplay between the host and pathogen is reflected in other risk factors for *V. cholerae* infection and cholera, such as blood group O (linked to more severe cholera), hypochlorhydria (which lowers the infectious dosage required to trigger infection), and partial gastrectomy (Bavishi & Dupont, 2011; Glass et al., 1985; Harris et al., 2005).

4 Cholera: Diagnosis and Clinical Manifestations

Clinical diagnosis suggests that patients with severe acute watery diarrhoea are often ipso facto identified with cholera. The culture of a stool sample to isolate and identify *V. cholerae* serogroups remains the benchmark for the diagnosis of the infection in a laboratory. Instant testing using Crystal® VC dipsticks or darkfield microscopy could assist in diagnosis whenever faecal culture is not readily accessible (Centers for Disease Control and Prevention [CDC], 2022). Cholera frequently manifests as diarrhoea and stomach discomfort, with the potential to cause fluid losses of upto 1 L/ hr. The most acute condition caused by the pathogen can be cholera gravis. Just after patient ingest contaminated food or drink, the bacteria colonise the small intestine for 12–72 h before complications arise (Phillips, 1964). Due to high fluid volume loss and metabolic acidosis caused by these losses, there is a risk of circulatory collapse and mortality (Wachsmuth et al., 1994). The average rice water stool contains 10^{10}–10^{12} vibrios/l. Before start of the disease, symptomatic patients may shed vibrio, and will continue to do so for 1–2 weeks after that (Cash et al., 1974; Feachem, 1982). Patients who are asymptomatic will normally only excrete 10^3 vibrios/g of stool for 1 day (Mosley et al., 1968).

The percentage of symptomatic patients is also dependent on age and the endemic nature of the disease. In regions where the disease is endemic, such as the Ganges River Delta, children are at higher risk to be diagnosed with severe conditions (Glass et al., 1982). However, in epidemic transmission patterns, all age groups appear to be equally prone to symptomatic infections (Harris et al., 2008; Holmberg et al., 1984).

5 Cholera: Environmental Perspective

Vibrio cholerae ideally grows at temperature 30 °C, a 15% salt concentration, and pH 8.5 or higher (Chowdhury et al., 2017). The condition is widespread in regions where human vulnerability—specifically, lack of potable water, improper sanitary conditions—concur with the modifications in hydro-meteorological processes, culminating scenarios which are favourable for the growth and proliferation of *V. cholerae* in water bodies. Additionally, both natural and man-made disasters are frequently linked to severe cholera outbreaks. One recent instance is the greatest cholera outbreak in 2016, which occurred after Hurricane Matthew pelted Haiti's southwest coast with rain (Khan et al., 2017). The unexpected occurrence of excessive rains and high air temperatures, combined with lack of or destroyed WASH (Water, Sanitation and Hygiene) infrastructure, are associated with epidemic cholera.

The disease will then be more likely to spread in humans as a result of exposure to the cholera bacterium (Jutla et al., 2015). Seasonal outbreaks of cholera occurs every year in areas where the disease is prevalent (Baracchini et al., 2017; Colwell, 1996; Deen et al., 2020; Shears, 2001). Changes in the aquatic environment might affect how intense a cholera outbreak is. Throughout outbreaks, the actual figure of new infections often increases during the warmest month. Heat waves and rising sea temperatures have been related to long-term elevations in prevalence of several virulent Vibrio species, notably *V. cholerae*, in Northern Europe and along the Atlantic coast of the USA (Vezzulli et al., 2016). Additionally, changing environment, like rise in water temperature, could result in extending season of *V. cholerae* abundance, having a significant impact on public health (Baker-Austin et al., 2010). In environments with high temperatures, pathogens may spread and multiply more quickly (e.g., raw food, water or environment). Annual precipitation can have a direct impact on the propagation of cholera in two different ways: first, torrential downpours can increase the possibility of wastewater contaminating (both treated and untreated water), accelerating human transmission to the environment; and secondly, milder rainfall can also raise the pathogen density in aquatic media causing spread from environment to humans (Pascual et al., 2002). Insufficient precipitation during a drought can also result in cholera epidemics (Bryden, 1871). Droughts and water scarcity, which can result in polluted rainwater use, increased microbial load in restricted water resources, or struggle for limited water resources (Mora et al., 2018).

6 Prevention and Control of Cholera

Mechanisms which supply potable water and proper sanitary facilities were developed to restrict cholera outbreaks in 1800s. However, approximately 2 billion people continue to lack in receiving these services and continued to be under threat of contacting the disease. The supply of clean water and adequate sanitation are among the prime goals of the Sustainable Development Goals although accomplishing them might take a generation (Ryan, 2011). Throughout cholera outbreak, the fundamental interventions ought to be diagnosis, rehydration-based treatment, and the provision of safe water, along with adequate hygiene and proper meal planning (Meeusen et al., 2013). These objectives have been employed for years in cholera-risk areas without lessening the continued effects of the illness, indicating that other control measures, including immunisation, are necessary (Farmer et al., 2011; Ivers et al., 2010).

In October 2017, the 35 members of Global Task Force on Cholera Control (GTFCC) came to a landmark agreement to fight cholera through the deployment of "Ending Cholera—A Global Roadmap to 2030". The GTFCC signatories urge that all partners must abide by the declaration in an effort to assist cholera-affected countries and collaborate in their efforts to stem the disease transmission (World Health Organization, 2017). This comprehensive strategy is essential to cholera prevention, control and mortality reduction. It is crucial to apply a combination of interventions including collaboration, public engagement, oral cholera immunisation, water, sanitation and hygiene (WASH), and surveillance. Cholera can be life-threatening, but is usually simple to prevent if basic sanitation precautions are taken (WHO, 1992).

6.1 Implementation/Reinforcement of Surveillance Systems

Cholera surveillance at present is inconsistent or negligible. Countries must increase the capability and scope of their current monitoring systems (e.g., laboratory diagnostic tests). Early detection and prompt treatment is possible. It plays significant role in cholera prevention and control when data are regularly analysed and disseminated at the local and national levels.

6.2 Water, Sanitation and Hygiene Promotion (WASH)

WASH is widely acknowledged as main factor in the prevention of various infectious diseases. The foundational WASH program involves improvements in sanitation, a hand-washing facility containing detergent and water in every home, civic engagement to preserve WASH facilities, and awareness promoting hygiene practices. Government freewill, community engagement is necessary for strategic WASH efforts to be implemented successfully. To promote awareness regarding open defection, food and environment safety and hygienic behaviours, specific advertisements should be available. Schools should receive special consideration. Political leaders' involvement could aid in supporting WASH objectives and putting food and water safety legislation into effect.

6.3 Deployment of Oral Cholera Vaccine (OCV)

The OCV is viewed as an effective deployment for cholera control and prevention. OCV immunisation campaigns that are both proactive and reactive in cholera hotspots in several African and Asian nations have demonstrated encouraging results and must be continued. To deploy OCV as efficiently as possible, an economic feasibility of large cholera immunisation campaigns must be considered.

6.4 Social Mobilization and Health Promotion

Community mobilisation should be focused on communication and education that advance understanding of the disease, prevention, and available treatments for it to be effective. These should offer open communication of information and suitable instruction regarding transmission routes and countermeasures. Information can be circulated fast if media and schools are properly involved.

6.5 Collaboration

For cholera, inter-sectoral collaboration between national and international health authorities, civil society, and other stakeholders is necessary for preparation and response. Cholera epidemics frequently include international borders, highlighting the value of international collaboration in containing and halting the disease's spread (Fig. 2).

7 Treatment of Cholera

Cholera outbreaks are associated with poor facilities, a shortage of potable water, or socioeconomic upheavals. Thus, researchers concur that a multimodal, dynamic technique is essential to treat infectious diseases, and following therapeutic techniques can effectively help in combating cholera.

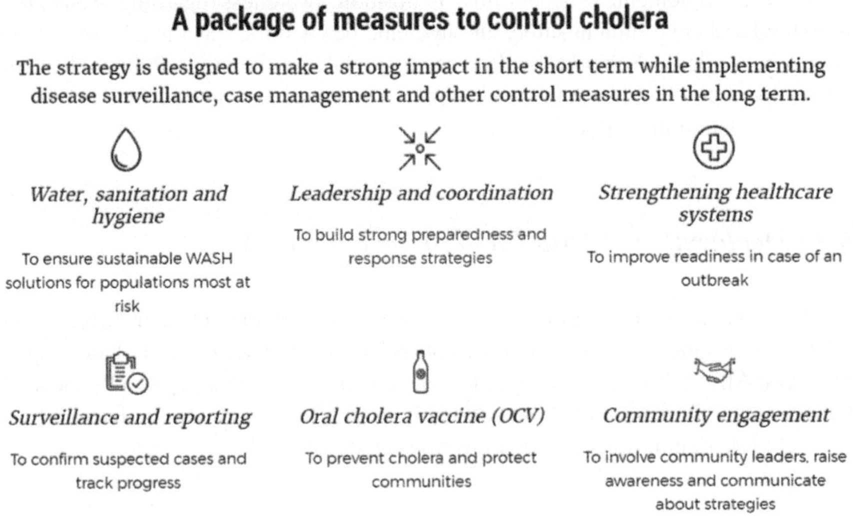

Fig. 2 Multi-sectoral interventions to control cholera. *Source* GTFCC

7.1 *Oral Rehydration Therapy*

Contemporary cholera control started with the invention of oral rehydration solution (ORS) in the late 1960s. The ORS systematically restored continued fluid losses using equimolar sodium and glucose contents to boost sodium absorption inside the small intestine (Guerrant et al., 2003). The death rate from cholera infections was greater than 50% before oral rehydration treatment (ORT). Conversely, ORT has cured the infection in thousands and saved numerous lives by supplementing the electrolytes and fluids lost throughout the disease (Kühn et al., 2014). Infections caused by *V. cholerae* are cured in more than 99% of patients receiving ORT, and the mortality rate decreased by over 97% (Milner et al., 2011). Considering *V. cholerae* infections force the intestinal mucosa to lose an abnormal quantity of critical electrolytes, the standard ORT advised by the WHO includes numerous key minerals such as sodium, chloride, potassium along with a carbon source in the form of glucose. There are still instances where ORT was unable to control the cholera symptoms. For instance, intravenous rehydration is necessary for severe dehydration. Due of ORT's partial efficacy, concurrent therapies including antibiotics and antimicrobials and vaccine may be required to maintain the reductions in cholera manifestations.

7.2 *Antibiotics/Antimicrobials*

Antibiotics are used to treat cholera infections by (1) decreasing the duration and extent of the infection and (2) preventing the spread. Antibiotics and ORT are used in combination to treat acute infections in patients who have severe dehydration (Hossain et al., 2018; Leitner et al., 2015). Doxycycline, azithromycin, and tetracycline are effective antibiotics for treating cholera. Adult symptom duration can be decreased from 4 to 2 days with several doses of 12.5 mg tetracycline given over the course of 3 days, and average stool volume can be decreased from 21 to 8 L. (Butler, 2017). Doxycycline, which comes 300 mg for adults and 6 mg for children, is as efficient as many tetracycline doses (Williams & Berkley, 2018). Similarly, a single dosage of 20 mg azithromycin can reduce vomiting frequency and allow passing of an average of 36 stools with volumes of around 5 L in less than 48 h (24 h earlier than with ciprofloxacin) (Leibovici-Weissman et al., 2014; Williams & Berkley, 2018). Tetracycline is appropriate for adults, however pregnant women and young children should take azithromycin. They are superior to ciprofloxacin and erythromycin (Orata et al., 2014). However, antibiotic resistance of cholera pathogen strains is a significant barrier to the utilisation of antibiotic therapy. To lessen the threat of cholera epidemics and pandemics, pharmaceutical companies need to invest more and continue their research into the creation of new drugs.

7.3 Nutritional Supplements

Nutritional therapy entail maintaining a high-energy diet immediately after the initial fluid loss is restored to avoid malnutrition in addition to resulting effect including hypokalemia and diabetes. Lactation must be encouraged for neonates in combination to ORS. In a clinical study, zinc supplementation shortened the duration of diarrhoea and the volume of faeces in children with cholera (Roy et al., 2008). Following diarrhoea in children, zinc supplementation decreased the likelihood of additional episodes of diarrhoea for several months (Bhandari et al., 2008). The WHO advises administering zinc to children under 5 years suffering from diarrhoea. Recommended dosage is 20 mg/day for 10 days for children between the age of 6 months and 5 years, and 10 mg/day for infants below 6 months (World Health Organization, 2005). Supplemental vitamin A is beneficial for children with diarrhoea in underdeveloped nations (Mayo-Wilson et al., 2011). However, antimotility medications and antiemetics have no proven benefits for treating cholera and may even make the infection worse or interfere with good oral rehydration by having sedative effects (Li et al., 2007).

7.4 Vaccines

The WHO advises oral cholera vaccines (OCVs), both in live-attenuated and inactivated oral whole cell as an interim precautionary measure in endemic regions or even during outbreaks since these have proven their efficacy when paired with other essential therapies, including antibiotics, ORT, plus basic healthcare (Harris, 2018). The OCVs primarily support mucosal immunity—mediated by antibodies, primarily IgA—against the pathogen. These antibodies target antigens like CT and O1-specific polysaccharide (Harris, 2018). The OCVs may also give unvaccinated people herd immunity, but more research is needed to determine how they affect unvaccinated children (Ali et al., 2019).

The WHO has already approved three oral cholera vaccines (OCV): Dukoral®, Shanchol™ and Euvichol-Plus®. All these vaccinations call for two doses to provide complete immunity. Dukoral (CTB-WC), which also contains recombinant CT (CTB) B subunits, has inactive/dead *V. cholerae* O1 (Cabrera et al., 2017; Wierzba, 2019). Dukoral is for travellers and has an effectiveness range of 55–88%; however, owing to its short shelf life, high price and needs for cold-chain circulation, it is not recommended for use by those living in endemic regions (Kirigia et al., 2009). Dukoral is only effective between the age 2 and 5 years although patients who have received the immunisation over the age of 5 years potentially acquire a 2-year protection against infection (Wierzba, 2019). Shanchol and Euvichol are made of inactivated *O1 Inaba, O1 Ogawa* and *O139* strains. CTB is not present in these vaccines (Wierzba, 2019). Based on a 5-year study, Shanchol has an effectiveness of roughly 65% protection (Hsiao et al., 2018). Shanchol and Euvichol may be used by all above the age of 1 year, while pregnant women should not (Russo et al., 2018). The Global OCV Stockpile currently has Shanchol and Euvichol-Plusvaccines available for mass vaccination. The Vaccine Alliance, or Gavi, is in favour of the stockpile (WHO, 2022).

7.5 Probiotics

Probiotics are being explored as a potential therapy or preventative measure for cholera infections. The gut microbiota is severely disrupted by *V. cholerae's* infection because of the excessive fluid retention and *V. cholerae* is the predominant bacterial species observed in the typical rice-water stools (Hay & Zhu, 2014). Hence, a novel measure for treating cholera infections is the restoration of the gut microbiome or the colony prevention using probiotic medication. It has been demonstrated that several bacterial species, including *Ruminococcus obeum* and a *Lactobacilli* strain, can eradicate or inhibit cholera pathogens (Hsiao et al., 2014; Kaur et al., 2018). Antibiotics and other medications, such as anti-inflammatory adjuvants, are frequently taken with probiotics; nevertheless, further research is necessary due to the drugs' unfavourable effects on the probiotics (Kaur et al., 2018). Probiotics that prevent cholera might reduce the spread of antibiotic-resistant *V. cholerae* strains by lowering antibiotic administration (Mao et al., 2018). Probiotics may also be a more effective treatment where it is impossible to prepare cold-chain vaccines and when clean water is scarce (Heikkilä et al., 2012).

The inherent issues of endemic or fragile territories, such as the economy, natural disasters, conflicts, national security and inadequate infrastructure present the greatest challenges in combating cholera. Due to these challenges, cholera must be treated utilising a multidimensional approach that may be modified to meet the unique needs of a specific outbreak. ORT has always been and will remain to serve as the first line of protection that can save cholera patients. ORTs are economical, simple to inject, and substantially decreases the severity of illness when paired with antibiotics. In addition, the WHO has prequalified Dukoral, Shanchol and Euvichol as modern cholera vaccines despite lack of the vaccines worldwide. The number of people who can get these immunizations could be significantly reduced during an outbreak. In addition, health surveillance is essential for directing preventive countermeasures in times of crisis, which helps to avert outbreaks. In order to reduce pandemic cholera, the world must improve ways for swiftly identifying epidemic strains, determine the most effective treatments for each strain and create novel treatments that are independent of clean water or cold-chain networks.

References

Ali, M., Nelson, A. R., Lopez, A. L., & Sack, D. A. (2015). Updated global burden of cholera in endemic countries. *PLoS Neglected Tropical Diseases, 9*(6), Article e0003832.

Ali, M., Kim, P., Zaman, K., & Clemens, J. (2019). Herd protection of unvaccinated adults by oral cholera vaccines in rural Bangladesh. *International Health, 11*(3), 229–234.

Baker-Austin, C., Stockley, L., Rangdale, R., & Martinez-Urtaza, J. (2010). Environmental occurrence and clinical impact of *Vibrio vulnificus* and *Vibrio parahaemolyticus*: A European perspective. *Environmental Microbiology Reports, 2*(1), 7–18.

Baracchini, T., King, A. A., Bouma, M. J., Rodó, X., Bertuzzo, E., & Pascual, M. (2017). Seasonality in cholera dynamics: A rainfall-driven model explains the wide range of patterns in endemic areas. *Advances in Water Resources, 108*, 357–366.

Barua, D. (1992). History of cholera. In *Cholera* (pp. 1–36). Springer.

Bavishi, C., & Dupont, H. L. (2011). Systematic review: The use of proton pump inhibitors and increased susceptibility to enteric infection. *Alimentary Pharmacology & Therapeutics, 34*(11–12), 1269–1281.

Bhandari, N., Mazumder, S., Taneja, S., Dube, B., Agarwal, R. C., Mahalanabis, D., & Bhan, M. K. (2008). Effectiveness of zinc supplementation plus oral rehydration salts compared with oral rehydration salts alone as a treatment for acute diarrhea in a primary care setting: A cluster randomized trial. *Pediatrics, 121*(5), e1279–e1285.

Bianchi, A. (1889). *Relazione e catalogo dei manoscritti di Filippo Pacini esistenti nella R.* Presso i principali librai.

Bryden, J. L. (1871). *Note on the epidemic connection of the cholera of Madras & Bombay with the cholera epidemics of the Bengal Presidency*. Office of the Superintendent of Government Print.

Butler, T. (2017). Treatment of severe cholera: A review of strategies to reduce stool output and volumes of rehydration fluid. *Transactions of the Royal Society of Tropical Medicine and Hygiene, 111*(5), 204–210.

Byrne, J. P. (Ed.). (2008). *Encyclopedia of pestilence, pandemics, and plagues*. Greenwood Press.

Cabrera, A., Lepage, J. E., Sullivan, K. M., & Seed, S. M. (2017). Vaxchora: A single-dose oral cholera vaccine. *Annals of Pharmacotherapy, 51*(7), 584–589.

Cash, R. A., Music, S. I., Libonati, J. P., Snyder, M. J., Wenzel, R. P., & Hornick, R. B. (1974). Response of man to infection with *Vibrio cholerae*. I. Clinical, serologic, and bacteriologic responses to a known inoculum. *Journal of Infectious Diseases, 129*(1), 45–52.

Centers for Disease Control and Prevention [CDC]. (2022). Diagnosis and detection. Retrieved from https://www.cdc.gov/cholera/diagnosis.html#:~:text=isolation%20and%20identification%20of%20vibrio,ideal%20for%20isolation%20and%20identification

Chowdhury, F. R., Nur, Z., Hassan, N., von Seidlein, L., & Dunachie, S. (2017). Pandemics, pathogenicity and changing molecular epidemiology of cholera in the era of global warming. *Annals of Clinical Microbiology and Antimicrobials, 16*(1), 1–6.

Colwell, R. R. (1996). Global climate and infectious disease: The cholera paradigm. *Science, 274*(5295), 2025–2031.

Deen, J., Mengel, M. A., & Clemens, J. D. (2020). Epidemiology of cholera. *Vaccine, 38*, A31–A40.

Farmer, P., Almazor, C. P., Bahnsen, E. T., Barry, D., Bazile, J., Bloom, B. R., & Weigel, J. L. (2011). Meeting cholera's challenge to Haiti and the world: A joint statement on cholera prevention and care. *PLoS Neglected Tropical Diseases, 5*(5), Article e1145.

Feachem, R. G. (1982). Environmental aspects of cholera epidemiology. III. Transmission and control. *Tropical Diseases Bulletin, 79*(1), 1–47.

Fredrick, T., Ponnaiah, M., Murhekar, M. V., Jayaraman, Y., David, J. K., Vadivoo, S., & Joshua, V. (2015). Cholera outbreak linked with lack of safe water supply following a tropical cyclone in Pondicherry, India, 2012. *Journal of health, population, and nutrition, 33*(1), 31–38.

Global Task Force on Cholera Control [GTFCC]. (2017). *Ending cholera a global roadmap to 2030*. Retrieved from https://www.gtfcc.org/wp-content/uploads/2019/10/gtfcc-ending-cholera-a-global-roadmap-to-2030.pdf

Glass, R. I., Becker, S., Huq, M. I., Stoll, B. J., Khan, M. U., Merson, M. H., & Black, R. E. (1982). Endemic cholera in rural Bangladesh, 1966–1980. *American Journal of Epidemiology, 116*(6), 959–970.

Glass, R. I., Holmgren, J. A. N., Haley, C. E., Khan, M. R., Svennerholm, A., Stoll, B. J., & Barua, D. (1985). Predisposition for cholera of individuals with O blood group possible evolutionary significance. *American Journal of Epidemiology, 121*(6), 791–796.

Guerrant, R. L., Carneiro-Filho, B. A., & Dillingham, R. A. (2003). Cholera, diarrhea, and oral rehydration therapy: Triumph and indictment. *Clinical Infectious Diseases, 37*(3), 398–405.

Haan, L. D., & Hirst, T. R. (2004). Cholera toxin: A paradigm for multi-functional engagement of cellular mechanisms. *Molecular Membrane Biology, 21*(2), 77–92.

Harris, J. B., Khan, A. I., LaRocque, R. C., Dorer, D. J., Chowdhury, F., Faruque, A. S., & Calderwood, S. B. (2005). Blood group, immunity, and risk of infection with *Vibrio cholerae* in an area of endemicity. *Infection and Immunity, 73*(11), 7422–7427.

Harris, J. B., LaRocque, R. C., Chowdhury, F., Khan, A. I., Logvinenko, T., Faruque, A. S., & Calderwood, S. B. (2008). Susceptibility to *Vibrio cholerae* infection in a cohort of household contacts of patients with cholera in Bangladesh. *PLoS Neglected Tropical Diseases, 2*(4), Article e221.

Harris, J. B., LaRocque, R. C., Qadri, F., Ryan, E. T., & Calderwood, S. B. (2012). Cholera. *Lancet,* 379(9835), 2466–2476. https://doi.org/10.1016/S0140-6736(12)60436-X

Harris, J. B. (2018). Cholera: Immunity and prospects in vaccine development. *The Journal of Infectious Diseases, 218*(suppl_3), S141–S146.

Hay, A., & Zhu, J. (2014). Microbiota talks cholera out of the gut. *Cell Host & Microbe, 16*(5), 549–550.

Hays, J. N. (2005). *Epidemics and pandemics: Their impacts on human history*. Abc-clio.

Heikkilä, J. E., Nybom, S. M., Salminen, S. J., & Meriluoto, J. A. (2012). Removal of cholera toxin from aqueous solution by probiotic bacteria. *Pharmaceuticals, 5*(6), 665–673.

Holman, S. R. (2015). *Beholden: Religion, global health, and human rights*. Oxford University Press.

Holmberg, S., Kay, D., Parker, R. R., Rao, N. U., Harris, J., Hargrett, N., & Blake, P. (1984). Foodborne transmission of cholera in Micronesian households. *The Lancet, 323*(8372), 325–328.

Holmgren, J. (1981). Actions of cholera toxin and the prevention and treatment of cholera. *Nature, 292*(5822), 413–417.

Hossain, Z. Z., Leekitcharoenphon, P., Dalsgaard, A., Sultana, R., Begum, A., Jensen, P. K. M., & Hendriksen, R. S. (2018). Comparative genomics of *Vibrio cholerae* O1 isolated from cholera patients in Bangladesh. *Letters in Applied Microbiology, 67*(4), 329–336.

Hsiao, A., Ahmed, A. M., Subramanian, S., Griffin, N. W., Drewry, L. L., Petri, W. A., & Gordon, J. I. (2014). Members of the human gut microbiota involved in recovery from *Vibrio cholerae* infection. *Nature, 515*(7527), 423–426.

Hsiao, A., Hall, A. H., Mogasale, V., & Quentin, W. (2018). The health economics of cholera: A systematic review. *Vaccine, 36*(30), 4404–4424.

Ivers, L. C., Farmer, P., Almazor, C. P., & Léandre, F. (2010). Five complementary interventions to slow cholera: Haiti. *The Lancet, 376*(9758), 2048–2051.

JSTOR Daily. (2020). *Anti-Asian racism in the 1817 cholera pandemic*. Retrieved from https://daily.jstor.org/anti-asian-racism-in-the-1817-cholera-pandemic/

Jutla, A., Aldaach, H., Billian, H., Akanda, A., Huq, A., & Colwell, R. (2015). Satellite based assessment of hydroclimatic conditions related to cholera in Zimbabwe. *PLoS ONE, 10*(9), Article e0137828.

Kaur, S., Sharma, P., Kalia, N., Singh, J., & Kaur, S. (2018). Anti-biofilm properties of the fecal probiotic lactobacilli against *Vibrio* spp. *Frontiers in Cellular and Infection Microbiology, 8*, 120.

Khan, R., Anwar, R., Akanda, S., McDonald, M. D., Huq, A., Jutla, A., & Colwell, R. (2017). Assessment of risk of cholera in Haiti following Hurricane Matthew. *The American Journal of Tropical Medicine and Hygiene, 97*(3), 896.

Kirigia, J. M., Sambo, L. G., Yokouide, A., Soumbey-Alley, E., Muthuri, L. K., & Kirigia, D. G. (2009). Economic burden of cholera in the WHO African region. *BMC International Health and Human Rights, 9*(1), 1–14.

Kirn, T. J., Lafferty, M. J., Sandoe, C. M., & Taylor, R. K. (2000). Delineation of pilin domains required for bacterial association into microcolonies and intestinal colonization by *Vibrio cholerae. Molecular Microbiology, 35*(4), 896–910.

Koch, R. (1884). An address on cholera and its bacillus. *British Medical Journal, 2*(1236), 453.

Kohn, G. C. (2007). *Encyclopedia of plague and pestilence: From ancient times to the present.* Infobase Publishing.

Kühn, J., Finger, F., Bertuzzo, E., Borgeaud, S., Gatto, M., Rinaldo, A., & Blokesch, M. (2014). Glucose-but not rice-based oral rehydration therapy enhances the production of virulence determinants in the human pathogen *Vibrio cholerae. PLoS Neglected Tropical Diseases, 8*(12), Article e3347.

Leibovici-Weissman, Y. A., Neuberger, A., Bitterman, R., Sinclair, D., Salam, M. A., & Paul, M. (2014). Antimicrobial drugs for treating cholera. *Cochrane Database of Systematic Reviews, 6.*

Leitner, D. R., Lichtenegger, S., Temel, P., Zingl, F. G., Ratzberger, D., Roier, S., & Schild, S. (2015). A combined vaccine approach against *Vibrio cholerae* and ETEC based on outer membrane vesicles. *Frontiers in Microbiology, 6*, 823.

Li, S. T. T., Grossman, D. C., & Cummings, P. (2007). Loperamide therapy for acute diarrhea in children: Systematic review and meta-analysis. *PLoS Medicine, 4*(3), Article e98.

Lippi, D., & Gotuzzo, E. (2014). The greatest steps towards the discovery of *Vibrio cholerae. Clinical Microbiology and Infection, 20*(3), 191–195.

Longini, I. M., Jr., Yunus, M., Zaman, K., Siddique, A. K., Sack, R. B., & Nizam, A. (2002). Epidemic and endemic cholera trends over a 33-year period in Bangladesh. *The Journal of Infectious Diseases, 186*(2), 246–251.

Lugo, R. V. (2008). Epidemia y Sociedad: Efectos del Cólera Morbo en Puerto Rico y en Costa Rica a Mediados Del Siglo Xix. *Diálogos Revista Electrónica, 9*, 218–242.

Mao, N., Cubillos-Ruiz, A., Cameron, D. E., & Collins, J. J. (2018). Probiotic strains detect and suppress cholera in mice. *Science Translational Medicine, 10*(445), eaao2586.

Mayo-Wilson, E., Imdad, A., Herzer, K., Yakoob, M. Y., & Bhutta, Z. A. (2011). Vitamin A supplements for preventing mortality, illness, and blindness in children aged under 5: Systematic review and meta-analysis. *BMJ, 343.*

McGrew, R. E. (1960). The first cholera epidemic and social history. *Bulletin of the History of Medicine, 34*(1), 61–73.

Meeusen, R., Duclos, M., Foster, C., Fry, A., Gleeson, M., Nieman, D., & Urhausen, A. (2013). Prevention, diagnosis and treatment of the overtraining syndrome: Joint consensus statement of the European College of Sport Science (ECSS) and the American College of Sports Medicine (ACSM). *European Journal of Sport Science, 13*(1), 1–24.

Mekalanos, J. J., Rubin, E. J., & Waldor, M. K. (1997). Cholera: Molecular basis for emergence and pathogenesis. *FEMS Immunology & Medical Microbiology, 18*(4), 241–248.

Milner, S. M., Green, W. B., III., Asuku, M. E., Feldman, M., Makam, R., Noppenberger, D., & Van Loon, I. N. (2011). From cholera to burns: A role for oral rehydration therapy. *Journal of Health, Population, and Nutrition, 29*(6), 648.

Mohareb, A. M., & Ivers, L. C. (2019). Disease and famine as weapons of war in Yemen. *The New England Journal of Medicine, 380*(2), 109–111.

Mondiale de la Santé, O., & World Health Organization [WHO]. (2021). Weekly epidemiological record. *Relevé épidémiologique hebdomadaire, 96*(37), 445–460.

Mora, C., Spirandelli, D., Franklin, E. C., Lynham, J., Kantar, M. B., Miles, W., & Hunter, C. L. (2018). Broad threat to humanity from cumulative climate hazards intensified by greenhouse gas emissions. *Nature Climate Change, 8*(12), 1062–1071.

Morales, R., Delgado, G., & Cravioto, A. (2008). Vibrio cholerae—genomics and molecular biology. incom

Mosley, W. H., Ahmad, S., Benenson, A. S., & Ahmed, A. (1968). The relationship of vibriocidal antibody titre to susceptibility to cholera in family contacts of cholera patients. *Bulletin of the World Health Organization, 38*(5), 777.

Mukerjee, S. (1963). Preliminary studies on the development of a live oral vaccine for anti-cholera immunization. *Bulletin of the World Health Organization, 29*(6), 753.

Mulhern, O. (2020). *Pandemic Map: Cholera.* Retrieved from https://earth.org/data_visualization/pandemic-map-cholera/

Nemmara, V. V., & Thompson, P. R. (2018). Development of activity-based proteomic probes for protein citrullination. *Activity-Based Protein Profiling*, 233–251.

Newton, A. E., Heiman, K. E., Schmitz, A., Török, T., Apostolou, A., Hanson, H., & Mintz, E. D. (2011). Cholera in United States associated with epidemic in Hispaniola. *Emerging Infectious Diseases, 17*(11), 2166.

Ng, Q. X., De Deyn, M. L. Z. Q., Loke, W., & Yeo, W. S. (2020). Yemen's cholera epidemic is a one health issue. *Journal of Preventive Medicine and Public Health, 53*(4), 289–292.

O'Connor, K. A., Cartwright, E., Loharikar, A., Routh, J., Gaines, J., Fouché, M. D. B., & Mahon, B. E. (2011). Risk factors early in the 2010 cholera epidemic, Haiti. *Emerging Infectious Diseases, 17*(11), 2136.

Orata, F. D., Keim, P. S., & Boucher, Y. (2014). The 2010 cholera outbreak in Haiti: How science solved a controversy. *PLoS Pathogens, 10*(4), Article e1003967.

Pascual, M., Bouma, M. J., & Dobson, A. P. (2002). Cholera and climate: Revisiting the quantitative evidence. *Microbes and Infection, 4*(2), 237–245.

Phillips, R. A. (1964). Water and electrolyte losses in cholera. In *Federation proceedings. Federation of American societies for experimental biology* (Vol. 23, No. 3; Pt. 1, pp. 705–12). Wash.

Pollitzer, R. (1959). *Cholera.* World Health Organization.

Pollitzer, R., & Burrows, W. (1955). Cholera studies: 4. Problems in immunology. *Bulletin of the World Health Organization, 12*(6), 945.

Rabbani, G. H., & Greenough III, W. B. (1999). Food as a vehicle of transmission of cholera. *Journal of Diarrhoeal Diseases Research*, 1–9.

Reen, F. J., Almagro-Moreno, S., Ussery, D., & Boyd, E. F. (2006). The genomic code: Inferring Vibrionaceae niche specialization. *Nature Reviews Microbiology, 4*(9), 697–704.

Reeves, P. R., & Lan, R. (1998). Cholera in the 1990s. *British Medical Bulletin, 54*(3), 611–623.

Reliefweb. (2021). *Cholera situation in Yemen*, April 2021. Retrieved from https://reliefweb.int/report/yemen/cholera-situation-yemen-april-2021

Richterman, A., Sainvilien, D. R., Eberly, L., & Ivers, L. C. (2018). Individual and household risk factors for symptomatic cholera infection: A systematic review and meta-analysis. *The Journal of Infectious Diseases, 218*(suppl_3), S154–S164.

Rosenberg, C. E. (2009). *The cholera years: The United States in 1832, 1849, and 1866.* University of Chicago Press.

Roy, S. K., Hossain, M. J., Khatun, W., Chakraborty, B., Chowdhury, S., Begum, A., & Chowdhury, R. (2008). Zinc supplementation in children with cholera in Bangladesh: Randomised controlled trial. *BMJ, 336*(7638), 266–268.

Russo, P., Ligsay, A. D., Olveda, R., Choi, S. K., Kim, D. R., Park, J. Y., & Excler, J. L. (2018). A randomized, observer-blinded, equivalence trial comparing two variations of Euvichol®, a bivalent killed whole-cell oral cholera vaccine, in healthy adults and children in the Philippines. *Vaccine, 36*(29), 4317–4324.

Ryan, E. T. (2011). The cholera pandemic, still with us after half a century: Time to rethink. *PLoS Neglected Tropical Diseases, 5*(1), Article e1003.

Sack, R. B., Siddique, A. K., Longini, I. M., Jr., Nizam, A., Yunus, M., Islam, M. S., & Colwell, R. R. (2003). A 4-year study of the epidemiology of *Vibrio cholerae* in four rural areas of Bangladesh. *The Journal of Infectious Diseases, 187*(1), 96–101.

Sack, D. A., Sack, R. B., & Chaignat, C. L. (2006). Getting serious about cholera. *New England Journal of Medicine, 355*(7), 649–651.

Seas, C., Miranda, J., Gil, A. I., Leon-Barua, R., Patz, J., Huq, A., & Sack, R. B. (2000). New insights on the emergence of cholera in Latin America during 1991: The Peruvian experience. *The American Journal of Tropical Medicine and Hygiene, 62*(4), 513–517.

Sharma, N. C., Mandal, P. K., Dhillon, R., & Jain, M. (2007). Changing profile of *Vibrio cholerae* 01, 0139 in Delhi & its periphery (2003–2005). *Indian Journal of Medical Research, 125*(5), 633.

Shears, P. (2001). Recent developments in cholera. *Current Opinion in Infectious Diseases, 14*(5), 553–558.

Silva, A. J., & Benitez, J. A. (2016). *Vibrio cholerae* biofilms and cholera pathogenesis. *PLoS Neglected Tropical Diseases, 10*(2), Article e0004330.

Smallman-Raynor, M. R., & Cliff, A. D. (2004). Impact of infectious diseases on war. *Infectious Disease Clinics, 18*(2), 341–368.

Snow, J. (1855). Further remarks on the mode of communication of cholera; including some comments on the recent reports on cholera by the General Board of Health. *Medical times and Gazette, 11*, 84–88.

Snowden, F. M. (1995). *Naples in the Time of Cholera, 1884–1911*. Cambridge University Press.

Subba, R. M., & Howard-Jones, N. (1978). Original observations of Filippo Pacini on *Vibrio cholera*. *Bulletin of the Indian Institute of History of Medicine Hyderabad, 8*, 32–38.

Taylor, R. K., Miller, V. L., Furlong, D. B., & Mekalanos, J. J. (1987). Use of phoA gene fusions to identify a pilus colonization factor coordinately regulated with cholera toxin. *Proceedings of the National Academy of Sciences, 84*(9), 2833–2837.

Tulchinsky, T. H. (2018). John Snow, cholera, the broad street pump: Waterborne diseases then and now. *Case Studies in Public Health*, 77–99. https://doi.org/10.1016/B978-0-12-804571-8.00017-2

Vezzulli, L., Grande, C., Reid, P. C., Hélaouët, P., Edwards, M., Höfle, M. G., & Pruzzo, C. (2016). Climate influence on Vibrio and associated human diseases during the past half-century in the coastal North Atlantic. *Proceedings of the National Academy of Sciences, 113*(34), E5062–E5071.

Wachsmuth, K., Blake, P. A., & Olsvik, Ø. (1994). *Vibrio cholerae* and cholera. ASM Press.

Waldor, M. K., Colwell, R., & Mekalanos, J. J. (1994). The *Vibrio cholerae* O139 serogroup antigen includes an O-antigen capsule and lipopolysaccharide virulence determinants. *Proceedings of the National Academy of Sciences, 91*(24), 11388–11392.

Wendt, E. C. (1885). *A treatise on Asiatic cholera*. W. Wood and Company.

Wierzba, T. F. (2019). Oral cholera vaccines and their impact on the global burden of disease. *Human Vaccines & Immunotherapeutics, 15*(6), 1294–1301.

Williams, P. C., & Berkley, J. A. (2018). Guidelines for the management of paediatric cholera infection: A systematic review of the evidence. *Paediatrics and International Child Health, 38*(supp 1), S16–S31.

World Health Organization [WHO]. (1992). *Emerging and other communicable diseases, surveillance and control. WHO guidance on formulation of national policy on the control of cholera*. Retrieved from https://apps.who.int/iris/bitstream/handle/10665/65507/whocddser9216rev1.pdf?sequence=1&isAllowed=y

World Health Organization [WHO]. (2005). The treatment of diarrhoea: A manual for physicians and other senior health workers (No. WHO/FCH/CAH/05.1). World Health Organization.

World Health Organization [WHO]. (2017). Ending cholera, a global roadmap to 2030. In *Ending cholera a global roadmap to 2030* (pp. 32–32).

World Health Organization [WHO]. (2022). *Cholera*. Retrieved from https://www.who.int/news-room/fact-sheets/detail/cholera

Yahr, T. L. (2006). A critical new pathway for toxin secretion? *New England Journal of Medicine, 355*(11), 1171.

Yoshifumi, T. (2010). *Vibrio cholerae* and cholera toxin: From Calcutta to Kolkata. *Science and Culture, 75*, 149–152.

Zhang, P., Li, F., Liang, W., Li, J., Kan, B., & Wang, D. (2014). The seventh pandemic *Vibrio cholerae* O1 El Tor isolate in China has undergone genetic shifts. *Journal of Clinical Microbiology, 52*(3), 964–967. https://doi.org/10.1128/JCM.03121-13

Chapter 7
Coronavirus Disease (COVID-19)

1 Introduction

Coronavirus disease 2019 (COVID-19), an extremely contagious viral infection associated with Severe Acute Respiratory Syndrome Coronavirus 2 (SARS-CoV-2), has had a disastrous impact on the world's demography, culminating about 6,520,263 deaths globally as of September 29, 2022, and also emerged as the most devastating global health catastrophe since the influenza pandemic during 1918 (World Health Organization [WHO], 2022a). CoVs are enveloped (E-protein), single-stranded, positive-sense RNA viruses belonging to the Coronaviridae family, Nidovirales order (Fan et al., 2019; ur Rehman et al., 2021). The virus has crown-like spikes on its surface, hence the name "Corona," which implies crown (Pal et al., 2020). The CoV family has four subgroups: alpha (α), beta (β), gamma (γ) and delta (δ) (Cowley, 2016). Bats, palm civets, livestock and other animals are the main carriers of virus (Islam et al., 2021). Prior to severe acute respiratory syndrome coronavirus (SARS-CoV) outbreak in Guangdong, China in 2002, it was thought that these pathogens could only spread infections amongst animals. Eventually, the Middle Eastern nations had an outburst of the Middle East respiratory syndrome coronavirus (MERS-CoV) (Mahalmani et al., 2020). Both SARS-CoV and the MERS-CoV share 80% genomic homology, respectively (Muralidar et al., 2020). All these outbreaks, i.e. MERS-Co, SARS-CoV and SARS-CoV-2 resulted in severe pneumonia with fatality rates of 2.9%, 9.6% and 36%, respectively (Hui & Zumla, 2019; Nassar et al., 2018; Wang et al., 2020a, 2020b, 2020c, 2020d, 2020e).

SARS-CoV has indeed been widely documented and is among the viruses thought to be the cause of respiratory illnesses. However on December 8, 2019, an epidemic outbreak with respiratory infections was recorded in Wuhan, the major metropolitan region in the Chinese province of Hubei. The infestation was initially thought to be a pneumonia-related with an unknown aetiology. However, the Chinese Health Authority eventually informed WHO China Country Office that the respiratory infection was actually caused by a novel CoV identified as 2019-nCoV (WHO, 2020a).

M. H. Fulekar et al., *Pandemic Over the Centuries*,
https://doi.org/10.1007/978-3-031-94270-9_7

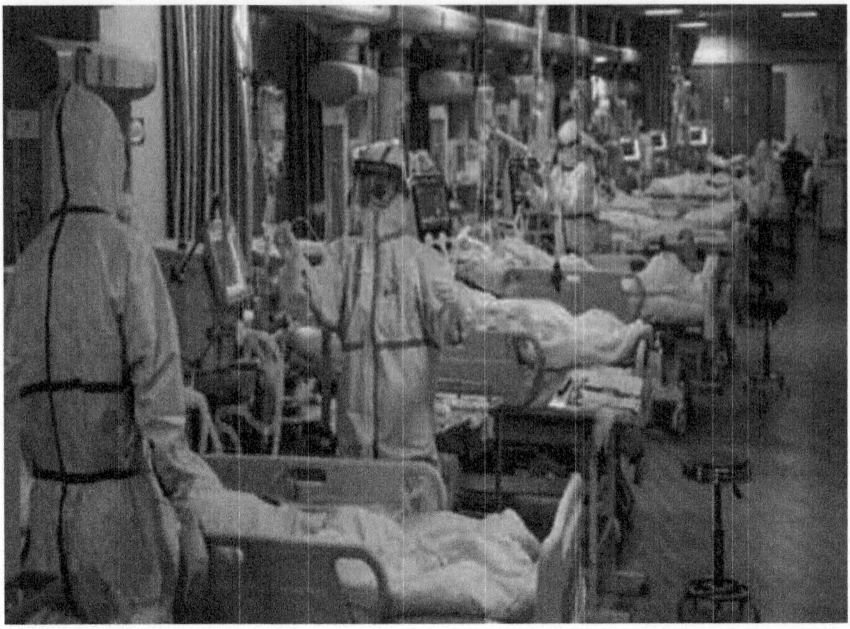

Fig. 1 Doctors attending COVID-19 patients. (*Source* Google)

Thereafter, the virus spread so widely and quickly that the WHO declared the outbreak to be a global emergency and termed it coronavirus disease-2019 (COVID-19), whereas this new 2019-nCoV got re-named as SARS-CoV-2 (Yang et al., 2020a, 2020b, 2020c) (Figs. 1, 2).

COVID-19 was declared a Public Health Emergency of International Concern (PHEIC) by the WHO on January 31, 2020; and subsequently declared a global pandemic on March 11, 2020 (Adibi et al., 2022; Rauf et al., 2020). SARS-CoV-2 is spread by close contact with infected persons at a distance of less than 1–2 m (often 3–6 ft.) through saliva and liquid products from a patient, as well as through coughing or sneezing (Ayenigbara, 2020). Additionally, if left unattended, the intensity of COVID-19 patients' coughs and sneezes can drive the SARS-CoV-2 to a further distance of 4.5–8.2 m (15–27 ft.), which is significantly dangerous for transmission of infection (Bourouiba, 2020). Additionally, SARS-CoV-2 can infect people whenever contaminated objects come in contact with their eyes, nose or mouth for at least 3 days. SARS-CoV-2 can also be discovered on any surfaces (WHO, 2020b). The transmission of SARS-CoV-2 could also occur when there are no obvious signs or symptoms of infection (asymptomatic), particularly 1–2 days prior to symptom onset, as one patient can potentially transmit the infection to 2 or more persons as was evident in some transmission cases in european countries (European Centre for Disease Prevention & Control, 2020).

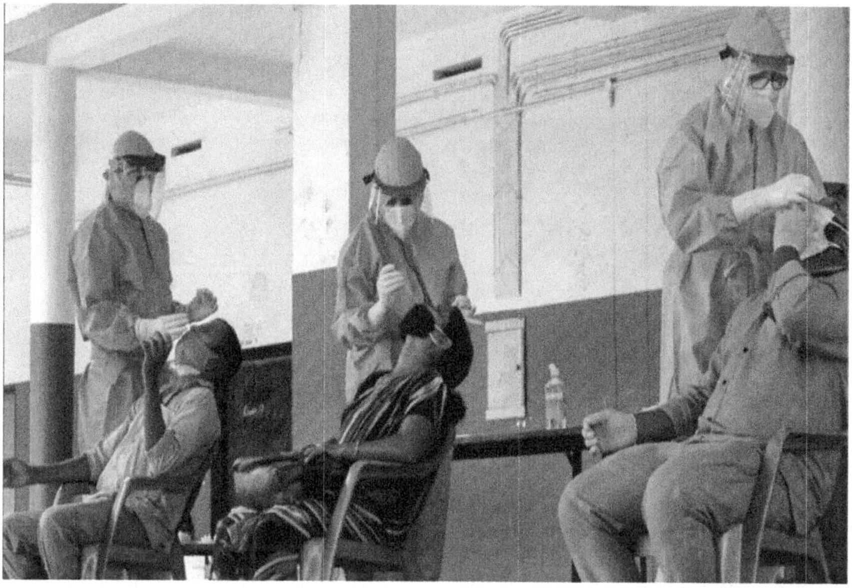

Fig. 2 Individual getting tested for SARS-COV-2. (*Source* Google)

COVID-19 has a varying infectivity time period, however, it lasts between 2 and 14 days (Velavan & Meyer, 2020). The disease's clinical features vary from asymptomatic scenarios to milder symptoms like fever, cough, sore throat, headache and nasal congestion, and serious forms like pneumonia, respiratory failure requiring respiratory support, multi-organ dysfunction, sepsis and mortality (WHO, 2020c; Zaim et al., 2020). Evidence from China revealed that individuals with inherent illnesses, like high blood pressure, diabetes, chronic pulmonary disease, cardiovascular disease and cancer, had considerably greater mortality rates than those who had any past record (Team, 2020). The COVID-19 virus has been spreading throughout the world for the past 2 years, wreaking havoc in nations such as USA, India, Brazil, Russian Federation, France, the United Kingdom, Italy, Spain, Argentina, Colombia, Germany, Mexico, Poland, Iran and Turkey, all having millions of confirmed COVID-19 cases (ur Rehman et al., 2021). There has been 771,191,203 confirmed cases of COVID-19, including 6,961,014 deaths throughout the world as of 4th October 2023 (Fig. 3).

SARS-CoV-2, like other RNA viruses, is susceptible to evolutionary change with the emergence of mutations over time, culminating in mutant forms that may possess distinctive characteristics from its parental lineages. Several SARS-CoV-2 subtypes have been identified throughout pandemic. However, only a small number were recognized to be variants of concern (VOCs) by the WHO attributed to their implications on public health around the world (Cascella et al., 2022). Five SARS-CoV-2 VOCs: Alpha (B.1.1.7); Beta (B.1.351); Gamma (P.1); Delta (B.1.617.2); and Omicron (B.1.1.529) have been detected since the beginning of the pandemic

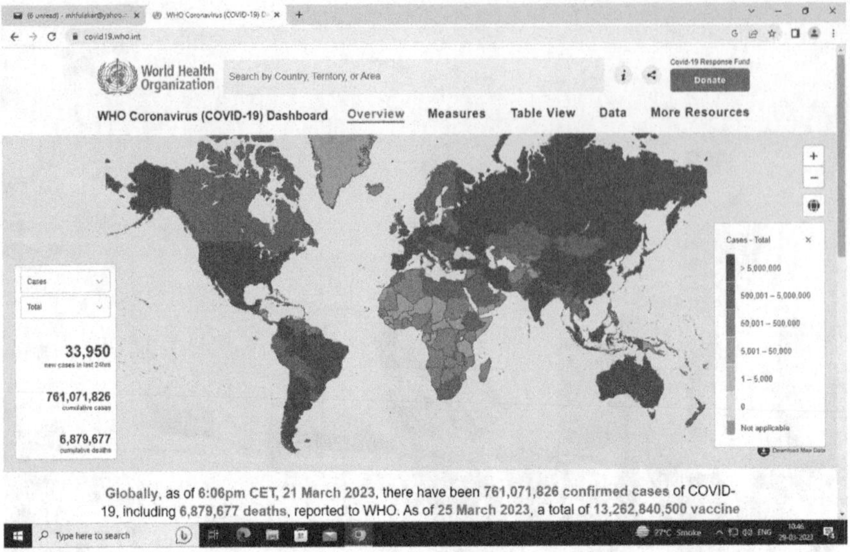

Fig. 3 COVID-19 world map. (*Source* WHO)

according to the WHO's epidemiological data, as of December 11, 2021 (Centers for Disease Control & Prevention [CDC], 2022a).

Although the pandemic is still looming, there are signs that it could be transforming to an endemic form (Biancolella et al., 2022). The prevailing Omicron strain may result in this (Adam, 2022). Extraordinary measures are being taken across the world to control the pandemic, which has collapsed healthcare systems and caused long-lasting socioeconomic and geopolitical consequences. A detailed knowledge of SARS-CoV-2 as well as the control of COVID-19, as a result of significant advancements in medical research, has aided in curtailing the further spread of this virus. Given the incredible pace with which vaccines for COVID-19 safeguard have been developed, as well as vigorous global mass vaccination initiatives, involving vaccine boosters, the introduction of these novel SARS-CoV-2 variants threatens to undo the enormous achievement accomplished to-date in preventing the transmission of this virus (WHO, 2022b).

2 SARS-COV-2 Epidemiology

2.1 Origin of SARS-CoV-2

The genesis of the pathogenic virus, SARS-CoV-2 has been the focus of extensive study ever since the first reporting of unusual pneumonia in Wuhan, Hubei Province, China (Jiang et al., 2020; Zhang et al., 2020). Numerous signs of such past zoonotic outbreaks are evident in the origin of SARS-CoV-2. The SARS-CoV that infected people in Foshan, Guangdong province, China, in November 2002 as well as in Guangzhou, Guangdong province, in 2003 clearly resembles it (Xu et al., 2004). Both of these SARS-CoV featured in market places which sold living animals, especially civets and raccoon dogs (Guan et al., 2003), which were also being sold alive in Wuhan markets in 2019 (Xiao et al., 2021) and are identified as being predisposed to SARS-CoV-2 infections (Freuling et al., 2020).

The Huanan market in Wuhan had been a significant and first site of SARS-CoV-2 infection, according to epidemiological statistics (Courtier-Orgogozo & de Ribera, 2022). The virus is currently believed to have primarily originated from bats, before spreading via one or more intermediate hosts, perhaps pangolins, to infect humans (El-Sayed & Kamel, 2021). The National Health Commission of China issued information regarding the outbreak and advised viral pneumonia on January 12, 2020 (Zhou et al., 2020a, 2020b). The virus was recognised as a new CoV from the sequence-based assessment of the patient specimens (WHO, 2021a). The pathogen was designated as SARS-CoV-2 by the International Committee on Taxonomy of Viruses (ICTV) on February 11, 2020 due to the virus' genetic similarity to the SARS CoV that caused the 2003 outbreak (WHO, 2020d). On March 11, 2020, WHO labeled the infection as "COVID 19" and declared it a worldwide pandemic, adhering to governing guidelines of the World Organization for Animal Health (OIE) and the Food and Agriculture Organization (FAO) of the United Nations (WHO, 2020e).

The 73rd World Health Assembly (WHA) (the regulatory body of the the WHO) voted on a resolution (WHA 73.1) in May 2020 demanding the WHO Director General to investigate the origin of SARSCoV-2 as well as the pathway of human transmission (WHO, 2021b). In July 2020, WHO deployed a delegation to China to draft a work plan and start gathering and evaluating data. A group of 34 professionals, including 17 from China and 17 from other nations and organisations, travelled to Wuhan between January 14 and February 10, 2021, to gather and analyse both new and existing data. The WHO published a report on the expedition on March 30, 2021, summarising the outcome of site inspections and interviews with both local and national specialists in China. The study reported following four theories for the origin of SARS-CoV-2 and evaluated how likely each was to be the reason of infection (WHO, 2021c).

1. **Introduction through an intermediate host (a "likely to-very-likely" hypothesis)**: The virus was carried by an intermediary host species that was infected by an animal reservoir host and then spread to humans. The existence of any intermediate hosts is unknown.
2. **Direct zoonotic spillover (a "possible-to-likely" hypothesis)**: A possible route through which SARS-CoV-2 might have spread from an animal reservoir host to a human. Given the significant genetic resemblance between SARS-CoV-2 and CoVs discovered in particular bat species found in China and other parts of South Asia, researchers believe that bats are a logical reservoir host.
3. **Introduction through cold/food-chain products (a "possible" hypothesis)**: SARS-CoV-2 might have transmitted through contaminated food, possibly

including frozen, imported items, which people then consumed. SARSCoV-2 has been isolated from frozen food, packaging and cold-chain items.
4. **Introduction through a laboratory incident (an "extremely unlikely" hypothesis)**: Laboratory staff might have unintentionally contracted and subsequently transmitted SARS-CoV-2 whilst investigating CoVs in bats.

The WHO report stated that the possibility that the virus was purposely unleashed had been ruled out by other scientists, thus, it did not take this into consideration. In the end, the researchers were unable to pinpoint the origin of SARS-CoV-2 and suggested additional research.

2.2 *Structure of SARS-CoV-2*

SARS-CoV-2 genome (30 kb in size) encodes a large, non-structural polyprotein (ORF1a/b), which is further proteolytically fragmented to produce 15/16 proteins, 4 structural proteins [spike (S), nucleocapsid (N), membrane (M) and envelope (E) proteins] and 5 accessory proteins (ORF3a, ORF6, ORF7, ORF8 and ORF9) (Li et al., 2020a, 2020b, 2020c; Mariano et al., 2020; Redondo et al., 2021). The structural proteins are crucial for SARS-CoV-2 assembly and infection (Satarker & Nampoothiri, 2020). The transmembrane protein known as the spike or S glycoprotein, is located in the virus's outermost surface and has a molecular weight of roughly 150 kDa. By targeting the angiotensin-converting enzyme 2 (ACE2) produced in cells of the lower respiratory tract, S protein creates homotrimers that protrude from the viral surface and aids in the attachment of enveloped viruses to host cells. The host cell's furin-like protease splits this glycoprotein into S1 and S2 components. With the receptor binding domain (RBD) make-up, part S1 determines the host virus spectrum and cellular tropism, whereas S2 drives virus fusion in transmitting host cells (Astuti, 2020). In order to generate monoclonal antibody (mAb) medications and also to direct the design and development of vaccinations, it is important to comprehend the structure and function of the S protein (Li et al., 2020a, 2020b, 2020c). The structural element of CoV that is found in endoplasmic reticulum-Golgi area and is structurally bound to the virus's nucleic acid material is the nucleocapsid known as N protein (McBride et al., 2014). The protein is engaged in mechanisms pertaining to the viral genome, including viral replication cycle and the biological response of host cells to virus infection since the protein is attached to RNA. Additionally, N protein is highly phosphorylated, which may alter its structural make-up and increase its affinity for viral RNA (Hatmal et al., 2020). The membrane protein, also known as the M protein, is another crucial component of this virus. It is the most structurally complex protein and affects how the virus's envelope is shaped. All other structural proteins may bind to this protein (Thomas, 2020). By stabilising the N protein-RNA complex within the inner viral genome, binding to M protein aids in the stabilisation of N proteins and facilitates the completion of viral assembly (Schoeman & Fielding, 2019). The final part of the SARSCoV is the envelope or E protein, which

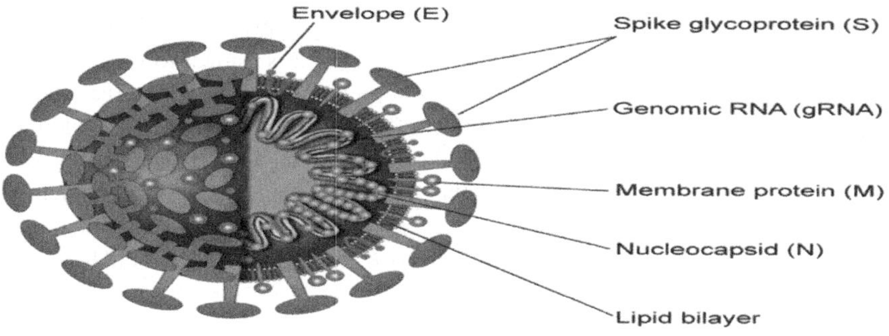

Fig. 4 Structure of SARS-COV-2. (*Courtesy* Beig Parikhani et al., 2021)

is the smallest protein and essential for the growth and maturation of the virus (Cao et al., 2021) (Fig. 4).

2.3 Classification of SARS-COV-2

The International Committee on Virus Taxonomy (ICT) has classified CoVs belonging to the Coronaviridae family and the Nidovirales order (de Groot et al., 2011; Jaiswal & Saxena, 2020). According to Tekes and Thiel (2016), the family is further distinguished into two subfamilies, i.e., Orthocoronavirinae and Torovirinae. The subfamily Orthocoronavirinae contains 4 genera: (a) αCoVs (14 subgenera and 19 species): HCoV-229E and HCoV-NL63 infecting humans; (b) βCoVs (5 subgenera and 14 species): HCoV-OC43, SARS-CoV, HCoV-HKU1 and MERS-CoV infecting humans; (c) γCoVs (3 subgenera and 5 species): infecting whales and birds; and (d) δCoVs(3 subgenera and 7 species): isolated from pigs and birds (Wang et al., 2018). SARS-CoV-2 is a member of the βCoVs genera, which also includes the two incredibly deadly viruses MERS-CoV and SARS-CoV (Abdelrahman & Wang, 2020; Sobti et al., 2021).

Infection with αCoVs and βCoVs causes respiratory issues in human and gastrointestinal problems in animals (Khan et al., 2021a, 2021b, 2021c). According to genomic analysis, the likely gene origins of αCoVs and βCoVs are bats and rodents. On the other hand, avian species appear to be the genetic origins of γCoVs and δCoVs. Multiple host species are susceptible to respiratory, intestinal, hepatic, and neurological ailments brought on by members of this huge virus family (Hughes et al., 2022). These viruses can overcome species barriers for unknown causes and infect people with symptoms varying from the common cold to more serious conditions like MERS and SARS (Dhama et al., 2020).

The SARS-CoV-2 single-stranded RNA genome has 29891 nucleotides, which code for 9860 amino acids (Chan et al., 2020). In fact, there is a significant degree of similarity (96%) between human SARS-CoV-2 and the CoV RaTG13 of bats (*Rhinolophus affinis*) (Ye et al., 2020). Acute severe pneumonia is the main symptom of both SARS-CoV and SARS-CoV-2, which is largely transmitted and enter through the respiratory tract (Chan et al., 2020). SARS-CoV-2 is known to have a greater histotropic range of infection than SARS-CoV, including the intestinal tract, kidney, neurological system and other areas (Chen et al., 2020; Hu et al., 2021). A lengthy gestation period (5.4 median incubation days) is projected for SARS-CoV-2 virus, which can reach as long as 13.7 days in 95% of symptomatic patients and sometimes even prolonged in certain asymptomatic patients. SARS-CoV-2 also exhibits high transmissibility, with a R0 value approximated at 2.3, however, that could be as high as 5.7 (Yang et al., 2020a, 2020b, 2020c).

2.4 *Variants of SARS-COV-2*

Viruses such as SARS-CoV-2 are constantly evolving as alterations in the genetic code (induced either by mutations or recombination) throughout genomic replication (Centers for Disease Control & Prevention, 2021a, 2021b, 2021c, 2021d). Moreover, a variant is a viral genotype with one or more mutations in its genetic sequence. Because SARS-CoV-2 is an RNA virus in origin, it is prone to mutations at a fast pace, which accelerates the emergence of multiple variations (Li et al., 2020a, 2020b, 2020c). The potential of RNA-dependent RNA polymerase (RdRp) to switch between strands and its error-prone behavior are the causes of the high genetic variation of SARS-CoV-2 (Lazarevic et al., 2021). A powerful SARS-CoV-2 strain that can infect humans eventually evolved as a result of the recurrent recombination of several coronavirus strains (Lai & Cavanagh, 1997). It is possible that the change in ACE-2 binding affinity is what causes the spike protein mutations, which mostly affect RBD, to transform a very deadly variant into highly infectious one (Zhao et al., 2004). Mutations may make the virus increasingly contagious, worsen the illness, affect diagnosis, treatments or immunizations.. Whenever these variants enhance the threat to public health, it is referred as "variants of concern." Ever since outbreak of COVID-19 pandemic, the Global Initiative on Sharing Avian Influenza Data (GISAID), Next strain and Pango sequencing database have been tracking virus mutations or variants (WHO, 2021d).

SARS-CoV-2 variants are classified into three groups by the CDC and WHO, viz. variants of interest (VOI), variants of concern (VOC) and variants of high consequences (VOHC). These are primarily described as variants linked to lowered treatment or vaccine effectiveness, enhanced trnsmission, and also serious infections. As the pandemic advanced, these were modified, and on September 21, 2021, a new category, i.e. Variants Being Monitored (VBM) was introduced (Feumba et al., 2022).

2.4.1 SARS-CoV-2 Variants Being Monitored (VBM)

VBM variants are those that have been linked to much more clinical complications or enhanced transmissions but are no longer found or are circulating at extremely low levels, and wherever evidence suggests that there is a probable or definite influence on authorised or licensed therapeutic interventions. Such variants do not present a serious or immediate public health threat (WHO, 2020f).

Whenever the frequency of occurrence of Variant of Interest or a Variant of Concern have significantly and consistently decreased over time or when additional data indicates that the variant does not significantly endanger public health, the variant may be relegated to this list (CDC, 2021a). The WHO does not identify any circulating variants that are being monitored as of August 2, 2022. A detailed list of COVID-19 VBM in past can be accessed from https://www.cdc.gov/coronavirus/2019-ncov/variants/variant-classifications.html?.

2.4.2 SARS-CoV-2 Variants of Concern (VOCs)

A variant that is associated with an indication of a rise in disease transmission, potentially serious ailment (like increase in hospitalised patients or mortalities), a notable decline in neutralisation of antigen by antibodies produced during earlier infection or vaccination, a decrease in the efficacy of therapeutic interventions or immunisations or diagnostic detection malfunctions (Aleem et al., 2022). Concerning variants may necessitate taking one or perhaps more relevant public health measures, like notifying WHO in accordance with the International Health Regulations, making local or regional efforts to control transmission, increasing screening or conducting research to ascertain the efficacy of vaccines and treatments against the variant. The WHO started allocating Greek alphabets to VOIs and VOCs in May 2021. This was to make them simpler to recall and to avoid the stigma attached to identifying them by the nation where they were first discovered, which might also discourage nations from disclosing its occurrence and is not always indicative of where they originally appeared. It is noteworthy that several Greek letters, such "Nu" and "Xi" were not allocated to a variant. According to the WHO, these letters were omitted because "Nu" might be mistaken for "new" quite readily and "Xi" is a popular family surname. Additional factors could be the creation of novel diagnostics, the modification of vaccines or treatments, depending on the traits of the variant (Gavi & The Vaccine Alliance, 2021a).

Alpha (B.1.1.7 Lineage)

Alpha (B.1.1.7 or GRY) was the first of the variants that gained widespread attention. Infections spiked in December 2020, the same year that Alpha initially appeared in Great Britain during November 2020 (Katella, 2022). It has 17 mutations that distinguish it from the classic Wuhan strain, 8 of which have been discovered in the spike protein, the virus's key to contracting living cells. One mutation is called N501Y that increases the spike protein's capacity to link to cellular receptors and makes the virus more infectious. Further, it has the P681H mutation, whose role is undetermined yet has repeatedly appeared unexpectedly, as well as the D614G mutation, that is predicted to enhance viral replication. According to estimates, Alpha is between 43 and 82% more pathogenic than the classic Wuhan strain (Gavi & The Vaccine Alliance, 2021a). The most typical Alpha variant symptoms include fever, cold, body pains and running nose. The WHO declared the alpha to be a Variant of Concern on December 18, 2020. No vaccination is believed to be 100% successful in avoiding this dreadful virus, regardless of the fact that it is developed in conformity with the original strain. Yet, some vaccines act as a line of defence against it. In context of the Alpha variant, vaccinations from companies like Pfizer, Biotech, AstraZeneca-Oxford, Johnson & Johnson, Novavax and Moderna function as a barrier and limit serious hospitalisation and fatal circumstances. However, according to Dr. Jenny Harries, Chief Executive of the UK Health and Security Agency, the Alpha had "disappeared," and had been replaced by the Delta (Sky news, 2021).

Beta (B.1.351 Lineage)

The second wave of COVID-19 outbreak was caused by variant B.1.351, also known as Beta variant or GH501Y.V2 with numerous spike mutations, and was initially discovered in South Africa in October 2020 (Aleem et al., 2022). On December 18, 2020, it was classified as a VOC. Although there is little proof that Beta is linked to more severe disease, it is considered to be about 50% more transmissible than earlier forms. The majority of the Beta variant symptoms were severe body aches, fever and loss of smell (Gavi & The Vaccine Alliance, 2021a).

The spike protein of the variant has 9 mutations (L18F, D80A, D215G, R246I, K417N, E484K, N501Y, D614G and A701V), 3 of which are present in the RBD and augment the binding ability for the ACE receptors (Cascella et al., 2022). According to reports, the K417N mutation has a lower ability to be neutralised by convalescent sera, post-vaccination sera and monoclonal antibody treatment.

The Oxford/AstraZeneca vaccination was 82% effective in preventing hospitalisation or death by either Beta or Gamma versions, according to recent factual data from Canada. Numerous commonly used vaccinations, such as Pfizer/BioNTech and Moderna vaccines, are thought to effectively defend it after two doses (Gavi & The Vaccine Alliance, 2021a).

Gamma (P.1 Lineage)

Gamma variant (or GR/501Y.V3) of the P.1 lineage was first noticed in Brazil in December 2020 and was deemed a variant of concern in January 2021. In contrast to other variants, patients who have recovered from COVID-19 had only 54–79% chances of protection against reinfection from the Gamma variant. This suggests that the Gamma variant is indeed 1.7–2.4 times more transmissible than other VOCs (Gavi & The Vaccine Alliance, 2021a). The fact is that COVID-19 vaccinations currently in use appear to be effective against it. The Gamma variant exhibited no symptoms that set it apart from other variants. Each of the manifestations were somewhat comparable to older COVID-19 versions.

It carries the mutations E484K, N501Y and D614G along with various additional severe mutations. Additionally, it carries the K417T mutation, which is linked to enhanced adherence to human cells and may facilitate the virus propagation, and the H655Y mutation, whose significance is undetermined. This E484K mutation has an immunological escape mechanism that allowed it to infiltrate the body and cause serious illness (Salehi-Vaziri et al., 2022). But after examining the impact of several vaccinations and their types against this variant, it was found that the Pfizer and Moderna vaccines had significantly worked as a barrier for this variant. Because no cases were found in the past weeks/months, the WHO declared the Alpha, Beta and Gamma variants as de-escalated variants in March 2022.

Delta (B.1.617.2)

The devastating second wave of COVID-19 in India was caused by the Delta form, which was discovered for the first time in May 2021 (Gavi & The Vaccine Alliance, 2021b). At first, the Delta variant was thought to be a variant of interest. Nonetheless, the WHO classified this type as a VOC in May 2021 as a result of its global spread. The Delta variant of SARS-CoV-2 is thought to be 40–60% more contagious than the Alpha and nearly twice as contagious as the classic Wuhan strain. According to data from Scotland, it is nearly twice likely to result in hospitalisation than Alpha. Similar to earlier strains, the symptoms included fever, running nose, headache and body aches. This strain also causes coughing and loss of smell. Additionally, some researchers assert that the Delta variant is much more deadly than other strains as it can cause additional body ailments and disabilities like gangrene, hearing loss, gastrointestinal problems and blood clots that kill the affected tissues (Gavi & The Vaccine Alliance, 2021a).

Delta has the D614G and a number of other mutations that are absent from other VOCs. These mutations include the L452R, which is considered to augment infection rate and may aid the virus in escaping immune cell destruction (Goher et al., 2022); T478K, which might also aid the virus in avoiding immune system detection (Jhun et al., 2021); and the P681R mutation, which is linked to an increased ability to cause severe symptoms (Liu et al., 2022a, 2022b). Despite claims that immunisations may be slightly less efficient than in the past at preventing infection with the Delta strain, they continue to offer excellent protection against severe disease. Delta variant was omitted from the list of VOCs on June 7, 2022, due to a 12-week non-detection period.

Omicron (B.1.1.529 Lineage)

Omicron form was initially discovered in South Africa on November 23, 2021, after South African researchers notified the WHO of an abrupt rise in COVID-19 cases in Gauteng province, which occurred at the same time (Viana et al., 2022). Because the Chinese surname Xi and the English word "new" are similar, the WHO omitted the Greek letters Nu and Xi that came before it (Indian Express, 2021). There are many mutations in Omicron, some of which are unsettling. As a result, on November 26, 2021, the WHO immediately identified variant as VOC, and now it is the only VOC in circulation. Although no unique symptoms have been noted in its patients, exhaustion and aches have been observed as prevalent symptoms, and the loss of smell, that was previously produced by other variants, has not yet been identified (Gavi & The Vaccine Alliance, 2021a).

The Omicron form has the greatest mutations of all the VOCs so far, making it more transmissible and somewhat resistant to the immunity developed by COVID-19 vaccines (Araf et al., 2022). The Omicron variant's genome sequencing data showed that the spike protein, which the SARS-CoV2 protein uses to recognise host cells, has more than 30 mutations (Wu et al., 2022). The data assessment of these mutations reveals the possibility of increased transmission through immune response evasion. The N501Y mutation improves the ACE2 receptor's binding ability, which is a significant factor in enhanced transmission (Kazybay et al., 2022), and when combined with Q498R, the binding ability increases, making it simpler for the Omicron type to enter the host. Additionally, the Omicron form has a very significant risk of reinfecting patients who already contracted COVID-19, demonstrating a stronger transmissibility (Pulliam et al., 2022). Near the furin cleavage site (FCS), Omicron variant mutations H655Y and N679K exist that can promote spike cleavage and boost the contagiousness of the virus (Hossain et al., 2022). On the contrary, P681H can increase the spike protein cleavage, which multiplies transmissibility (Wu et al., 2021). Additionally, the "S gene target failure" in the novel Omicron causes false negative results in polymerase chain reaction tests, which paves the way for the virus to spread more quickly globally (Torjesen, 2021). As Omicron spreads globally in few days and cases sharply rising, there is growing concern about its transmissibility. In the USA, a 2.5% increase in Omicron variant capacity was seen in just 2 weeks, according to the CDC data. In a number of nations, the version sparked the fourth wave of the COVID-19 (Foxnews, 2021). Unfortunately, until May 2022, this variant was already discovered in 190 different nations. Omicron subvariants identified in a number of different ways, are continuously emerging. The Omicron variant is now related with 310 Pango lineages (Gangavarapu et al., 2022). The two other sublineages are now known as BA.2 (or B.1.1.529.2) and BA.3, while the "standard" sublineage is known as BA.1 (or B.1.1.529.1 or B.1.1.529.3). Recently, BA.4 and BA.5 have been found in a number of countries (Rana et al., 2022).

The COVID-19-vaccinated patients had the Omicron variant of SARS-CoV2, which suggested immune invasion of the novel variant and called for improved vaccinations. The COVID-19 vaccines made by Pfizer, AstraZeneca, Moderna and Johnson & Johnson are the four most widely used COVID-19 vaccines and the

Table 1 Variants of concern (VOCs) in past and circulating

WHO label	Pango lineage	GISAID clade	Earliest documented samples	Date of designation
Alpha	B.1.1.7	GRY	United Kingdom	VOC: 18-Dec-2020 Previous VOC: 09-Mar-2022
Beta	B.1.351	GH/501Y.V2	South Africa	VOC: 18-Dec-2020 Previous VOC: 09-Mar-2022
Gamma	P.1	GR/501Y.V3	Brazil	VOC: 11-Jan-2021 Previous VOC: 09-Mar-2022
Delta	B.1.617.2	G/478K.V1	India	VOI: 4-Apr-2021 VOC: 11-May-2021 Previous VOC: 7-Jun-2022
Omicron	B.1.1.529	GR/484A	Many countries	VUM: 24-Nov-2021 VOC: 26-Nov-2021

Source World Health Organization

Omicron variant is naturally resistant to neutralisation by plasma from both convalescent patients and those who received the vaccines (Hoffmann et al., 2022). With two doses of the vaccine, the efficiency of immunological protection is reduced by a factor of 34. Even those who had been vaccinated (33–40% immune protection) and boosted with mRNA-based vaccines (80–85% immunological protection) had serum that dramatically lowered neutralising antibody efficacy against the Omicron variant (Liu et al., 2022a, 2022b). A third booster vaccine is advised because (according to Agrawal and colleagues) "Omicron represents a formidable threat to the prevailing two-dose immunisation strategy presently utilised by many nations globally " (Sharma et al., 2022). Numerous COVID vaccinations have proven to be secure and effective when administered as a booster dose. In the Cov-Boost study, seven vaccinations (made by AstraZeneca, Curevac, Johnson & Johnson (Janssen), Moderna, Novavax, Pfizer and Valneva) were used as boosters after two doses of either the AstraZeneca or Pfizer vaccines. The study showed that all vaccinations (except from Curevac, which was stopped from using), enhanced the immune response, although the number of antibodies varied significantly depending on the mix of vaccinations (Chavda & Apostolopoulos, 2022) (Table 1).

2.4.3 Variant of Interest (VOI)

A variant with particular genetic markers that have been linked to alterations in receptor binding, decreased neutralisation by antibodies raised in response to prior infection or vaccination, decreased treatment effectiveness, a possible impact on diagnosis or a predicted rise in transmissibility or disease severity (CDC, 2022a).

Depending on VOI, one or more adequate public health measures, such as improved sequence surveillance, improved laboratory characterisation or epidemiological investigations, may be necessary to determine the degree of the disease, transmission of the virus, the effectiveness of therapeutic strategies, and whether or not vaccines that are currently approved or authorised provide protection.

The WHO has so far identified eight VOIs since the start of the pandemic, including Epsilon (B.1.427 and B.1.429), Zeta (P.2), Eta (B.1.525), Theta (P.3), Iota (B.1.526), Kappa (B.1.617.1), Lambda (C.37) and Mu (B.1.621). The WHO does not have any circulating VOI listed as of August 2, 2022 (Aleem et al., 2022).

2.4.4 Variant of High Consequence (VOHC)

A variant of high consequence (VOHC) contains strong evidence that medical counter or precautionary measures are substantially less efficient than to previously circulating variants (CDC, 2022a). A VOHC would call for reporting to the WHO under the International Health Regulations, the announcement of prevention or containment measures, and suggestions for new medications and vaccines. No SARS-CoV-2 variants have been classified as VOHC yet (Nasereddin et al., 2022).

2.5 *Transmission of SARS-COV-2*

SARS-COV-2 is spread through two routes: respiratory transmission and direct contact or fomite transmission (World Health Organization, 2020e). Moreover, there have also been theories about potential transmission through other channels discussed below.

2.5.1 Respiratory Transmission (Aerosol/Droplet)

Respiratory transmission is the primary way SARS-CoV-2 spreads. Increasing evidence suggests that contagious viruses can be found in aerosols (or droplets) in expiratory breath specimens, and is highly probable that the virus might well be transmitted at a distance through aerosols under certain conditions, such as during aerosol-generating processes, whilst also speaking or indoor with poor ventilation (Jayaweera et al., 2020). Greater droplet and aerosol ejection from the mouth cavity and respiratory system is brought by coughing and sneezing. These droplets being released by COVID-19 patients contain virus that could infect humans if inhaled, consumed or in contact with mucous membranes (Southest AIDS Education and Training Center [SEAETC], n.d.). The virus survives in aerosols for at least 3 h and on stainless steel and plastic surfaces for 48–72 h (CDC, 2021b). The most deadly means for healthcare personnel to acquire COVID-19 is through respiratory transmission.

2.5.2 Direct Contact and Fomites Transmission

Direct contact transmission can occur when a person comes into contact with an object or surface that is infected with a virus and contacts the disease by touching mouth, nose or eyes (Belser et al., 2013). One reason there are so many nosocomial infections is that healthcare workers dealing with COVID-19 patients are more susceptible to infection through this form of transmission (Wang et al., 2020a, 2020b, 2020c, 2020d, 2020e). When they do therapies on the respiratory tract, provide dental care, hemodialysis or manage an intensive care unit (ICU), nosocomial infections can occur through aerosols (Li et al., 2021). Utilisation of high-flow nasal oxygen or endoscopy is routinely necessary for the management of respiratory diseases and disorders; however, these procedures have the potential to release COVID-19 virus-laden aerosols into the surroundings and infect unprotected people. Along with the social distance practise, certain other measures should also be taken (Cook et al., 2020), viz. frequently washing hands with soap and water or an alcohol-based hand rub and avoiding touching eyes, nose or mouth with infected hands.

2.5.3 Faecal-Oral Transmission

As stated earlier, gastrointestinal issues are frequent signs of COVID-19 disease. The RNA virus is present in certain SARS-CoV-2 patients' faeces. After fever has subsided, the COVID-19 virus clears the respiratory tract in 2 weeks, but the faeces can continue to test positive for viral RNA for upto 4 weeks (Danchin et al., 2020). Recent studies have revealed that SARS-CoV-2 in faecal samples from COVID-19 patients can be another method of virus transmission in addition to cells that are infected by the COVID-19 virus through the surface ACE-2 enzyme (ACE2) (Xu et al., 2020). Further evidence indicating the virus transmitted by the faecal-oral route was found to reveal that high expressions of ACE2 are present in stomach glandular, colon and ileum absorptive enterocytes, duodenal and rectal cells (Rahman et al., 2020). Poorly maintained restrooms put users at risk of becoming infected via touch. Even though it is still controversial, sharing bathrooms with people who have COVID-19 may help virus to spread via the faecal-oral pathway (Huang et al., 2020).

2.5.4 Transmission Through Surgical Operations

The respiratory tract secretions and tissues are where the COVID-19 virus primarily lives. If these individuals need surgery, there are chances for the medical professionals who treat them to get affected. This especially happens when it comes to organ transplantation. Organ recipients are especially at risk of contacting COVID-19 because they must be placed in an immunocompromised state for the surgery (Hussein & Kareem, 2020). With increasing distance from the source and lengthening of time after exhale, the chance of infection lowers. The proportion of virus to which an individual is subjected in the environment or by contacting a surface tainted by virus is determined by two main processes (CDC, 2021c). The amount of virus in the air is decreasing as larger, heavier respiratory droplets which includes the virus drop to the ground or other surfaces due to gravity, and the very small respiratory droplets and aerosol particles that are still in the airstream gradually mix with and become diluted within the expanding volume and streams of air they encounter. The initial flaring of breaths and thermal layering can also affect how homogeneous this mixing is. Environmental elements like temperature, humidity, and UV radiation have an effect on the gradual deterioration of viral viability and infectivity over time (e.g., sunlight).

2.6 SARS-COV-2: Life Cycle

SARS-CoV-2 is a Nidovirus family member that can be acquired by humans and animals like bats. The virus progresses through six phases to complete its life cycle: entry, translation of the viral replication machinery, replication, translation of the viral structure proteins, virion assembly and virus release (Pampel, 2022).

2.6.1 Entry of Virus

SARS-CoV-2 can seize the cell in two different ways: through fusion of the plasma membrane and endosomes. Both times, the SARS-CoV-2 spike proteins (S1, S2) facilitate adherence to host cell membrane and activate ACE-2, an entrance receptor present in a number of organs including heart, lungs, kidneys, and gastrointestinal system (Hoffmann et al., 2020). The spike protein is activated by cathepsin L when virions are brought into endosomes. Alternately, type II transmembrane serine protease (TMPRSS2) that is located on the surface of human host near to the ACE-2 receptor can break the spike protein between S1 and S2 domains, which starts the fusion of viral membrane with plasma membrane (Jackson et al., 2022). The virus can enter cells when the S proteins are activated, altering their shape. The plasma membrane fusion entrance is more effective for viral replication because host cell antiviral immunity is less likely to be triggered by it (Tang et al., 2020).

2.6.2 Translation of Viral Replication Machinery and Replication

This virus releases mRNA, which is prepared to be converted into protein. This virus's genome is supplemented by roughly 14 open reading frames (ORF), each of which encodes a spectrum of structural and nonstructural proteins important to its viability and pathogenicity (Astuti, 2020). The coronavirus genomic RNA encodes nonstructural proteins (NSPs), and this process first translates into ORF1a and ORF1b to produce two large overlapping polyproteins, pp1a and pp1ab. These polyproteins are then cleaved by the papain-like protease (PLpro, Nsp3) and 3C-like protease (3CLpro, Nsp5) to form functional NSPs like helicase or the RNA replicase–transcriptase complex (RdRp) (Shin et al., 2018).

2.6.3 Translation of Viral Structure Proteins, Virion Assembly and Release of Virus

The subgenomic proteins are transcribed into structural and accessory proteins including Membrane (M), Structural (S) and Envelope (E) proteins, which are then coated in the endoplasmic reticulum and transferred to the endoplasmic reticulum-Golgi intermediate compartment (ERGIC) (Chakravarti et al., 2021). During this process, the already duplicated genome programme can enter the ERGIC and immediately link the N protein to the nucleocapsid form (Astuti, 2020). The nucleocapsids in this compartment will combine with a number of other structural proteins to create tiny wallet vesicles that will be exocytosed out of the cell (Fehr & Perlman, 2015) (Fig. 5).

3 SARS-CoV-2 Clinical Manifestations

SARS-COV-2 patients might exhibit a variety of symptoms ranging from asymptomatic to mild to moderate symptoms, severe to critical illness (Gavriatopoulou et al., 2020). Almost majority of patients will experience symptoms within 11.5 days of infection, while the median incubation period for SARS-CoV-2 is anticipated to be 5.1 days (Lauer et al., 2020). The National Institutes of Health (NIH) published recommendations that categorise COVID-19 into five different types based on the severity of the presenting condition, which encompasses disease manifestations, laboratory and radiographic abnormalities, hemodynamics, and organ function (Cascella et al., 2022).

(a) **Asymptomatic or presymptomatic infection**: Patients who tested positive for SARS-CoV-2 but did not exhibit any COVID-19-like clinical signs.
(b) **Mild illness**: Patients who exhibit any COVID-19 symptoms, e.g., fever, cough, sore throat, fatigue, headache, muscle spasms, nausea, vomiting, diarrhoea, anosmia or dysgeusia, but not abnormal chest imaging or breathing difficulties.
(c) **Moderate illness**: Patients who have oxygen saturation (SpO_2) levels under 94% in room air and either clinical signs of lower respiratory tract disease or radiologic proof of it.
(d) **Severe illness**: Patients with SpO_2 less than 94% on room air (PaO_2/FiO_2) less than 300 with marked tachypnea and respiratory frequency greater than 30 breaths/min or those with lung infiltrates greater than 50%.

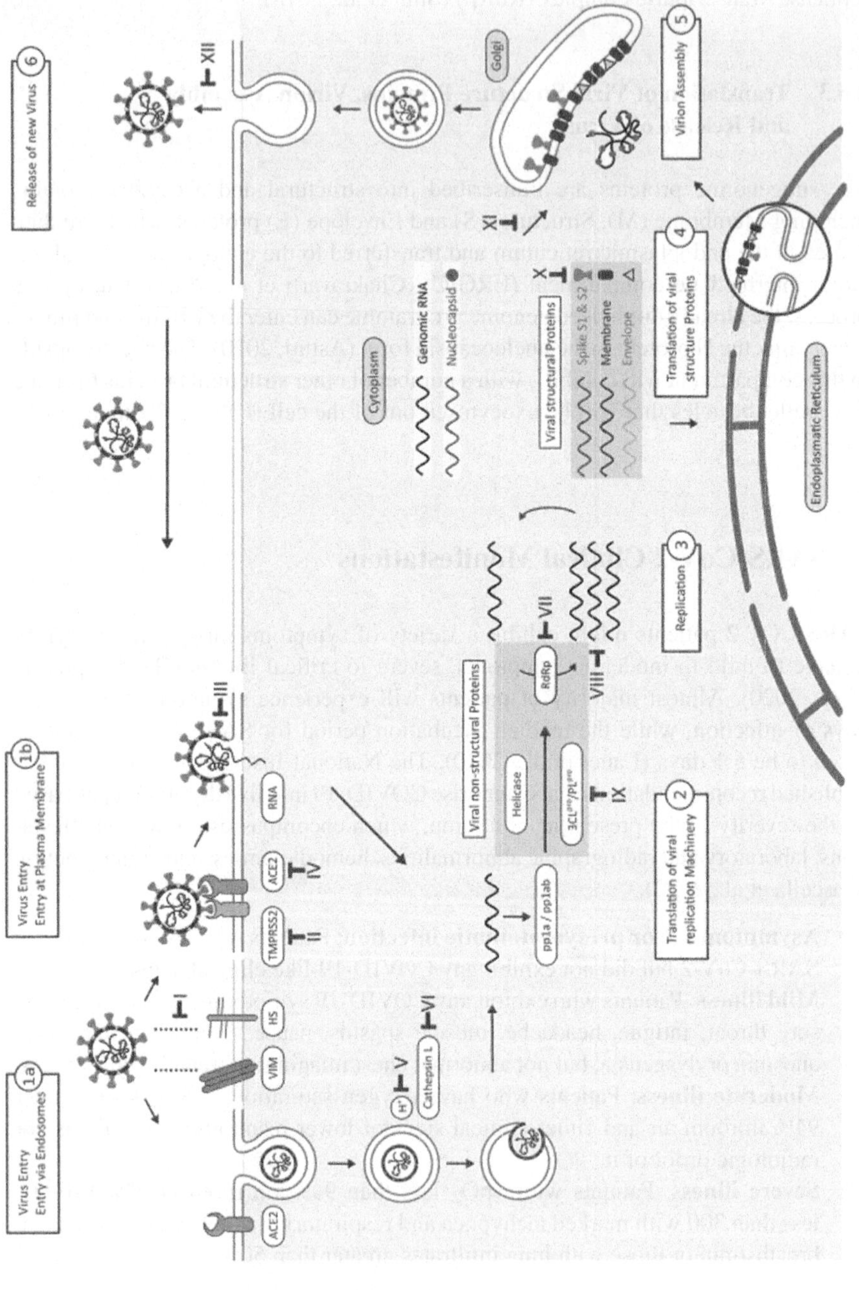

Fig. 5 SARS-CoV-2 Life Cycle. (*Source* https://www.antibodies-online.com/resources/18/5410/sars-cov-2-life-cycle-stages-and-inhibition-targets/?utm_source=free-download&utm_medium=sarscov2-replication-cycle&utm_content=text

(i) **Critical illness**: Patients who suffer from multiple organ dysfunction, septic shock or severe respiratory failure. Acute respiratory distress syndrome (ARDS), which typically develops around a week after symptoms appear, can cause patients with severe COVID-19 disease to become dangerous. ARDS is defined by serious new-onset respiratory failure or worsening of an existing respiratory condition. A specific set of clinical and ventilatory requirements must be met, including the use of chest imaging techniques such as chest radiograph, CT scan or lung ultrasound that shows bilateral opacities (lung infiltrates >50%). The Berlin classification divides ARDS into three categories according to the level of hypoxia, with PaO_2/FiO_2 or P/F ratio serving as the standard variable.

(ii) **Mild ARDS**: 200 mmHg $< PaO_2/FiO_2 \leq$ 300 mmHg in patients not receiving mechanical ventilation or in those managed through non-invasive ventilation (NIV) by using positive end-expiratory pressure (PEEP) or a continuous positive airway pressure (CPAP) $\geq$ 5 cmH_2O.

(iii) **Moderate ARDS**: 100 mmHg $< PaO_2/FiO_2 \leq$ 200 mmHg

(iv) **Severe ARDS**: $PaO_2/FiO_2 \leq$ 100 mmHg.

When PaO_2 is not available, a ratio of $SpO_2/FiO_2 \leq 315$ is suggestive of ARDS.

In addition, the following COVID-19 manifestations are well documented.

3.1 Pulmonary Manifestation

The most typical organ system impacted by SARS-CoV-2 is the pulmonary system. The pulmonary signs of COVID-19, which include coughing, breathlessness, phlegm formation, respiratory distress and ARDS, have routinely been described in several previous studies (Huang et al., 2020; Rodriguez-Morales et al., 2020). The extent of lung involvement by SARS-CoV-2 infection can vary from mild pneumonia or no symptoms (81%) to severe illness-related hypoxia (14%), critical disease brought by shock, respiratory failure, and multiorgan failure (5%) or death (2.3%) (Gavriatopoulou et al., 2020). It was demonstrated that SARS-CoV-2 infests respiratory, olfactory, and paranasal sinus epithelia with tropism to ciliated mucosa, and the adjacent tissues, and can describe why anosmia and dysgeusia are common symptoms in patients infected. The nasopharyngeal and oropharyngeal tissues are the main entry and (early) replication sites of SARS-CoV-2 (Wolfel et al., 2020). In roughly 10% of hospitalised cases, intensive care treatment with invasive mechanical ventilatory support is necessary (Docherty et al., 2020). Alveolar damage and interstitial inflammation are caused by SARS-CoV-2 infection (Dockrell et al., 2022). The most common cause of death in COVID-19 patients is pulmonary manifestation, which often manifests as peripheral lung ground-glass opacities on computed tomography (CT) (Raptis et al., 2020).

3.2 Cardiovascular Manifestation

The development of cardiovascular complications, like myocardial injury, heart failure, and arrhythmias, has been linked to poor sustenance between many COVID-19 patients, while the underlying cardiovascular comorbidities, such as hypertension, diabetes and particularly cardiovascular disease, have indeed been related to negative consequences. (Guan et al., 2020; Huang et al., 2020). Additionally, obesity is linked to worsen cardiovascular complications (Sanchis-Gomar et al., 2020). Magnetic resonance imaging (MRI) cohort-study by Wei et al., 2020 reported that upto 78% of COVID-19 patients on clinical admittance had acute cardiac complicity, defined as elevated cardiac injury markers like troponin, arrhythmia, lowered ejection fraction and ventricular dyskinesia. These occurrences are consistent with those linked to the Beta coronavirus SARS-CoV, which causes severe lung illness.

Along with the aforementioned, a non-negligible number of COVID-19 patients also appear to be in a hyperinflammatory condition, which is characterised by substantial increase in inflammatory cytokines and other markers of inflammatory process (Wu et al., 2020). The endothelial cells of the underlying atherosclerotic lesions may become more susceptible to disruption due to the circulating cytokines' stimulation of macrophages and leucocyte adhesion molecule production, which raises the probability of a clinically obvious acute coronary syndrome (Bonow et al., 2020). Type 2 myocardial infarction, which is the subsequent result of an imbalance between myocardial oxygen supply and demand, can also cause myocardial damage. Additionally, hypotension, a common clinical symptom of both sepsis and the cytokine storm syndrome, might decrease the myocardial oxygen supply (Libby, 2020). A Kawasaki-like syndrome, which is defined by circulatory dysfunction and macrophage activation syndrome, is another typical manifestation of COVID-19 in the cardiovascular system that has now been identified (Viner & Whittaker, 2020).

3.3 Gastrointestinal and Liver Related Manifestations

The 10.6% of SARS patients exhibited gastrointestinal (GI) signs, with diarrhoea as the primary clinical symptom (Leung et al., 2003). The most common GI symptoms are nausea, vomiting, diarrhoea and abdominal discomfort (Jin et al., 2020; Perisetti et al., 2020). 20–25% of COVID-19 patients had diarrhoea and abdominal pain, which occasionally came before respiratory symptoms, according to researchers (Lin et al., 2020).

Liver morphological changes are quite frequent in both living and dead COVID-19 patients, however, they are typically non-specific. Steatosis, which is primarily macrovesicular and common (30–70%), is more likely to be an existing feature and/or the outcome of hypoxia and pharmacological side effects than a virus-induced disease (Nardo et al., 2021). Other alterations include minor lobular hepatitis, centrilobular congestion and necrosis, platelet-fibrin microthrombi (17–70% of instances), and cholestasis. These alterations are likely connected to the cytokine storm, general hypoxia in COVID-19 patients, and septic sequelae (Evert et al., 2021; Zhao et al., 2021).

3.4 *Haematological Manifestations*

Hematological indicators of disease development in COVID-19 patients include changes in immune cell population and coagulation factors (Słomka et al., 2020). Adult COVID-19 infected with most prevalent symptom appear to have lymphocytopenia, which may be brought by the degeneration of lymphoid organs (Ludvigsson, 2020). Platelets play an important role in controlling inflammatory reactions. According to a meta-analysis, the intensity of COVID-19 is correlated with lower platelet counts, which can lead to thrombocytopenia through enhanced platelet disintegration or lower platelet generation (Lippi et al., 2020). High serum levels of biomarkers such procalcitonin, CRP, and ferritin have also been identified as poor prognostic indicators (Lippi & Plebani, 2020; Zhou et al., 2020a, 2020b). More recently, it has been discovered that patients with severe COVID-19 may have elevated cortisol levels at presentation, which may represent the severity of their disease (Tan et al., 2020). Additionally, hospitalised COVID-19 patients frequently have blood hypercoagulability, particularly those with severe illness (Deng et al., 2020; Levi et al., 2020; Zhou et al., 2020a, 2020b). In severe individuals, elevated D-dimer values are frequently seen (Lippi & Favaloro, 2020). The development of thrombosis, endothelial damage, and hemorrhagic consequences are all serious risks for such patients (Maier et al., 2020).

3.5 *Neurological Manifestations*

SARS-CoV-2 enters the central nervous system (CNS) through the blood or nerves. CNS related neurological symptoms include encephalitis, convulsions, and febrile seizures (Desforges et al., 2019). According to researchers, almost one quarter of infected patients experienced a variety of neurological problems that affected their skeletal muscles, peripheral nervous system and CNS (Helms et al., 2020; Mao et al., 2020). As per epidemiological studies, SARS-CoV-2 may only need a 7-day incubation in order to infect and harm medullary neurons. Acute cerebrovascular dysfunction, altered blood pressure, viral encephalitis, and infectious toxic encephalopathy are some of the significant CNS symptoms linked to COVID-19 (Miller et al., 2020). Encephalopathy, myopathy, autonomic neuropathy, and polyneuromyopathy connected to critical illness frequently develop in patients hospitalised to the intensive care unit (ICU) (Vanhorebeek et al., 2020). Another possibility is an indirect injury by the effects of acute coronavirus infection on the host immune system, which could partially account for the occurrence of Guillain-Barré Syndrome (GBS) cases, transverse myelitis or acute disseminated encephalomyelitis in COVID-19 and other virus epidemics (Toscano et al., 2020).

An indirect injury related to host immune response effects after acute coronavirus infection could also be possible, explaining (to some extent) the occurrence of Guillain-Barré syndrome (GBS) cases, transverse myelitis or acute disseminated encephalomyelitis in patients with COVID-19 and other virus epidemics (Toscano et al., 2020). Neurological symptoms caused by systemic illness especially in severely ill patients could also justify neurological manifestations of COVID-19 infection (Mao et al., 2020). Patients admitted to ICU often develop encephalopathy, myopathy, autonomic neuropathy and polyneuromyopathy related to critical illness (Vanhorebeek et al., 2020).

3.6 Renal Manifestations

Acute kidney injury and increased disease severity are both associated with renal injury, has been a widely reported COVID-19 related renal presenting feature (Braun et al., 2020; Pei et al., 2020). Dialysis, kidney transplant and chronic kidney disease patients are particularly susceptible to COVID-19, which has a significant severity and fatality rate (Jager et al., 2020). Acute tubular injury to the kidneys is typically present, followed by collapsing glomerulopathy or, more frequently, unspecific or secondary, localised segmental glomerulosclerosis (Ferlicot et al., 2021). Although noticed, microthrombi or thrombotic microangiopathy are difficult to directly link to the SARS-CoV-2 infection (Tiwari et al., 2021).

3.7 Dermal Manifestations

Numerous studies describing skin related seen in COVID-19 infected have indeed been documented. Based on specific datasets, upto 20% of COVID-19 patients experience skin abnormalities, which can range from papulosquamous lesions to urticarial, vesicular and purpuric lesions (Guarneri et al., 2021). Although preliminary findings, much more research is required on COVID-19 dermal-related symptoms.

3.8 Endocrine Organs

In COVID-19, thyroid dysfunction is common and can range from thyrotoxicosis (15–20%) to hypothyroidism and nonthyroidal disease syndrome (NTIS) (Chen et al., 2021a, 2021b; Khoo et al., 2021). The understanding of how SARS-CoV-2 affects various endocrine organs, including the pituitary gland, parathyroid, adrenals, gonads, endocrine pancreas, and diffuse neuroendocrine system, is still limited (Parolin et al., 2021).

4 Mucormycosis and SARS-COV-2: An Epidemic Within Pandemic

SARS-CoV-2 has unleashed a catastrophic pandemic that the entire world is currently experiencing. The second wave of COVID-19 was difficult for India, moreover sudden appearance of Covid-19 Associated Mucormycosis (CAM) infections increased the concern. Medical pactitioners treating COVID-19 patients need to be more mindful of mucormycosis because the present pandemic continues to be a major public health concern globally and also because the two illnesses could significantly increase morbidity and mortality if they coexist. According to some studies, there could be 140 instances of the disease for every 1 million people in India (Banerjee et al., 2022; Rocha et al., 2021). Due to this, the GoI officially designated it an illness that must be reported, and numerous state governments classified it as an epidemic (Rocha et al., 2021). Despite being less common than bacterial infections, fungal infections have a relatively significant death rate (Garbee et al., 2017). The main aspect which seems to be promoting Mucorales spore germination in COVID-19 patients is a perfect setting of lower oxygen level, elevated blood sugar, acidic medium, increased iron levels (ferritin), and reduced phagocytic activity of white blood cells due to compromised immunity owing to virus and steroid mediated treatment (Pippal et al., 2021). Dexamethasone, a systemic corticosteroid, can decrease death in patients who have most serious disease presentations, but when combined with immunological and other clinical variables, it also put patients at risk for secondary fungi infestations like aspergillosis, invasive candidiasis, and mucormycosis (Arastehfar et al., 2020). If detected early, mucormycosis can be managed; nonetheless, if not treated promptly, it can harm vital organs like brain, eyes and upper section of the nasal cavity. In such situation, the patient needs urgent surgical surgery to prevent deadly infections from spreading to other important organs. The non-specific clinical and radiological characteristics of pulmonary and disseminated mucormycosis can overlap with signs assumed to be related to COVID-19, making the diagnosis of CAM difficult and leading to missed or delayed diagnoses (Cornely et al., 2019).

Baker in 1957 first proposed the term "Mucormycosis" (alias Black Fungus); an angioinvasive fungal infection brought on by fungi from several genera, including *Rhizopus, Apophysomyces, Rhizomucor, Lichtheimia, Cunninghamella, Mucor* and *Saksenaea* (Roden et al., 2005). To be specific *Rhizopus arrhizus*, a member of the order Mucorales and the subphylum *Mucoromycotina*, is the most prevalent cause of mucormycosis. Mucorales are abundant in nature and can be found in soil, decomposing plant waste, food and dirt (Reid et al., 2020). Exposure of spores containing Mucorales from the consumption of infected food products or through injured skin and wounds causes the mucormycosis (Richardson, 2009). Mucormycosis-causing fungi are frequently diagnosed in low- and middle-income nations, including India and can be identified based on their phenotypic traits such as growth rate, colony morphology, and reproductive structures (Skiada et al., 2018). According to a previous projection for 2019–2020, the prevalence of mucormycosis ranged from 0.005 to 1.7 per million people worldwide, however it is approximately 80 times more prevalent (0.14 per 1000) in India than in developed countries (Prakash & Chakrabarti, 2019; Skiada et al., 2020). Another way, India has the highest rate of mucormycosis in the entire world. India has the second-highest prevalence of diabetes mellitus (DM) and, until recently, was the world's diabetes capital, according to (Federation, 2019). According to a recent analysis of 275 cases of CAM, 233 of which were from India and 42 from the rest of the world, DM was the most common underlying risk factor for CAM in India compared to other countries (Muthu et al., 2021). Before the COVID-19 pandemic, the mortality rate from mugormycosis was 50%. However, it has increased to 85% during the current pandemic in India (Aranjani et al., 2021). Mucormycosis symptoms and indicators include facial swelling, fever, headache, sinus or nasal congestion, and black scars on the nose or inside the mouth. Different body parts, including sinuses, brain and lungs, are impacted by mucormycosis (Zegarelli, 1993).

Patients with CAM are infected through a number of routes. The immune system is significantly impacted by COVID-19 infection by inflammation, an increase in neutrophil cell number, a decrease in lymphocyte cell quantity, and primarily by its influence on CD4+ and CD8+ T lymphocytes. Neutrophils increase in quantity after SARS-CoV-2 infection. An essential part of mucormycosis infection in COVID-19 hosts is played by CD4+ and CD8+ T lymphocytes (John et al., 2021). Because of adverse effects, such osteomyelitis, systemic infection, and cavernous sinus thrombosis, CAM has significant death risk. Radiographic imaging, biopsies and standard blood tests are all used to diagnose mucormycosis. The recommended course of action for mucormycosis is to first eliminate risk factors, have the affected tissue or organ surgically removed, and then administer intravenous antifungal medications like liposomal Amphotericin B (Koehler et al., 2020).

4.1 *Types of Mucormycosis*

A mucormycosis infection is classified (according to anatomical site) as: (i) rhino-orbital cerebral mucormycosis (ROCM), (ii) pulmonary, (iii) cutaneous, (iv) gastrointestinal, (v) disseminated and (vi) miscellaneous, which includes uncommon, rare forms like endocarditis, osteomyelitis and peritonitis (Alloush et al., 2022).

4.1.1 Rhino-Orbital Cerebral Mucormycosis (ROCM)

The most prevalent form of mucormycosis among immunocompromised individuals with uncontrolled diabetic mellitus, hematologic dyscrasias, retroviral illness, steroid therapy, and malnutrition is known as Rhino-orbital cerebral mucormycosis (ROCM) (Jeong et al., 2019). High blood sugar is the most aetiology of ROCM, and *Rhizopus* grows well in this environment (Singh et al., 2021). The fungus is capable of causing both biological and mechanical harm to the interior lamina of blood vessels, particularly the arteries, lymphatics, and veins. In immunocompromised patients, the infection begins to grow into several hyphae as soon as spores penetrate the sinuses. The fungus expands and invades the vascular and neural zones throughout for the next several hours. Thrombosis and nerve dysfunction are brought by invagination. Blood vessels, cartilage, bone, neural and perineural regions, and frequently meninges are all involved in disease presentation. Palate eschars and nasal turbinates are also harmed by necrosis in the palate region. The sinus infection erodes the bone and gradually engulfs the orbital components. As a result, the infection enters the brain via retro-orbital channel or the frontal lobes via ethmoid sinuses. Cranial nerve palsies may develop as the infection develops from the sphenoid sinuses to the next cavernous sinus. The patient's condition may continue to deteriorate due to the widespread thrombosis in the carotid artery, cavernous sinus, and juglar veins. The identification of the impacted can be aided by CT and MRI scans (Therakathu et al., 2018).

4.1.2 Pulmonary Mucormycosis

Recent reports on COVID-19 carriers with fungal infections include cases of candidiasis and pulmonary aspergillosis. Due to spore inhalation caused by hematogenous or lymphatic dissemination in immunocompromised patients, infection in the pulmonary areas begins. Mucorales immediately target arteries, veins and lymphatics after entering through the respiratory system, where they produce thrombosis and infarction (Helenglass et al., 1981). If pulmonary mucormycosis is not identified and treated within a timeframe, it can be aggressive and fatal (Iqbal et al., 2017). Healthy hosts have polymorphonuclear and mononuclear phagocytes that are equipped and may kill Mucorales by producing cationic defensins and oxidative metabolites in the pulmonary system. Neutrophils use an oxidative burst to kill hyphae whereas macrophages aid in defence by preventing spore sprouting in healthy persons (Guarner & Brandt, 2011). The immunocompromised host succumbs to the coinfection by COVID-19 because it lacks efficient phagocytes and macrophages to combat the spreading fungus in the pulmonary areas. The use of continuous corticosteroids during COVID-19 therapy and pre-existing comorbidities like diabetes have recently been linked to higher morbidity and death in COVID-19 patients with pulmonary mucormycosis (Al-Tawfiq et al., 2021; Hussain et al., 2021).

4.1.3 Cutaneous Mucormycosis

An invasive fungus belonging to the *Glomeromycota phylum* and *Mucormycotina subphylum* is the source of the uncommon fungal infection known as *Cutaneous mucormycosis* (Castrejón-Pérez et al., 2017). Mucormycosis can present itself clinically and geographically in many different ways. Vascular invasion and irregularly shaped broad, non-septate, twisted, ribbon-like, and thick-walled fungal hyphae are essential features of *C. mucormycosis* (Umbert & Su, 1989). Given the rising number of patients whose immune systems have been impaired as a result of COVID-19 therapy, is becoming more widespread. The skin is most severely damaged on arms and legs. A biopsy and molecular diagnostic tests are advised.

4.1.4 Gastrointestinal Mucormycosis

The stomach, colon, and ileum are the three most typical sites for gastrointestinal mucormycosis infestations (Spellberg, 2012). Malnutrition, particularly in newborns and toddlers, is a risk factor (Mohta et al., 2011). The fungus generally enter the gastrointestinal tract via ingestion (Spellberg et al., 2005). Depending on the location affected, gastrointestinal mucormycosis symptoms can vary. The most typical symptoms are non-specific abdominal pain and distention coupled with nausea and vomiting. Hematochezia and fever can also happen. It is frequently believed that the patient has had an intra-abdominal abscess (Spellberg, 2012).

4.1.5 Disseminated Mucormycosis

Disseminated mucormycosis (DM) develops when the infection travels through bloodstream to some other parts of the body. Although the brain is the organ that is most frequently affected, the illness could also impact the spleen, heart and skin (CDC, 2021d). The generic nature of DM symptoms makes diagnosis challenging (Gupta et al., 1987). DM affects people who have significant immunosuppression; it accounts for 1.6% of all invasive fungal infections and is more common in immunosuppressed people with risk factors (Pozo Laderas et al., 2015). DM manifests itself in a variety of ways, depending on the host in addition to the position and severity of vascular invasion and tissue injury in the affected organs (Petrikkos et al., 2012).

4.2 *Immunocompromised Conditions in COVID-19 Patients Favouring Mucormycosis*

Critical care units, ventilators, corticosteroids, and interleukin antagonists like tocilizumab were used in the treatment of severe COVID-19 infections because they lessen cytokine burst and constriction of inflamed airways. The host's immunity is hampered by unregulated high blood sugar levels and the careless use of immunosuppressive medications like corticosteroids and interleukins (IL), which raises the risk of numerous opportunistic infections by mucormycosis.

Three hypotheses concerning how immunological and inflammatory processes may be involved in CAM include: (a) COVID-19 dramatically reduces T cell (CD4+ and CD8+) accessibility, which allows opportunistic fungal infections to spread by opening the door for substantial lymphopenia. The fungus, which is perpetually present in the air, enters immune-weakened host cells, creates fungal-specific T cells, and releases IL-4, IL-10, IL-17, and IFN-γ. The host cell is further damaged as a result of the pro-inflammatory cytokines' stimulation of CD4+ T cells. The hyphae of fungi obstruct the secretion of various immunomodulatory molecules like RANTES (regulated upon activation, normal T cell expressed and secreted) and IFN-γ secreted by NK cells and seen in the early stage of infection; (b) enhanced pro-inflammatory markers in patients with severe disease; and (c) notable damage to pulmonary tissues by COVID-19 helps the invasive fungi, particularly the air-borne or those that enter through the respiratory tract (Bhatt et al., 2021). Besides, Hyperferritinemia triggered by excessive IL-6 release and macrophage activation boosts the amount of free iron inside cells. Endothelitis, an inflammation and endothelial cell death, results as a consequence of this. In addition, the virus's mimicking of hepcidin also causes ferritin to be produced independently of an inflammatory response (Jose et al., 2021).

Therefore, COVID-19-induced immunosuppression elevates the likelihood of opportunistic pathogen, impacts the endothelium and alveoli, facilitating fungal invasion, and raises blood sugar levels owing to the acute diabetes-like state caused by pancreatic impairment. Additionally, elevated ferritin amounts sustain spore germination, and elevated body temperature is ideal for thermotolerant fungi. The imbalance of ACE-2 expression in multiple body organs, immunosuppression, and the development of a microenvironment system raises the risk of coinfection in COVID-19 (Pandiar et al., 2021).

4.3 *Clinical Identification and Diagnosis of Fungus*

Diagnosing the fungi responsible for the infections is crucial. The Smith and Krichner criteria form the basis of the reference standard for diagnosing mucormycosis clinically in the laboratory. The following symptoms are necessary for microbiological identification: (a) black necrotic turbinates that can be mistaken for dried flecked blood; (b) nasal discharge that is imbued with blood and causes facial pain; (c) soft peri-orbital and peri-nasal swelling with discoloration and induration; (d) ptosis of the eyelid, proptosis of the eyeball, and complete ophthalmo (Smith & Kirchner, 1958).

Differentiating mucormycosis from other fungal infections requires use of factors such as the diameter of the hyphae, presence of septa, pigmentation, and the branching angle (Ghosh et al., 2019). A few fundamental diagnostic procedures, including direct

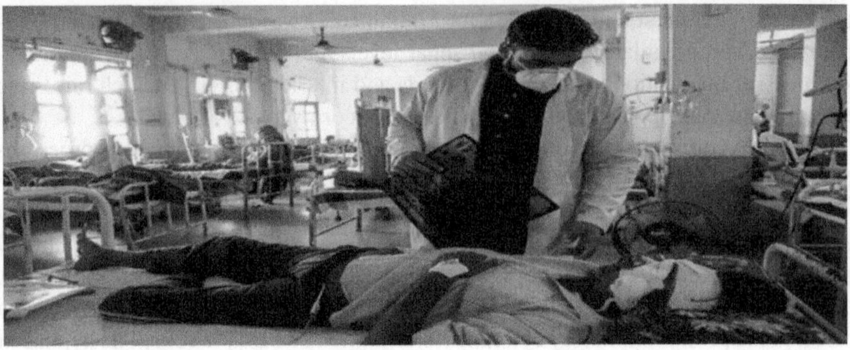

Fig. 6 COVID associated mucormycosis patient receiving medical care at hospital. (*Source* BBC)

microscopy, fungal culture, histopathology, susceptibility testing, antigen testing, antibody testing, and molecular assays (DNA and RNA) of the suspected specimen patient, serve as the basis for the diagnosis of a fungal infection. To diagnose mucormycosis, physical examination of the patient's nasal airways as well as other signs and symptoms are required (Fig. 6).

4.4 Risk Factors Associated with Mucormycosis

4.4.1 Diabetes Mellitus

Immunocompetent hosts' phagocytes are capable of killing Mucorales spores by creating oxidative metabolites. This immune response may be altered by uncontrolled diabetes brought by diabetic ketoacidosis (DKA), which could result in decreased granulocyte phagocytosis and altered polymorphonuclear leukocyte response (Saidha et al., 2021). Additionally, the ketone reductase enzyme found in Mucorales flourishes in high glucose environments and DKA, which leads to a disappointing prognosis (Bayram et al., 2021). According to estimates, India, which has a population of 1.38 billion, has 77 million diabetes patients in the 20–79 age group as of 2021, and the figure is expected to rise to 124.8 million by 2045 (Down to Earth, 2021). When diabetes coexists with a systemic immunological change by COVID-19, immunity is lowered and the risk of secondary infections is elevated (Lim et al., 2021).

4.4.2 Immunomodulator Use

Another important risk factor is the improper use of corticosteroids, mAb against the IL-6 receptor like tocilizumab and sotrovimab, and JAK inhibitors like baricitinib and tofacitinib to inhibit the overactive immune response in the treatment of moderate to severe COVID-19 infections. 71.4% of cases reported in a systematic study had received corticosteroids, according to some researchers (Patel et al., 2021; Salama et al., 2021; Tandon & Pandey, 2021). In a study conducted during the first wave of COVID-19, more than 60% of Indian patients received the incorrect treatment and 75% of patients received corticosteroids (Patel et al., 2021). Those with severe and critical COVID-19 during the early stages of the pandemic, the Indian authorities suggested a larger methylprednisolone dose (about 70–140 mg). More immunosuppressed patients may have been affected by this, leaving them open to fungal infection. Later, it was lowered to 32 mg/day recommended by the World Health Organization. In addition, even milder patients took corticosteroids out of anxiety without receiving medical advice (Rocha et al., 2021). Immunosuppressant usage on a regular basis, together with other risk factors, creates the perfect setting for invasive fungal infections. Furthermore, corticosteroids may interfere with glucose homeostasis, causing unchecked hyperglycemia and the acceleration of DKA. Reduced cell-mediated immune responses (phagocytic activity), both brought on by corticosteroids and a hyperglycemic condition, make ketoacidosis a perfect setting for the germination of fungal spores, which worsens the situation and leads to invasive mucormycosis (Artis et al., 1982). Additionally, improper broad-spectrum antibiotic treatment in COVID-19 patients may raise the chance of CAM (Banerjee et al., 2022). The healthy sinonasal microbiome is disrupted by both broad-spectrum antibiotics and antifungal prophylaxis, potentially opening the door for opportunistic infections. Additionally, macrolide medicines like azithromycin prevent IL-6 and hence interfere with the antimicrobial immune response (Thakar & Lal, 2021).

4.4.3 Iron and Zinc Supplements

A hyperferritinemic condition and high cytokine release are two characteristics of severe COVID-19 infection. High amounts of intracellular iron are caused by excessive ferritin production and decreased extracellular iron transport (Perricone et al., 2020). Because of the release of iron from binding proteins (notably transferrin) in the presence of acidosis, patients with DKA also exhibit higher serum levels of free iron (Artis et al., 1982). Increased amounts of reactive oxygen species, which harm tissues and release free iron into the bloodstream, are produced as a result of higher intracellular iron levels. Renal dysfunction and hypoxia, which are common in COVID-19 patients, might make the acidic condition even worse (Chan et al., 2021). There is evidence that diabetics have optimal conditions for fungal reproduction due to low levels of dialysable inhibitory factor and elevated serum free iron levels. Thus, the availability of free iron (which is necessary for cell growth and development) makes the acidic milieu (which is the overall effect of a number of factors) an ideal environment for fungal proliferation (Artis et al., 1982). Studies have shown that zinc deficiency causes biological stress and hinders the growth of fungi by acting as transcription factor for the biological processes necessary for fungi (Eide, 2011). The COVID-19 era has seen a growth in the use of micronutrient supplements, particularly zinc as potential immune boosters, which may be a factor in the rise of mucormycosis cases.

4.5 Prevention and Treatment of Mucormycosis

Mucormycosis spores are abundant in the environment. It is challenging to prevent inhaling the fungus spores. There is no vaccination available yet to stop mucormycosis. There may be some measures to reduce the risk of acquiring mucormycosis for those with compromised immune systems. First, stay away from dusty areas like construction or demolition sites. N95 respirators should be used if they cannot be ignored. Additionally, direct contact with flooded areas and water-damaged structures following hurricanes and other natural disasters must be averted. Whenever handling soil, moss or manure, use of gloves is advised. Also, cleanliness must be practised in order to lower the risk of skin infection. Monitoring and management of underlying medical conditions, particularly diabetes mellitus, is required. Treatment options for mucormycosis can be categorised into several groups, including first-line therapy, salvage therapy and surgical treatment.

4.5.1 First-Line Therapy

Amphotericin-B liposomes are administered in high doses as first-line therapy. At such an early phase of infection, liposomal amphotericin-B treatment should be started (Chamilos et al., 2008). The preferred treatment is liposomal amphotericin-B as an initial dose of 5 mg/kg body weight. It must be continued for weeks until it takes a positive response and stabilise the disease. A unexpected rise in the need for amphotericin-B during the COVID-19 pandemic led to shortage of medication (Chamilos et al., 2008). Although amphotericin B liposomal formulations are less harmful, they are expensive therapeutic option (Sahu et al., 2021).

4.5.2 Salvage Therapy

Salvage therapy (sometimes called rescue therapy) is a treatment used when the condition does not improve with conventional treatment (Amfani & Ibrahim, 2017). If outcomes with liposomal amphotericin-B are inconsistent, doctor may recommend oral Posaconazole (300 mg delayed release tablets twice daily for 1 day followed by 300 mg daily) or Isavuconazole (200 mg 1 tablet three times daily for 2 days followed by 200 mg daily) (Darwish et al., 2022). The treatment must be continued until the clinical cure of infection-related signs and symptoms, the radiological resolution of active disease, and the elimination of risk factors that predispose to infection, such as hyperglycemia, immunosuppression, etc. (National Centre For Disease Control-Libya, 2022).

4.5.3 Surgical Treatment

Surgery is associated with higher survival rates because mucormycosis is difficult to treat due to low medication penetration of necrotic tissues. The efficient therapy of mucormycosis necessitates the surgical removal of mucormycosis-infected tissue (Sahu et al., 2021). Surgeons may undertake surgical procedures on SARS-CoV-2 infected patients using the barrier draping approach (Tsagkovits et al., 2021). Drawback is that surgery could be an expensive treatment option, which might be quite burdensome in low-income settings.

5 Prevention of SARS-COV-2

The WHO has emphasised that the key measures in managing serious diseases like COVID-19 include awareness, isolation, prevention, regulating the transmission, and treating the infected. To minimise the spread of SARS-COV-2, initial detection, quarantine, and treatment are required (Güner et al., 2020). There is currently no effective treatment, hence the majority of healthcare systems must rely on personal safeguard, hand wash, and social segregation for patients as well as barrier precautions for healthcare workers (Dixit et al., 2020). Although there are safe and efficient vaccines available that are useful for controlling COVID-19, it remains crucial to implement precautionary measures like personal protective equipment (PPE) use, hygiene practices, proper ventilation, social distancing, and preventing crowds while working to develop innovative remedies. Consequently, it is urgently necessary to design an effective response mechanism in public places to safeguard the general public, maintain society's normal functioning, and minimise economic losses throughout the pandemic (Pan et al., 2022). Following is a summary of significant COVID-19 prevention and control strategies in the community.

5.1 Administrative Management

A strong operational management system lowers the probability of COVID-19 infection. Managers and owners should bear the primary responsibility for COVID-19 prevention and management in public places. It is advised that they create an emergency plan, gather employee health data, organise shifts sensibly, and increase employee and consumer precautionary participation via videos, posters, brochures. Managers should implement infection prevention and control methods after thoroughly evaluating the risks at the workplace. To minimise employee infection risk, they should use the right balance of technical and administrative measures, safe work practices, personal protective equipment (PPE), and increasing employee immunisation rates (Occupational Safety and Health Administration [OSHA], n.d.). The CDC (2020) advises any person having fever, cough or other symptom should get medical attention as soon as possible. People who have an elevated body temperature should be prohibited from entering public places and those already there should be quickly confined to the emergency area if they have sudden suspicious symptoms. Optimal ventilation is essential. It is advised that wastes be categorised and properly disposed in approved waste treatment areas (Qualls et al., 2017). Employers should keep the number of workers assigned to each shift to a minimum and make sure workstations are at least 6 ft apart. To reduce exposure, changes to their jobs or workplaces should be taken into account. In workplaces and public areas, education and information about illness prevention and control must be disseminated (OSHA, n.d.). If confirmed cases are noticed in public places, terminal disinfection must be performed as directed by local authorities. Cleaning employees need to be safeguarded from exposure to the dangerous chemicals utilised in these duties. The PPE and Hazard Communication guidelines may be applicable in such circumstances.

5.2 Social Distancing

A unique strategy that solely depends on human behaviour pattern is social distancing. By limiting droplet and contact transmissions and potential aerosol transmission, maintaining social distance can reduce exposure and prevent SARS-CoV-2 transmission (Fiorino et al., 2020). The WHO advises keeping minimum of 1 m (3 ft.) distance between people to minimise the spread of disease by contaminated respiratory droplets (Varghese & John, 2020). Stay-at-home, digitisation of services (shopping for necessities, education and teleworking from home), expanding physical distance between workers on the job site, spreading out work schedules, reducing social interactions at work, etc., should all be promoted (Fong et al., 2020). Services that are not necessary should be temporarily suspended. Premises must be thoroughly cleaned before opening the public areas if shutdown is not avoidable. Minimise public gatherings like exhibitions, meetings at one location, and group dinners. It is advised that departmental stores, pharmacies, restaurants, etc. suspend in-store services and start sidewalk pickup or online ordering for home delivery.

5.3 Cleaning and Disinfection

SARS-CoV-2 will continue to circulate in public places if appropriate sanitation and disinfection procedures are not undertaken. According to reports, SARS-CoV-2 had been found on surfaces in the confinement area with suspected patients. A plethora of studies furthermore validated that hands or gloves of healthcare workers could get infected by touching surfaces that were contaminated with SARS-CoV-2, indicating the need for surface cleaning and disinfection (Guo & Wang, 2020; Saitoh et al., 2020). In densely populated public areas, the cleaning frequency should be properly regulated (Fig. 7).

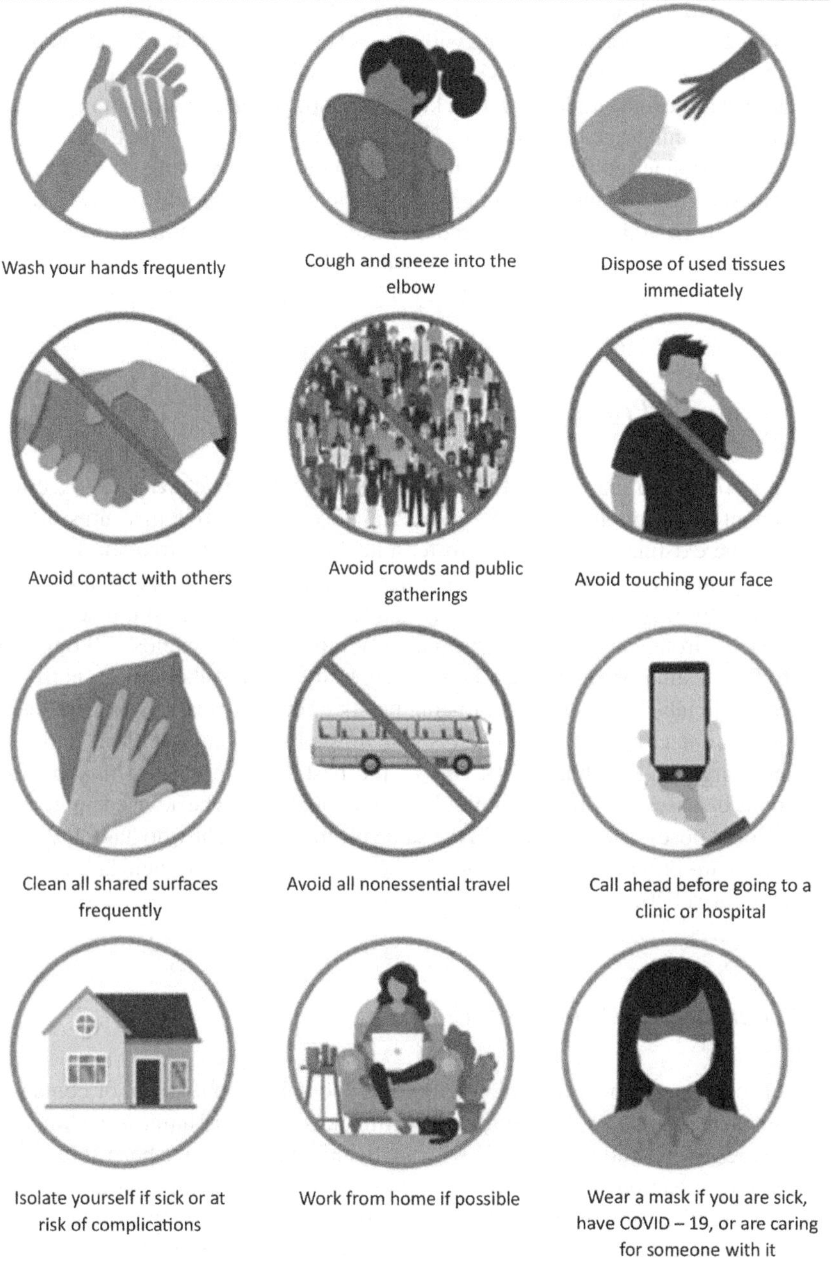

Fig. 7 Prevention of COVID-19. (*Source* https://www.medicalnewstoday.com/articles/coronavirus-prevention)

It is advised to disinfect surfaces by spraying with a solution containing 250–500 mg/L of available chlorine or wiping with alcohol or another reliable disinfectant. For surfaces that cannot be cleaned with bleach, 70% ethanol can be used. High-touch places such as bedside tables and door knobs must be sanitised daily with a common household cleaning agent comprising diluted bleach solution (1 part bleach to 99 parts water) (Güner et al., 2020). Whenever touching or sanitising objects, clothes or linen contaminated with body fluids, disposable gloves should be worn. Before discarding them with regular household waste, all used, contaminated disposable items should be put in a sealed bin. Bath and hand towels, sheets and clothes should all be washed in a machine at 60–90 °C with standard laundry detergent and water.

5.4 Personal Protective Measures

The spirit of dedication is reflected in the individual protective measures—a crucial component of infection management. It entails extra-legislative measures that can improve the existing condition and foster a healthy future. The frequency of infection and the spread of disease can be significantly reduced with proper hand hygiene, maintained by washing hands with soap and water. It is advised to utilise masks in high-risk environments where people gather or are somewhat enclosed (Fong et al., 2020). Self-testing products must be used correctly and results must be notified to the local officials if one suspects of being positive for the disease. When frequently coming in contact with infected objects, one must always keep sanitisers and disinfectant (Varghese & John, 2020). One of the best ways to prevent the spread of infectious diseases is by timely immunisation. People should be advised not to touch their mouth, noses or eyes. When in the same room as a patient who has COVID-19, those who are providing care for them at home must also have their faces covered (Güner et al., 2020).

5.5 Quarantine/Home Isolation

One of the first and most reliable methods for containing outbreaks of contagious diseases is quarantine (Wilder-Smith & Freedman, 2020). Quarantine is the limitation of activities or separation of people who are healthy but may have been subjected to a disease or infectious agent with the goal of tracking their symptoms and assuring the early identification of cases (Patel et al., 2020). Healthcare professionals analyse patients suspected of having COVID-19 infection to determine whether the condition may be handled at home and/or patients can receive symptomatic medication. Patients are advised to stay in a room with good ventilation that is not shared with other family members. To reduce the danger of transmission, patients must be contained to the house and should not be allowed to leave the room. The patient should have a single attendant, who is a family member. When helping the patient with needs, the caregiver must keep a proper distance. Until the patient has recovered completely, visitors not to be allowed. Everyone in family should maintain a safe distance from the patient. People living in the same home are to be suspected of having COVID-19, and should adhere to WHO quarantine recommendations. The WHO (2020g) advises quarantining contacts of patients with COVID-19 for 14 days after their last contact with the patient, as determined by the laboratory. Authorities that have confirmed the case should frequently check the patient and offer necessary medical care (Shereen et al., 2020).

6 Therapeutics for COVID-19

The treatments offered to COVID-19 patients depend on individual symptoms, and there is no specific known treatment for total recuperation in victims. In an attempt to treat COVID-19 effectively, scientists and medical professionals are evaluating a wide range of treatments, including vaccinations, immunosuppressants, monoclonal antibodies, and antiviral drugs (Zumla et al., 2020). However, testing their effectiveness in vitro and in vivo might require several months, and the outcomes of clinical trials might take even longer. WHO is coordinating sizeable clinical trials, in more than 100 nations worldwide, in an effort to reduce timeframe. Treatment for suspected and confirmed cases should take place in hospitals with specific designations, effective isolation, and safe environments.

As of 20/09/2022, the FDA has approved the use of specific antiviral drugs and monoclonal antibodies for treating mild to moderate COVID-19 to those who are more prone to become seriously ill, according to the CDC (2022b).

6.1 Antiviral Treatments

Antiviral treatments target particular virus components to prevent it from spreading throughout the body, reducing the risk of serious disease and death (CDC, 2022b). An antiviral treatment can be administered in one of two ways: orally, as a pill (oral antivirals) or intravenously, as an infusion (IV antivirals). Oral antivirals must be taken by the patient prescribed by an antiviral within the first 5 days when COVID-19 symptoms start to manifest. Within the first 7 days after symptoms start, patient must receive IV antivirals. There are two medications available: Paxlovid™ (Pfizer) and Lagevrio (molnupiravir)

6.2 Monoclonal (mAb) Antibodies

These antibodies aid the immune system in recognising and combating the virus more efficiently (CDC, 2022b). The mAb therapy can prevent the SARS-CoV-2 virus from infecting body's cells, hence reducing the amount of virus present there (COMBATCOVID, n.d.). The medication must be administered within the first 7 days of the onset of symptoms to those having high risk of experiencing serious COVID-19 symptoms. The mAb therapy is administered either through IV or a series of injections. Bebtelovimab has been approved by the FDA (2022a) for use in emergency for patients based on their age, medical history, and duration of symptoms.

6.3 Long-Acting Antibody (LAAB) Treatment

LAAB treatment may prevent some from contacting COVID-19 before coming into contact with the SARS-CoV-2 virus. Evusheld™ is the LAAB therapy and is a form of preventive medicine (PrEP). PrEP therapy may prevent the spread of COVID19 to those who have never been exposed to SARS-CoV-2 virus. After starting a PrEP regimen, if exposed to the virus, the regimen can aid in recovery from a COVID-19 infection. Evusheld is not a COVID-19 vaccination alternative. When exposed to the virus that causes COVID-19, Evusheld is to help those not having strong immune response (COMBATCOVID, n.d.).

Moreover, the following Table 2 provides a summary of the previously employed and currently in use treatment options for COVID-19 patients.

7 Vaccination for SARS-COV-2

In middle of COVID-19 pandemic, which began with the initial confirmed cases on December 31, 2019, it was clear that vaccines are crucial for boosting immunity, in an attempt to avoid serious complications of COVID-19 infection, lessening the pressure on health services, and minimising financial losses (Jeon, 2020). Globally researchers made incredible efforts and developed powerful vaccines targeting SARS-CoV-2 are now at an exponential rate to manage this deadly disease that already ravaged mankind (Cascella et al., 2022). Vaccines are therapeutic compounds that, when given, trigger an immune response (generation of antibody) against antigens such as bacteria, virus or even other pathogens. This protects the host from contacting the disease from aforementioned microorganisms (Ahmed et al., 2021). CD4+ helper T cells activate B-cells to create virus-specific neutralising antibodies and CD8+ cytotoxic T cells to identify and destroy virus-infected cells (WHO, 2020h). Many COVID-19 vaccines use the antigenic variants for S protein, assumed to be the main target for protective immune response (Lurie et al., 2020; Ni et al.,

Table 2 Therapeutics previously employed and currently in-use in treating COVID-19 patients

Type of treatment	Name	Description
Drug	Remdesivir	• Gilead Sciences discovered Remdesivir, an adenosine analogue antiviral prodrug to combat the Ebola virus outbreak in Western Africa. It has already shown antiviral effectiveness versus SARS-CoV-2 in laboratory (Rauf et al., 2020) • The active component of remdesivir (GS-441524) inhibits the synthesis of viral RNA via attaching to RNA-dependent RNA polymerase (Khan et al., 2021a, 2021b, 2021c) • FDA approved it for hospitalised adults (above age of 12 and weighing more than 40 kgs) and infants (under age of 12 and weight between 3.5 and 40 kg) (Salahshoori et al., 2021) • Recommended dosage: Three consecutive days of intravenous (IV) infusions at hospital (CDC, 2022b)
	Paxlovid	• Antiviral treatment including the co-packaging of two different drugs (nirmatrelvir and ritonavir) (USFDA, 2022a) • Paxlovid is an experimental drug for the treatment of mild-to-moderate COVID-19 patients aged 12 and above having weight above 40 kgs (USFDA, 2022a) • Drug has been granted Emergency Use Authorization (EUA) (The Hill, 2022) • Paxlovid may interact with hormonal contraceptives (e.g. pills, implants, intrauterine devices [IUDs] etc.) hence, alternate contraception is recommended (COMBATCOVID, n.d.) • Patients with severe liver or kidney problems should refrain from taking Paxlovid (Mishra, 2020)

(continued)

Table 2 (continued)

Type of treatment	Name	Description
	Molnupiravir (Lagevrio)	• Lagevrio is an antiviral medication that suppresses SARS-CoV-2 infectivity (AIFA, 2021) • This drug is given orally to manage COVID-19 symptoms that are mild to moderate • It must be administered within 5 days just after appearance of the initial COVID-19 symptoms (NPSMEDICINEWISE, 2022) • Lagevrio can be used by adults of 18 or above age • Lagevrio also not advised to be taken during pregnancy or while nursing. Additionally, after the final dose, supplementary contraceptive measures are needed for a brief period (COMBATCOVID, n.d.) • EUA has been granted (USFDA, 2022b)
	Favipiravir	• Favipiravir may prove to be an effective medication for the treatment of mild-to-moderate SARSCoV-2-infected infections (Agrawal et al., 2020) • It reduces viral RdRp reproduction by binding to it (Shannon et al., 2020) • Adults may take favipiravir as suggested dosage of 1800 mg orally twice a day on day one, followed by 800 mg orally each day for maximum of 14 days (Joshi et al., 2021)
	Chloroquine or Hydroxychloroquine	• Chloroquine is used to treat malaria. Another analogue of chloroquine is hydroxychloroquine (Das et al., 2020) • During the early phase of COVID 19, medication was offered as an antiviral (Gasmi et al., 2021) • COVID-19 Treatment Guidelines Panel advises against treating COVID-19 to hospitalised patients with chloroquine or hydroxychloroquine, with or without azithromycin (National Institutes of Health [NIH], 2022) • Not advised for usage in patients who are not hospitalised. Chloroquine is an antimalarial drug. Hydroxychloroquine is an analogue of chloroquine (Das et al., 2020)

(continued)

Table 2 (continued)

Type of treatment	Name	Description
	Lopinavir/Ritonavir	• FDA-approved HIV combo medicine that was offered as antiviral against SARS-COV-2S during initial stages (Meini et al., 2020) • May prevent SARS-CoV-2 protease from cleaving proteins, that could prevent virus from replicating (Alanagreh et al., 2020) • According to WHO (2020m) it is currently not recommended for the treatment of COVID-19 in both hospitalised and non-hospitalised patients
	Ivermectin	• Anti-parasitic medication used globally against SARS-COV-2 earlier (Cascella et al., 2022) • Based on an in vitro investigation, it has shown its effectiveness in treating the virus by suppressing its proliferation (Pisoschi et al., 2022) • Not advised for use against SARS-COV-2 (USFDA, 2021)
Immune-based therapy	Convalescent plasma	• Convalescent plasma therapy includes providing plasma from sick people who healed people currently ill with the disease (Rauf et al., 2020) • The USFDA approved the use of convalescent plasma in hospitalised COVID-19 patients as an EUA (Lotfi, Rezaei and Hamblin, 2020) • It neutralises the virus and improves patient's immunity (Rojas et al., 2020) • FDA promotes plasma donation from COVID-19 recovered patients (for at least two weeks). According to estimates, one donation can produce enough plasma for four patients (Bloch & Feldweg, 2021; USFDA, 2022c)
	Tocilizumab	• Tocilizumab, a mAB against the interleukin-6 receptor alpha, has also been approved for use in treating a number of rheumatological conditions (Cascella et al., 2022) • The medication has been tested in hospitals in China with severe COVID-19 patients (Rauf et al., 2020) • Yet not approved by USFDA

(continued)

Table 2 (continued)

Type of treatment	Name	Description
Anti-SARS-CoV-2 antibody products	Bamlanivimab	• Spike protein neutralizing mAB which prevents entry of SARS-CoV-2 into host cells • The mAB attaches to several overlapping epitopes in RBD of S structural protein of SARS-COV-2 (COVID-19 Treatment Guidelines Panel, n.d.) • Non-recommended for use
	Baricitinib	• Patients with moderate-to-severe rheumatoid arthritis (RA) are currently advised to use Baricitinib orally which selectively inhibits Janus kinase (JAK) 1 and JAK 2 • Due to its inhibition activity on SARS-CoV-2 endocytosis in vitro and on the intracellular signalling cascade of cytokines that induce the late-onset hyperinflammatory state which culminates into serious illnesses, baricitinib was assumed to be a viable treatment for COVID-19 (Cascella et al., 2022) • Non-recommended for use
	Bebtelovimab	• Neutralising antibody medication that functions by attaching to S protein of SARS-COV-2 • USFDA has granted an EUA to the drug for treating SARS-COV-2 patients • Drug is administered intravenously for approximately 30 seconds following which patients must be monitored for around an hour (USFDA, 2022d)
Corticosteroids	Dexamethasone	• Dexamethasone, a corticosteroid with anti-inflammatory and immunosuppressive properties, is prescribed for many diseases (WHO, 2021e) • It was suggested as a helpful treatment for cytokine release syndrome (CRS) in SARS-COV-2 patients (Sharun et al., 2020) • Not advised to administer corticosteroids to COVID-19 patients owing to increased risk of vascular necrosis and diabetes (Zhang et al., 2021)

(continued)

Table 2 (continued)

Type of treatment	Name	Description
Adjunctive therapy	Vitamin C	• With anti-inflammatory, immunomodulatory, antioxidative, antithrombotic, and antiviral effects, vitamin C is a water-soluble vitamin (Tsoupras, 2022) • No concrete proof that vitamin C works better against COVID-19. It is believed to reduce inflammation and vascular damage in COVID-19 patients (Milani et al., 2021)
	Vitamin D	• By altering the transcription of immune-response genes in macrophages, T cells, and dendritic cells, vitamin D controls inflammatory processes (Fernandez, Ramrez-Meja and Urcuqui-Inchima, 2022) • Vitamin D is anticipated to help avert, cure or reduce SARS-CoV-2 symptoms like severe pneumonia (Ali, 2020)
	Zinc	• Maintenance of natural tissue barriers like respiratory epithelium, preventing pathogen invasion, and ensuring a healthy immune system function depend on zinc (Wessels et al., 2020) • Zinc insufficiency can be included to list of risk factors for COVID-19 infection and harmful COVID-19 development. • Zinc supplementation could be advantageous for prophylaxis. But there is no clinical recommendation for its use (Kumar et al., 2020)

2020). Numerous companies globally use various platforms and techniques to create vaccines. A number of vaccinations are being considered for their possible contribution for controlling COVID-19. The 68% of world population has got at least one dose of COVID-19 vaccine and globally, 12.728 billion doses have been given, and as of September 28, 2022, a total of 3.22 million are being administered daily according to "Our World in Data" dashboard.

7.1 Vaccine Development: Phases of Clinical Trials

A potential vaccine patient goes through a complex development procedure (Bisht & Dubey, 2020). The development process consists of two phases: (a) preclinical trials, in which the target antigen is recognised and the vaccine safety and effectiveness are assessed in laboratory (in vitro) and on animals (experimental) (in vivo) and (b) clinical trials involving humans (National Centre for Immunisation Research and Surveillance [NCIRS], 2020; Singh & Mehta, 2016). Phases in clinical trials are discussed as follows.

7.1.1 Clinical Trial Phase I

The vaccine is tested in humans first in a small cohort of 20-100 persons. The vaccine undergoes testing on humans for the first time in order to assess its safety, establish an appropriate dose ratio and produces the best immune reaction, and shows vaccination-related side effects. Investigators first check the vaccine on adults, and if meant for yougsters, then steadily reduce the age of the test group till they reach the desired age (International Federation of Pharmaceutical Manufacturers & Associations [IFPMA], 2019). Phase I requires at least 1 year to execute (NCIRS, 2020). Phase II clinical studies will begin if the phase I clinical trials show encouraging results (COVID19 VACCINE TRACKER, 2022). Not every person who completes phase I clinical trials proceeds to phase II.

7.1.2 Clinical Trial Phase II

Objectives of this phase are to investigate the candidate vaccine's immunology, suggested doses, vaccination regimen, and route of administration (IFPMA 2019). The NCIRS (2020) stresses the importance of ensuring that the potential vaccine promotes both humoral and cellular immune reaction against target antigen. These trials are randomised, well-controlled, and involve a control group; few of the participants may be members of high-risk groups for the condition (Scheppler et al., 2021). The vaccined candidates moving on to phase III clinical trials must have substantial phase II outcomes.

7.1.3 Clinical Trial Phase III

Clinical trials in phase III are crucial as they can offer the most convincing proof of the safety and efficacy of the vaccine. Thousands to sometimes lakhs of qualified participants take interest in phase III studies and are allocated randomly to either obtain the vaccination or go unvaccinated (COVID19 VACCINE TRACKER, 2022). Researchers examine whether the vaccinated people shields from the disease in comparison to those who do not receive it. To ascertain whether the vaccination is successful at preventing the disease, the pathogen must be circulating during the trial (NCIRS, 2020). Occasionally Phases II and III of clinical trials are merged to create phase II/III clinical trials that accomplish the goals of both phases (COVID19 VACCINE TRACKER, 2022).

7.1.4 Approval and Licensure

Following positive Phase III study, vaccine investigator will submit a report to the relevant competent authority for evaluation and approval. The concerned authority will continue to regulate vaccine manufacturing after licensure, conduct facility inspections, and examine the vaccine company's testing regime (IFPMA, 2019) (Fig. 8).

Earlier, it took years or even decades for vaccines to develop and undergo testing before there was sufficient proof of their efficacy and safety. Due to inadequate funding, resource defficiency which lenghthened the schedule, and the reality that only a few businesses and investigators were focusing on vaccine development; the slow progress of the vaccine development cycle became standard (COVID19 VACCINE TRACKER, 2022). The COVID-19 pandemic was so critical that the potential vaccinations had to be developed quickly. Development of the COVID-19 vaccine was given highest priority by several nations, businesses, research institutes and health organisations, who invested their most of the time, money and effort in vaccine development. Phase I clinical trials started in March 2020 for the first vaccine candidate (Moderna), 2 months after the SARS-CoV-2 genome was sequenced and made public (Deplanque & Launay, 2021). Clinical studies in Phase II commenced before Phase I ended. Phase I and Phase II clinical studies for COVID-19 vaccine were frequently combined in order to advance the trial timelines. The dose, safety and immunogenicity measures were assessed, but the scientific approach remained intact. Additionally, Phase III clinical trials started before Phase II trials were finished. Phase II and Phase III trials were occasionally combined (NCIRS 2020). The duration of developing a vaccine has been reduced to roughly 10 months attributable to clinical trial phases that crossover and are integrated, international cooperative efforts, funding and manufacturing pre-planning in the urgent need of a vaccine that is both safe and effective. In many nations, COVID-19 vaccinations are being distributed with EUA label. The NCIRS (2020) explains that because there are not much safety data, the vaccine will not be fully registered until after extensive safety surveillance, which again will take several years. As of September 19, 2022, the 227 vaccine candidates have since entered various stages of clinical development. Out of them, 47 vaccines have been approved, 70 are in Phase II of their trials, 89 are in Phase III

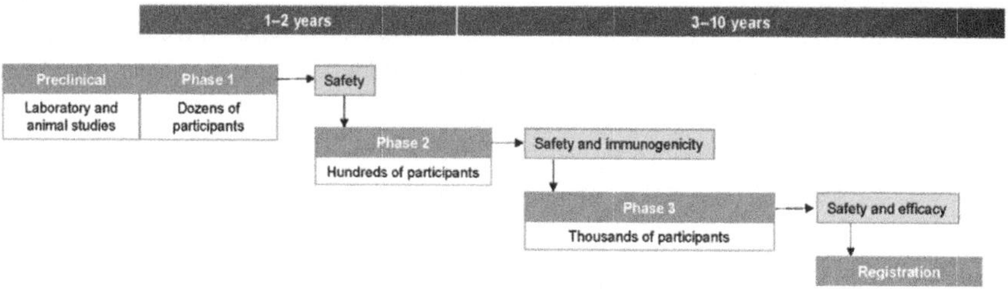

Fig. 8 Conventional pathway of vaccine development. (*Source* National Centre for Immunisation Research and Surveillance)

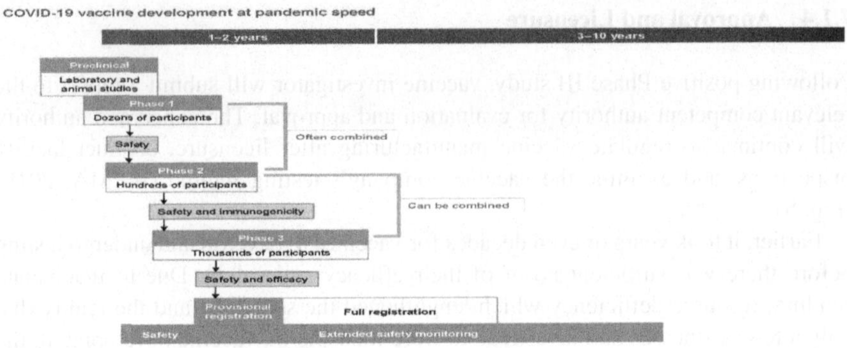

Fig. 9 SARS-COV-2 Vaccine development pace. (*Source* National Centre for Immunisation Research and Surveillance)

and 57 vaccines are in Phase I of their trials. However, only 11 vaccines have received Emergency Use Listing (EUL) designation from WHO (COVID 19 Vaccine Tracker, 2022). Details of the same can be accessed from https://covid19.trackvaccines.org/vaccines/ (Fig. 9).

7.2 Vaccine Development Platforms

Candidates for COVID-19 vaccine are available in a range of formulations, including conventional whole-pathogen vaccines and modern generation vaccines (Wang et al., 2020a, 2020b, 2020c, 2020d, 2020e). Live attenuated virus, inactive virus, protein subunit and virus-like particles are examples of common vaccine delivery systems (Murphy & Weaver, 2016). Recombinant viral vector, nucleic acid-based and antigen-presenting cells are among the next-generation platforms (Jafari et al., 2022). Since recombinant protein-based development seems to be simpler to scale up utilising defined industrial capacity, some platforms offer significant advantages such as nucleic acid-based with antigen manipulation versatility and viral vectors provides powerful immune response, excellent protein expression, and protracted reliability (Chauhan et al., 2020).

The major vaccination platforms that have advanced to Phase III trials are divided into two categories, viz. component viral vaccines and whole virus vaccines described as follows.

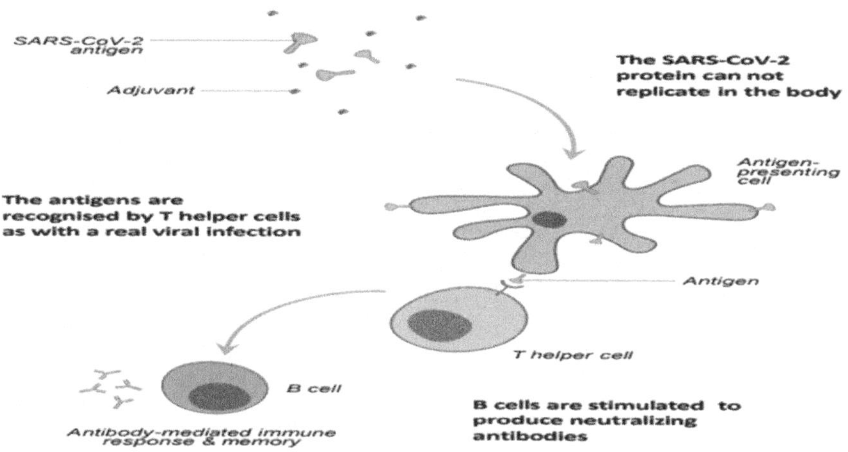

Fig. 10 Mechanism: subunit vaccines. (*Source* World Health Organization Coronavirus Update 73 (2022d))

7.2.1 Component Viral Vaccines

Protein Subunit

A subunit vaccine uses synthesised peptides or recombinant antigenic proteins, that are essential for triggering a prolonged immune reaction that is beneficial and/or curative (Wang et al., 2020a, 2020b, 2020c, 2020d, 2020e). The RBD of the SARS-CoV-2 S protein is used as an antigen in the subunit vaccine to trigger a potent immune reaction against virus (Jafari et al., 2022). These vaccinations are considered to be quite safer following injection because they contain no active viral component (Hammoud & Myronchenko, 2021). The main drawback of such vaccine is that it primarily induces humoral immune reactions and perhaps offers limited protection against virus infection owing to low immunogenicity of the proteins (Jeyanathan et al., 2020). In order to build immunological memory, these vaccinations need to be administered repeatedly in addition to an adjuvant in the composition (Fig. 10).

Virus-Like Particles (VLPs)

Virus-like particles, which generate particles naturally built from numerous viral structural proteins have gradually evolved as vaccine delivery vehicles (Jafari et al., 2022). In terms of size, shape and antigen presentation, they closely approximate its native virus (Zepeda-Cervantes et al., 2020). Furthermore, VLPs can be modified to contain many antigens of interest, which is a crucial characteristic for creating effective and comprehensive vaccinations (Prates-Syed et al., 2021). They have a greater immunogenicity, are safer and more efficient. Importantly, VLPs exhibit

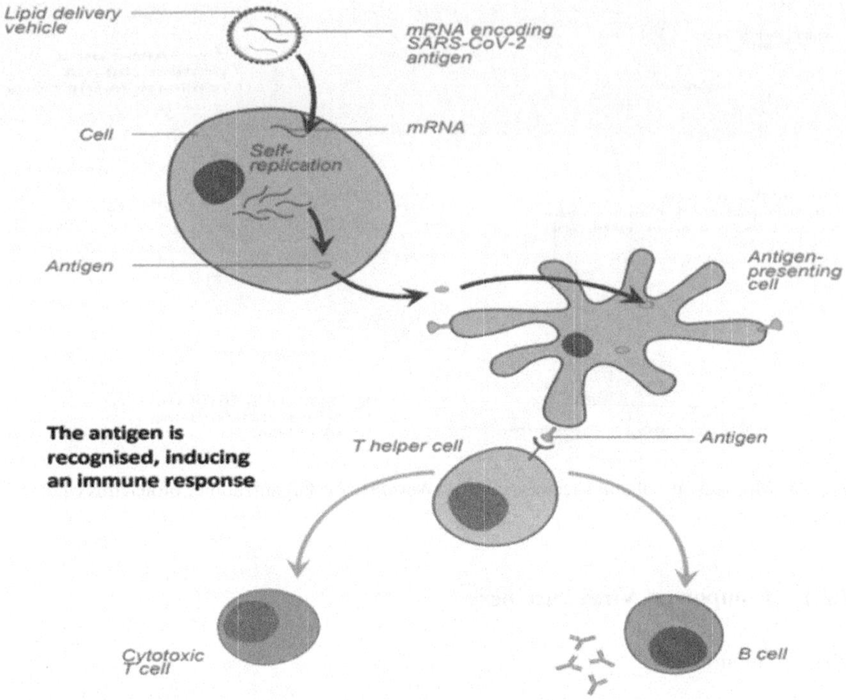

Fig. 11 Mechanism: RNA vaccine. (*Source* World Health Organization Coronavirus Update 73 (2022d))

exceptional adjuvant properties and have the capacity to stimulate both cellular and humoral immune reactions (Jafari et al., 2022).

RNA-Based

The mRNA-based vaccines are often more appealing than other vaccine techniques because of their minimal cost and quick production (Pardi et al., 2018). The mRNA vaccine development is often a simple procedure. Engineered microorganisms are propagated and cultivated to obtain the necessary genetic sequences when the target antigen from the emergent pathogen is detected, its gene sequence is optimised and the plasmid is created (Fang et al., 2022). This type of vaccine has many benefits as it addresses both cellular and humoral immune reactions, has a strong safety profile, does not integrate the human genome, does not produce antivector immunity, and does not increase the risk of viral infection (Kowalzik et al., 2021) (Fig. 11).

However, this platform has some shortcomings. For vaccine persistence to be maintained, preservation at quite cold temperature is necessary. This type of vaccine is not suitable for low-income countries owing to the high manufacturing cost (Frederiksen et al., 2020). Lack of systematic methods for identifying the primary physicochemical mechanisms that lead to the finalised mRNA vaccine degradation (Crommelin et al., 2021). Since this platform is still quite new, little is known about how it will scale up (Krammer, 2020).

DNA-Based

DNA vaccines consist of a genetically modified plasmid that encodes the pathogen's particular antigen and a vector that delivers the plasmid to the host cell (Cui, 2005). Plasmid DNA vaccines have a number of benefits, including the ability to target and stimulate both humoral and cellular immune responses, versatility for multivalency and room-temperature preservation of the ultimate vaccine. These advantages make plasmid DNA vaccines perfect for quick responses all through pandemics (Silveira et al., 2017; Xu et al., 2014). Moreover, such vaccines have few significant disadvantages: (a) relatively low reactivity, requiring multiple doses of the vaccine to provide the best protection (Qin et al., 2021) and (b) possible incorporation in cellular chromosomes may increase the risk of cancer (Doerfler, 2021).

Replicating and Non-replicating Viral Vectors

These vaccines either use bioengineered viral structures or replication-deficient viral vectors (Lundstrom, 2020); Zhu et al., 2020). Adenoviruses represent the most prevalent replication-incompetent or defective viral vectors now in use for transporting and delivering a chosen, double-stranded DNA segment of the RNA of SARS-CoV-2 which encodes the virus S-protein. This segment is originally produced on a plasmid. The vectors penetrate body cells following administration, however, they are unable to proliferate inside of cells. The genetic material that has been intended to be transmitted escape the vectors and passed through to the nucleus, where it gets preserved but could not combine with the body DNA. It is then translated into spike proteins, which start to gather on the exterior of infected cells, after being read and translated into mRNA, which leaves the nucleus (Lundstrom, 2020; Wold & Toth, 2013). When the immune system detects S-proteins or their fragments, it begins to send alerts, produce specific neutralising antibodies, activation of t cells (CD4+ and CD8+) and memory B- and T-cell cells (Alberts et al., 2002; Cano & Lopera, 2013). These vaccine vectors have a hard protein covering that aids in safeguarding the genetic material they contain. Due to this, the vaccine does not need to be kept frozen and can be kept in a refrigerator at a temperature between 2 and 8°C for at least 6 months. Furthermore, these vaccinations do not need adjuvants to work and can trigger powerful immunogenicity (Abdulla et al., 2021). Nonetheless, several disadvantages connected with their use are:

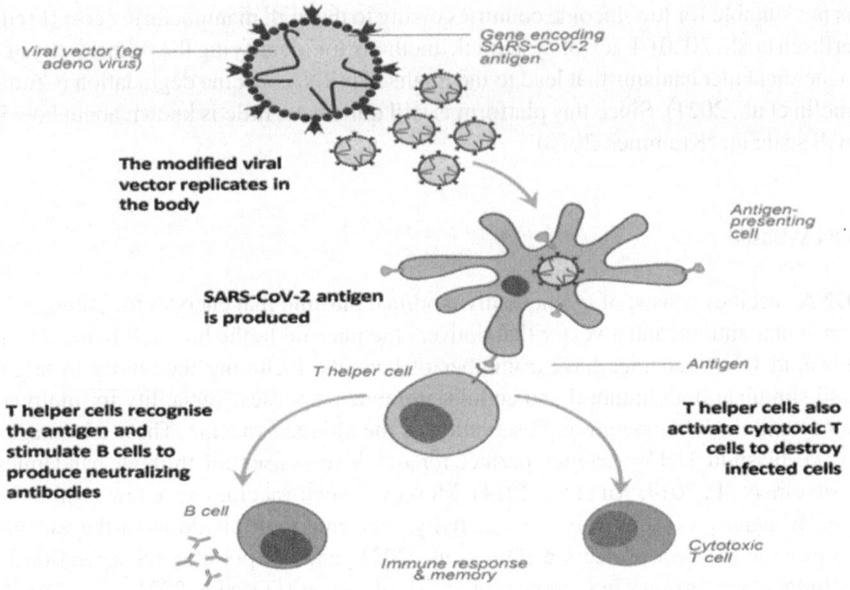

Fig. 12 Mechanism: viral vector vaccines. (*Source* World Health Organization Coronavirus Update 73 (2022d))

(a) The produced vaccine candidate would be useless if there was already immunity to the viral vector used (Nayak & Herzog, 2010). But, this might be reduced by prepping with a different kind of vaccination, like a DNA vaccine (Yang et al., 2003) or by switching up the way the prime and booster dosages are administered (Pandey et al., 2012).
(b) The viral vector itself triggers an immunological reaction that subsequently prevents the use of the same vector in following vaccinations (Nayak & Herzog, 2010).
(c) Strong reactivity in vaccine recipients suggests that some viral vectors may cause genetic toxicity; but, this can be reduced by employing hybrid vectors (Huang & Kamihira, 2013).
(d) Possibility of virus recombination while the product is being produced (Fig. 12).

7.2.2 Whole Virus Vaccines

Inactivated

The genetic information of the viruses has indeed been damaged in inactivated virus vaccinations to inhibit its ability to cause disease (Rodrigues et al., 2015). Larger

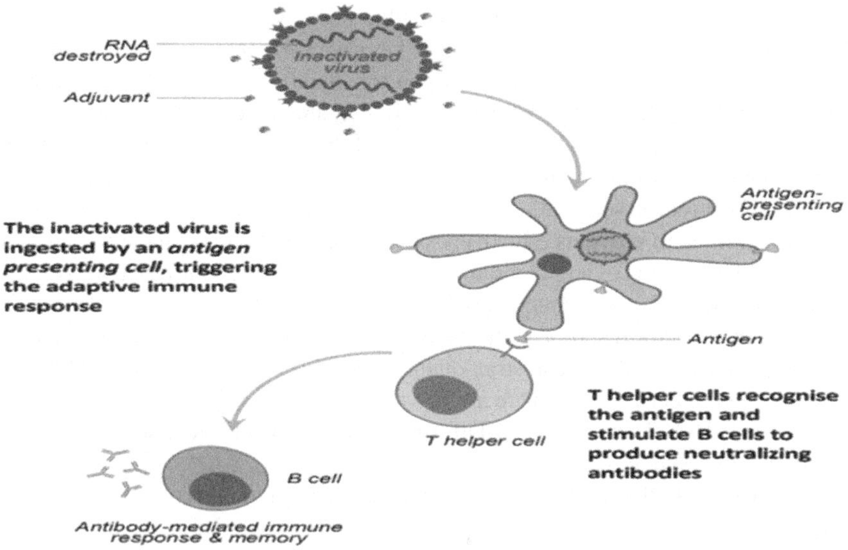

Fig. 13 Mechanism: inactivated virus vaccines. (*Source* World Health Organization Coronavirus Update 73 (2022d))

doses are required since an inactivated virus cannot multiply inside the body (Robert-Guroff, 2007). In order to achieve wide immune responses, inactivated vaccines are injected intramuscularly after being adjuvanted with alum or certain oils.

This is because the vaccine encourages targeting of a variety of different virus proteins, such as S, M and N in SARS-COV-2 (Krammer, 2020). The benefits of these vaccinations include: safer and more reliable than LAVs. It already has the infrastructure and technology needed for development. Also, possesses a history of SARS-CoV and other disease testing. When used with adjuvants, it can make them more immunogenic (Kaur & Gupta, 2020). However, there are three difficulties with the formulation and manufacture of this kind of vaccination. Requiring booster doses in order to retain immunity. It is necessary to handle large quantities of viruses while ensuring the integrity of the immunogenic particles. Additionally, it is difficult to produce billions of doses quickly, which makes the pandemic responses exceptionally disrupted (Frederiksen et al., 2020) (Fig. 13).

Live-Attenuated

Live-attenuated virus vaccines are created by using a non-virulent strain of the virus or by creating a biologically weaker version of the virus, for whom the restricted rounds of replication can elicit immune reactions comparable to those stimulated by a natural infection but also are insufficient to trigger illness (Dai et al., 2019). By subjecting and modifying the virus to unfavourable conditions, including low temperature, proliferation in a non-susceptible hosts, the virus can be attenuated. Codon de-optimization or gene deletion of key genes for enhancing innate immune recognition can also be accomplished by genetic alteration using RG system (Broadbent et al., 2016; Talon et al., 2000). The use of live-attenuated SARS-CoV-2 vaccine candidates has several benefits, including the ability to target and stimulate strong cellular and mucosal immunity, which is required for safeguarding without the use of adjuvants (Frederiksen et al., 2020). However, there are certain downsides to this kind of vaccine: To prove its effectiveness and safety, LAV must undergo thorough supplementary testing. The potential for nucleotide substitution during viral replication can lead to post-vaccination production of recombinants. Large-scale vaccine manufacture is a slow response to the pandemic since the manufacturing t procedures are labor-intensive and demand strict quality control (Frederiksen et al., 2020).

As of September 19, 2022 there are 11 vaccines Granted Emergency Use Listing (EUL) by WHO. The vaccines, platform of development and manufacturer are mentioned in Table 3.

Most COVID-19 vaccinations are intended to be administered in two doses. The second vaccination mimics the immune system's own defences. Immunological memory is primed by the first dose and solidified by the second. The immune system requires time to respond to the first vaccination dose and to develop memory cells that will detect the pathogen when it is contacted again. When opposed to a small timeframe, a longer time between the first and second dose may trigger a higher immunological response (WHO, 2020i).

8 SARS-CoV-2: Environmental Perspective

The COVID-19 pandemic had an impact on many facets of human life and the world economy (Klemeš et al., 2020). It is the biggest pandemic of the 21st century and has caused the largest quarantine and lockdown in human history. This once-in-a-century phenomena paradoxically had some beneficial effects for the health of the planet (Khan et al., 2021b). Consider how lower levels of criteria pollutants like carbon monoxide (CO), nitrogen oxide (NO_2), methane (CH_4), sulphur oxide (SO_2), black carbon (BC), and particulate matter (PM10 and PM2.5) were the result of the closure of industrial sites, power stations, and reduced transit and shipments (Gautam, 2020; Somani et al., 2020). Noise pollution was also decreased as a result of fewer commercial operations and public transit (Zambrano-Monserrate et al., 2020). The lockdown reduced air pollution by 30% (Muhammad et al., 2020). COVID-19 had a noticeable impact on the improvement of water bodies, like Ganga River in India (Lokhandwala & Gautam, 2020). However, these beneficial results were only temporary because of the lockdown especially during the early stages of coronavirus. These advantageous environmental impacts disappeared as varied activities returned to their regular levels. On the other hand, COVID-19 has had a number of negative effects, including an increase in the production of biomedical waste and the threat of

Table 3 Vaccines approved by World Health Organization granted emergency use listing (EUL)

Vaccine	Manufacturer	Platform of development	Effective date
Covovax (Novavax formulation)	Serum Institute of India Pvt. Ltd. (SIIPL), India	Protein Subunit	17 December 2021
Nuvaxovid	Novavax CZ a.s., Czech Republic	Protein Subunit	20 December 2021
Moderna Spikevax	Moderna Biotech, Spain	RNA	30 April 2021
Pfizer/BioNTech Comirnaty	BioNTech Manufacturing GmbH, Germany	RNA	31 December 2020
CanSino Convidecia	CanSino Biologics Inc., People's Republic of China	Non-replicating Viral Vector	19 May 2022
Janssen (Johnson & Johnson) Jcovden	Janssen–Cilag International NV, Belgium	Non-replicating Viral Vector	12 March 2021
Oxford/AstraZeneca Vaxzevria	AstraZeneca/SK Bioscience Co. Ltd, Republic of Korea	Non-replicating Viral Vector	15 February 2021
Covishield (Oxford/ AstraZeneca formulation)	Serum Institute of India Pvt. Ltd, India	Non-replicating Viral Vector	15 February 2021
Covaxin	Bharat Biotech International Limited, India	Inactivated	03 November 2021
Sinopharm (Beijing) Covilo	Beijing Institute of Biological Products Co., Ltd. (BIBP), China	Inactivated	07 May 2021
Sinovac CoronaVac	Sinovac Life Sciences Co., Ltd., China	Inactivated	01 June 2021

Source COVID19 Vaccine Tracker

SARS-COV-2 spreading due to the identification of the virus in wastewater samples (Sangkham, 2020).

8.1 Role of Climatic Factors in COVID-19

A major factor influencing the transmission of viruses like SARS is the climate (Petersen et al., 2020). Particularly, climatic factors like temperature, humidity and wind speed have a significant impact on the persistence and propagation of several viruses (Watson et al., 2006). Temperature change, moisture and wind velocity were found to have an impact on the incidence and lethality of COVID-19 cases in Wuhan,

China. The chance of developing COVID-19 and deaths from it is increased by 2.92 times for every unit decrease in the daytime temperature (Chen et al., 2021a, 2021b). On the contrary, a drop in the number of people dying from this virus was correlated with rise in temperature of 1 unit and absolute humidity (Ma et al., 2020). As a result, lesser humidity throughout the winter may also be linked to higher fatality from COVID-19.

Results from earlier studies revealed that many outbreaks of respiratory diseases have seasonal patterns (Leecaster et al., 2011). Cold weather reduces human immunity and creates the ideal circumstances for viral replication and survival, presenting new possibilities for viral invasion (Eccles & Wilkinson, 2015). During the winter and early spring, the first wave of the pandemic struck the Northern Hemisphere (China, Europe and the USA). There is evidence that SARS-CoV-2 can live on surfaces lengthier in temperate than in warm climates (Riddell et al., 2020). In tropical countries, COVID-19 cases were significantly lower, may be in part, because of the increased temperature (Ghosh et al., 2020). Interestingly, COVID-19 cases increased over the warmer months in some American states like Florida where airconditioners are used widely, implying that outdoor temperatures might not be a significant factor in SARS-CoV-2 transmission. This is in contrast to the summer months in Europe where COVID-19 cases greatly reduced having followed spring lockdowns and ongoing social distance measures. According to a study by Menebo (2020) conducted in Oslo, Norway, temperature and rainfall are linked to the number of regular instances of COVID-19 at highest and regular temperatures. Temperature is highly associated with COVID-19, while precipitation has a negative correlation.

8.2 Impact on Water Quality

Water pollution is a prevalent problem in developing countries like India and Bangladesh, where household and industrial pollutants are discharged into waterways without being treated (Hasan et al., 2019; Rai et al., 2022). In several regions of the world during the pandemic, water ecosystems displayed evidence of recovery owing to lockdown. The water quality of the "National River" Ganga in India improved, particularly in and near industrial hubs and populated areas. Enhanced oxygen levels and reduced biological oxygen demand, as well as lower levels of faecal coliform, total coliform and nitrate concentration, were indicators of the effect. The river's ability to self-clean was boosted with better water quality. The main industrial sources of pollution that harm aquatic ecosystems, like dumping of industrial waste, crude oil, heavy metals and plastic dropped or stopped entirely during lockdown (Häder et al., 2020; Rai & Fulekar, 2020b). Additional human activities including social and religious gatherings, waterway transportation, fishing and others were also forbidden, which reduced the amount of solid waste dumped along the river banks and the amount of trash that locals and visitors left behind (Pathak et al., 2021). Recent studies by the Indian Institute of Technology, Roorkee (India), demonstrated that the pandemic period showed that Ganga River water became fit for drinking

(Indian Express, 2020). The river (Bhagirathi at Gangotri, Rudraprayag, Devprayag, Raiwala-Uttarakhand, Rishikesh, Bijnor and Diamond Harbor in West Bengal, India) was classified as class "A" by the Central Pollution Control Board since it was deemed to be good for drinking following disinfection (Singh et al., 2022). After 2 months of lockdown, Venice (Italy) water also seemed cleaner and underwater wildlife had become more apparent, something that had not been seen in the city for long time. The activity on the waterways, sediment churning, and other sources of water pollution significantly decreased as the number of tourists decreased, leaving Venice canals with significantly cleaner water than they were prior to COVID-19 (Saadat et al., 2020).

Contrarily, the COVID-19 pandemic, which has harmed water quality: (i) there is a high likelihood that SARS-CoV-2 RNA can be conveyed and found in wastewater through the faeces of infected individuals (Ji et al., 2021); (ii) during the COVID-19 era, high levels of detergent usage and water consumption made it possible for many organic and metallic compounds to enter domestic waters, which led to decline in water quality (Chirani et al., 2021). According to the World Health Organization (2020j), sewage or drinking water have not yet been identified as a source of exposure for this virus. Policymakers advised implementing techniques for the sustainable water management and controlling new hazardous species that were found in water during COVID-19 (Miyah et al., 2022).

8.3 Impact of COVID-19 on Air Quality

A sharp decline in greenhouse gas (GHG) emissions has been caused by the closure of business, travel and industry. Vehicles and aeroplanes are believed to be the main sources of emissions, accounting for around 72% and 11% of the GHG emissions in the transportation sector, respectively (Henriques, 2020). Several cities in Asia, Europe and USA have reported lower air pollution levels, particularly for nitrogen dioxide (NO_2), particulates smaller than 2.5 mm (PM) and black carbon (CN) (Tobías et al., 2020). Due to the lockdown caused by COVID-19, the level of air pollution in New York (USA) decreased by upto 50% in 2020 compared to 2019 (Singh & Mishra, 2021). According to the European Environmental Agency (EEA), during COVID-19, NO_2 emissions decreased by up to 60% in the continent's biggest cities (Bruyninckx, 2020). During the nationwide lockdown, Delhi's (India) NO_2 level dropped upto 70%. The Central Pollution Control Board [CPCB], (2021) reports, air quality has increased relative to the standard. Ozone concentration was found to have increased above CPCB requirements whereas PM2.5, PM10, NO, NO_2, SO_2, CO and toluene concentrations were reduced. In the coastal regions of China, the concentrations of NO_2 and CO fell by 63.1 and 6.9%, respectively, while the concentration of O_3 increased by 37.9%. The considerable fall in NO_2 concentration can be linked to freighters' and heavy vehicles' reduced emissions, which resulted in a drop in NO concentration. The decrease in NO_2 during lockdown was credited for the rise in ozone levels (Wang et al., 2020a, 2020b, 2020c, 2020d, 2020e). In addition, the COVID-19 is said to have reduced air travel worldwide by 96% compared to same period last year (Wallace, 2020), which has a long-term impact on the environment. Reduced use of fossil fuels reduces GHG emissions, aiding in the fight against global warming.

8.4 *Impact of COVID-19 on Biomedical Waste Management*

Globally, an upsurge in the source of medical waste ever since COVID-19 pandemic, poses a serious risk to both the environment and humans (Rai & Fulekar, 2020a). Many infectious and biological wastes are produced from hospitals for specimen collection of the suspicious COVID-19 sufferers, testing, treatment of numerous patients, and sanitation purposes (Zambrano-Monserrate et al., 2020). When the first phase of lockdown began, the production of medical waste in Ahmedabad (India), jumped from 550–600 kg/day to roughly 1000 kg/day (Somani et al., 2020). The local waste management authorities are facing a serious problem as a result of the unexpected increase in hazardous material and how to dispose it properly. The SARS-CoV-2 virus can reportedly survive for 1 day on cardboard and upto 3 days on plastic and stainless steel, as per recently published (Van Doremalen et al., 2020). In order to prevent future infection and environmental contamination, which is a concern on global scale, waste produced by hospitals (needles, syringes, bandages, masks, gloves, used tissue, etc.) should be managed properly.

8.5 *Impact of Municipal Waste Management*

Food intake throughout the pandemic has significantly altered the daily lifestyle and dietary practices of the majority of people (Cosgrove et al., 2021); Scacchi et al., 2021). Additionally, this pandemic is causing the advent of other new waste sources, complicating the monitoring of municipal solid waste by organisations and governments (Mehta & Angadi, 2020; Tripathi et al., 2020). Frequent usage of single-use items and impulsive buying have been linked to high levels of plastic waste pollution in the environment (Silva et al., 2021). The need to packaging demands for the delivery and food take-out or healthcare is the source of such waste (Van Fan et al., 2021). During the COVID-19 pandemic, most of the nations dropped significantly recycling rate in order to prevent COVID-19 spread. However, a few scientists have found that most people mix COVID-19 protective tools with domestic wastes, which could end-up causing the virus to transmit (Acharya et al., 2021). For instance, waste recycling was reduced (upto 46%) in many American cities to prevent the spread of COVID-19 to those participating in such activities. Waste pickers gather recyclable waste from roadways and waste dumps and sell it so that it can be recycled (Rai et al., 2020). For underdeveloped nations, these waste pickers are important to the circular economy and waste management (Moreno-Sánchez & Maldonado,

2006). The world's environmental contamination rose as a result of improper waste management practices and garbage recycling (Somani et al., 2020).

8.6 *Impact of COVID-19 on Energy Consumption*

All transport and industrial sectors were shut during lockdown, which significantly reduced energy demand and consumption. This improved energy security was used by medical industry to produce products and personal protective equipment (Nundy et al., 2021). According to Corpus-Mendoza et al., 2021 the decrease in movement led to the shutdown and/or reduced functioning of transportation, which in-turn reduced electricity usage.

In conclusion, the nation-wide lockdown was seen to be an effective solution for preventing pandemic transmission owing to a scarcity of COVID-19 treatment during the early transmission phase. The lockdown reduced environmental pollutants, boosting the quality of the water and air. Positive signs were also reported from all over the world that COVID-19-induced lockdown was indeed enhancing the environment, and it simultaneously caused a significant drop in PM 2.5, NO_2 and CO concentrations, which led to a significant rise in O_3 concentration. This restoration of the damaged habitat shows that human-caused environmental destruction is reversible. Everyone was able to observe nature's recuperation in just 2–3 months. This is due to pay attention and respond. To ensure that this restoration is permanent, government officials and decision makers must take appropriate actions. Thorough research is required to determine the impact of implementing such a short-term shutdown as an alternate method for reducing pollution and how it will influence the economy.

9 COVID-19: Indian Scenario

India was the second-most populated country in the world and the second-worst impacted by COVID-19 to-date (Niranjan, 2020). The first wave in India started in March 2020 and continued throughout November 2020, whereas the second wave started in March 2021 and continued through the end of May 2021 (Zirpe et al., 2021). The first wave peaked in September 2020 with more than 90,000 confirmed cases/ day, and then steadily reduced in intensity with 10,000 cases/day in February 2021, according to "ourworldindata" (Garg et al., 2022). In India, there had been more than 15 million infections during the initial wave (Shil et al., 2022). Consequently, India followed Western nations, likely due to variation in seasonal timing. There were severe deficits of hospital beds, drugs and oxygen during the second wave, which caused massive havoc. As of September 23, 2022, SARS-COV-2 was linked to 4,46,00,000 confirmed cases and 5,28,000 fatalities (Ministry of Health and Family

Welfare [MoHFW], 2022). Therefore, an assessment of the nation's efforts to combat the pandemic since its inception is crucial.

9.1 First Wave of Pandemic

India announced its first COVID-19 on January 30, 2020 in Kerala. The case was a student who was hospitalised in isolation after arriving from Wuhan. There were 3 recorded incidents in Kerala as of February 3, 2020; each case originated in a separate city. They were deemed recovered by February 20, 2020 (WHO, 2020k). It is unclear whether the first COVID-19 cases in India were contacts of the first case or they had prior travel history. Within a month, the number of cases began to rise and by beginning of March, 2020 several states and union territories were already affected. MoHFW claims that limited public transmission of COVID-19 was first recorded on March 30, 2020, and that it is mostly linked to traveling and localized transmission of imported cases. To stop the spread, the MoHFW released travel advisory limitations, which were akin to those for previous pandemics like SARS, Ebola and bubonic plague. These restrictions included that all foreign visitors travelling to India enter self-quarantine for a period of 14 days. Furthermore, until April 15th, 2020 several travel visas were prohibited to avoid/reduce the rate and extent of transmission of disease, which resulted in reduction in the growth, severity and death (Khanna et al., 2020). MoHFW issued several preventive measures on March 16, 2020, including social distancing of 1 m (Karotia & Kumar, 2020).

On 19 March 2020, Prime Minister of India, Shri Narendra Modi, appealed all Indians to follow a 14-hour Janata curfew ("people's curfew") on 22 March, and to honour frontline workers by clapping or ringing bells at 5 pm outside their houses (Pandey, 2020). The curfew was used to assess if a nationwide lockdown was feasible. On March 24, 2020, India's first phase of a 21-day lockdown began with 519 confirmed cases and 9 fatalities (Pulla, 2020). As a result of lockdown, there was less movement in supermarkets and pharmacies, recreational and retail, transportation to station to parks and workplaces (Saha et al., 2020). On March 25, 2020, all incoming international flights were stopped (Joy, 2020). This suspension was until April 14, 2020. Indian Railways issued a directive declaring "Force Majeure" effective March 22, 2020 to April 14, 2020 (Hussain & Kareem, 2020). On April 2, 2020, the Indian government released the ArogyaSetu smartphone App through a public-private collaboration to allow consumers to evaluate their risk of contacting COVID while ensuring the security of consumers (Singh, 2020). The Prime Minister Office (PMO) approved INR 15,000 crores COVID-19 Emergency Response and Health System Preparedness Package; INR 7,774 crores for immediate response and rest medium-term support (1–4 years) (WHO, 2020l) (Fig. 14).

Later, Indian government announced a prolonged second phase lockdown until May 3, 2020, which was then extended till May 17, 2020 and finally till May 31, 2020 due to COVID-19 consultation with States (Saha et al., 2020). Mid-March 2020 saw the effective enactment of the Disaster Management Act of 2005 and

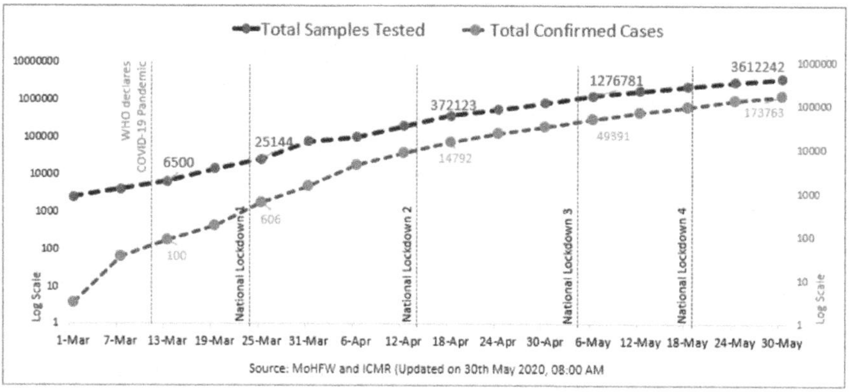

Fig. 14 Trend of lockdown in India. (*Source* Ministry of Health and Family Welfare and Indian Council of Medical Research)

the Epidemic Diseases Act of 1897. On April 16, 2020 districts were grouped into "Red" (hotspot), "Orange" or "Green" (little to no transmission) zones using a color-coded tier system based on incidence rates (Mave et al., 2021). In a directive issued on May 1st, 2020, the Union Ministry of Home Affairs (MHA), India authorised the movement of stranded migrant workers, pilgrims, tourists, students and other individuals by special trains run by the Ministry of Railways (MoR) (Ministry of Home Affairs MHA, 2020a). The lockdown was profitably used to expand the India's health infrastructure and system capacity. Total 3027 dedicated COVID hospitals, COVID health centres and 7013 COVID care centres were built promptly (Ghosh, n.d.).

According to the WHO (2020m) situation report, there were 3,46,759 deaths and 2,88,09,339 confirmed cases of COVID-19 in India from 3 January 2020 to 31 May 2021. Maharashtra state being most severely affected, accounting for 20.3% of all cases in India as of May 31, 2021. Due to COVID-19, the Maharashtra state reported 28.6% of all fatalities nationwide. However, effective recovery was achieved in more than 95% of the infections (Shil et al., 2022).

9.1.1 The Unlocks

The lockdown conditions were gradually relaxed. In order to combat COVID-19 the MHA issued new regulations for Unlock 1:0. These regulations went into effect on June 1, 2020, and remained until June 30, 2020 (MHA, 2020b). Beginning on June 8, this initial phase allowed hotels, restaurants, shopping malls and houses of worship to reopen. Large gatherings were forbidden, although interstate movement was authorised. All areas had night curfews in place from 9 p.m. to 5 a.m. and state governments were free to impose acceptable limitations on any activity.

Under the rules and directives of the MHA and the National Disaster Management Authority (NDMA), unlock 2:0 commenced on July 1, 2020. Lockdown restrictions were only put in place in confinement zones (MHA, 2020c). The majority of activities were allowed in all other regions. All areas had night curfew from 10 p.m. to 5 a.m. State borders continued to operate, but state governments were permitted to impose reasonable limits on all activities. Within the framework of the Vande Bharat Mission, Ministry of Home Affairs, certain international travel was permitted (MHA, 2020d). Until July 31, 2020 all metros, educational institutions and recreational facilities were closed. With stringent parameter control, "intense contact tracing, house-to-house surveillance and other therapeutic interventions", only necessary activities were allowed in confinement zones.

During Unlock 3.0 (1–31 August 2020), night curfews were relaxed, allowing gyms and yoga studios to reopen on August 5. Institutions of higher learning remained closed until August 31. All modes of intrastate and interstate transportation were allowed. Though West Bengal enforced lockdowns twice a week, Maharashtra and Tamil Nadu imposed lockdowns for the entire month due to large number of instances (MHA, 2020e).

The MHA issued regulations outlining what is allowed under Unlock 4.0. According to the MHA (2020f), "Lockdown shall stay in effect in the Containment Zones until September 30, 2020". However, some extra operations were permitted outside the confinement zones. Funeral/last rites and wedding ceremonies with a maximum of 50 persons were permitted. Events with upto 100 persons including those related to religion, politics, entertainment, sports and academia were allowed. Face masks were now mandatory in public areas, at work and while travelling. Preference for online instruction in schools was also announced in Unlock 5.0, however, States and Union Territories had discretion starting on 15 October. Lockdown remained in effect until 30 November 2020, only in the containment zones. After closed for long, movie theatres eventually reopened on October 15, 2020, with only 50% of their seating capacity. The Indian government decided to open all academic institutions, viz. schools, colleges and universities by January 2021 (MHA, 2020g).

9.2 *Second Wave of Pandemic*

The likelihood of a second wave was considered weak by experts who were studying the pandemic, who believed that India had already reached the "shore" (Chinnaswamy, 2021). Despite the fact it was only targeting adults over 45 years, India started its immunisation effort in mid-January 2021 whilst still relishing in this unprecedented success. At this point, India used two vaccines: an inactivated entire virus and an adenovector-based spike protein-generating vaccine, but had no idea that a terrifying second wave in early to mid-2021 would hit like tsunami. A disastrous second wave struck India in March 2021 (Times of India [ToI], n.d.). Paucity of hospital beds, oxygen cylinders and medications made the situation worse. Over

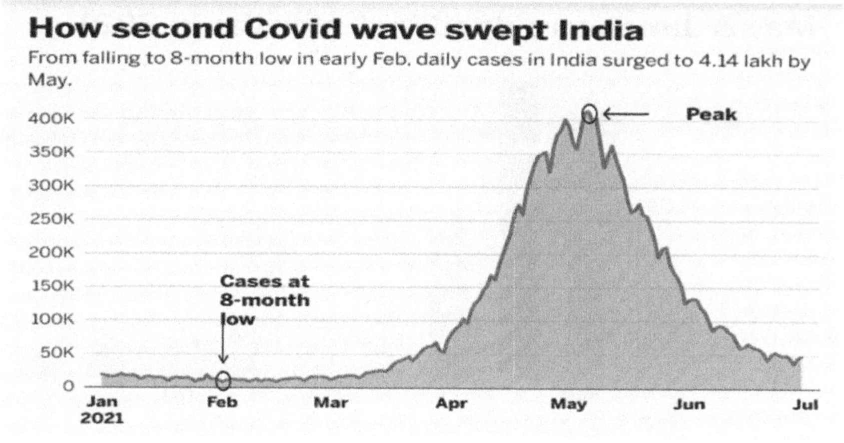

Fig. 15 Graph depicting peak of second wave in May, 2021. (*Source* Times of India, n.d.)

time, the public and the administration both began events carelessly and neglected the developing pandemic totally (Goel et al., 2021). Elevated infections were caused by large crowds (such as protests, rallies, elections and festival celebrations), disobedience with COVID protocols and slow immunisation (Srivastava et al., 2020). According to Times of India (n.d.) on April 9 India overtook Brazil to become the second worst-affected nation in the world (Fig. 15).

The country had a spike in incidents that began in mid-March 2021 and quickly increased in April. India recorded slightly over 4,00,000 new COVID-19 cases on a single day on May 1 after logging more than 3,00,000 new infections for 10 days (Samarasekera, 2021). According to estimates, the second wave killed 2 lakh people (ToI, n.d.). A new variant, now known as B.1.617, was discovered in the state of Maharashtra, as per the Indian health ministry report at the end of March, 2021. The increase is associated with B.1.617 and B.1.1.7, the variant that was initially discovered in the UK. The majority of cases in the nation were brought by the B.1.617 (Mlcochova et al., 2021). India's struggle to vaccinate its humongous population against SARS-CoV-2 was yet another factor contributing to this catastrophe.

Affected by the Delta variant, India recorded a startling 1.7 crore cases of Covid in just over 2 months, along with thousands of fatalities. The second wave had such an impact that, with the exception of April, the total number of cases for the next months fell far short of the 90 lakh infections reported in May, 2021. Unfortunately, there were an unacceptably high number of fatalities. Approximately 2.4 crore infections were recorded in India in 2021, more than twice reported in 2020, according to data of December 27, 2021. Given that India experienced approximately 1.8 lakh (122%) more fatalities that year than the previous, the gap in mortality rates is even more obvious. In 2021 the heartbreaking visuals of many cremation pyres and huge lines outside crematoriums came to represent India's Covid dilemma (ToI, 2021).

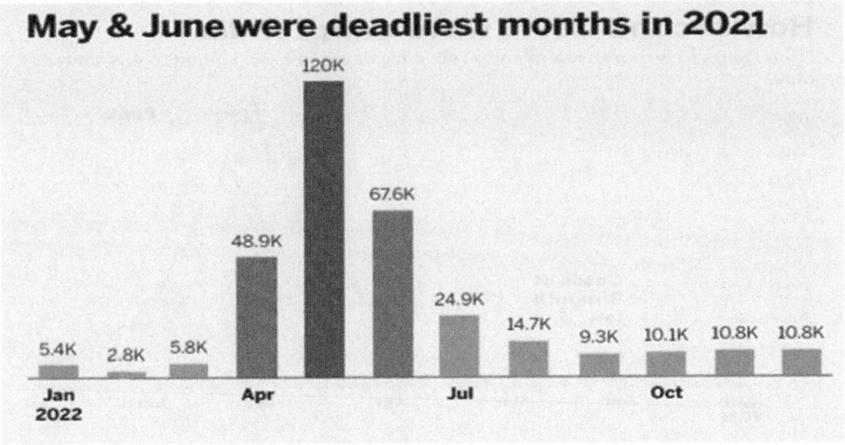

Fig. 16 Graph depicting deadliest months (May and June, 2021) of COVID-19 in India. (*Source* Times of India, n.d.)

Official statistics show that India's Covid death toll in May, 2021 was 43 times more than the country's lowest death toll in February, 2021. The two deadliest months of the pandemic, with over 1.2 lakh fatalities reported in May and another 67,000 in June, both occurred in India in 2021. The mortality decreased dramatically in July as the second wave subsided, and continued to decline in the months that followed as India's vaccination programme accelerated (ToI, 2021) (Fig. 16).

India began their vaccination programme in January 2021 with an emphasis on persons over 45 and frontline employees (Samarasekera, 2021). On May 1, 2021 all adults above 18 years were included for immunisation. The availability of vaccines was a problem; in early May, numerous states reported COVID-19 vaccine deficits. Just over 9% of India's population got one dose of COVID-19 vaccine as of May 3, 2021 and only 2% were completely vaccinated.

However, things became more organised after the Indian Government took control of procurement and introduced a liberalised policy on June 21, 2021. By October 21, 2021 India became the second nation in the world to provide its citizens more than a billion doses (ToI, 2021).

9.3 Third Wave of Pandemic

India reported between 7,000 and 9,000 new COVID-19 cases daily in the third week of November 2021 (News 18, 2022). The COVID-19 vaccination programme was speeding up, and there was more supply of vaccines than was in need. There was a sense of optimism that things will get back on track by early 2022. Then, in last week of November, 2021 Omicron (B.1.1.529)—the fifth and most recent variation of concern (VOC) of SARS-CoV-2 emerged, completely transforming the situation. The WHO reports that there is an evidence that suggests that Omicron has a sizeable substantial advantage over Delta. In nations where community transmission is known to occur, the Omicron version was spreading far more quickly than the Delta variant because of its superior transmissibility (Rana et al., 2022). On January 3, 2022 India recorded almost 37,000 new instances of COVID, a considerable increase from November and most of December when new cases each day were under 10,000 (Quartz India, 2022). This rise was mostly concentrated in large urban areas like Delhi, Mumbai and Kolkata. For instance, the number of new infections in Delhi increased from 54 on December 3, 2021 to 4099 on January 3, 2022. During same time, the test positivity rate rose from 0.06 to 6.46%. According to the Indian Government, the daily caseload peaked on January 20, 2022, when it reached 3,17,532. However, in 2 weeks, i.e. from January 21, 2022 to February 3, 2022 the number of daily Covid cases fell by 50% (from 3,47,254 to 1,72,433), signalling beginning of the end of the third wave in India. In March 2022, there were only 6,000 cases left. The governments had implemented control measures like curfews at night, limit on the size of meetings, closures of educational institutions and restrictions on the number of people allowed in theatres and restaurants during the third wave (Indian Express, 2022). In the third wave, the vaccination campaign of the Indian government greatly reduced the infection and mortality rates. The Oxford-AstraZeneca vaccine, locally known as Covishield, Covaxin by Indian company Bharat Biotech, Sputnik V produced in Russia and Corbevax are the four vaccinations currently used in India (BBC, 2022). However, so far, only Covishield and Covaxin have seen widespread usage. Notably, India launched a campaign to provide all adults with free booster doses (or precautionary doses) for 75 days. The period coincided with the nation's 75th anniversary of independence, which was celebrated on August 15, 2022, according to Business Standard. As of October 5, 2022, the nation had received 2,18,83,40,816 doses of vaccinations (MoHFW, 2022).

10 Conclusion

COVID-19 pandemic is still raging. However, there are signs that we might be moving towards the endemic phase. This trend might be caused by the widespread Omicron variant, which is currently dominating. With more resources at disposal, such as vaccinations, booster shots and treatments, the world is now in a better position to safeguard from COVID-19 catastrophic malady. And yet exuberant excitement needs to be restrained by a sense of realism and compassion for developing nations. In the fight against SARS-CoV-2, developed nations must remain attentive and support developing nations, with the goal of identifying novel variants of concern, looking into animal reservoirs and exchanging diagnostic equipment, monitoring and treatments.

References

Abdelrahman, Z., Li, M., & Wang, X. (2020). Comparative review of SARS-CoV-2, SARS-CoV, MERS-CoV, and influenza a respiratory virus. *Frontiers in Immunology,* 2309.

Abdulla, Z. A., Al-Bashir, S. M., Al-Salih, N. S., Aldamen, A. A., & Abdulazeez, M. Z. (2021). A summary of the SARS-CoV-2 vaccines and technologies available or under development. *Pathogens, 10*(7), 788.

Acharya, A., Bastola, G., Modi, B., Marhatta, A., Belbase, S., Lamichhane, G., & Dahal, R. K. (2021). The impact of COVID-19 outbreak and perceptions of people towards household waste management chain in Nepal. *Geoenvironmental Disasters, 8*(1), 1–11.

Adam, D. (2022). Will Omicron end the pandemic? Here's what experts say. *Nature,* 20–21.

Adibi, S., Griffin, P., Sanicas, M., Rashidi, M., & Lanfranchi, F. (Eds.). (2022). *Frontiers of COVID-19: Scientific and clinical aspects of the novel coronavirus* 2019. Springer Nature.

Agrawal, U., Raju, R., & Udwadia, Z. F. (2020). Favipiravir: A new and emerging antiviral option in COVID-19. *Medical Journal Armed Forces India, 76*(4), 370–376.

Ahmed, S., Khan, S., Imran, I., Al Mughairbi, F., Sheikh, F. S., Hussain, J., & Al-Harrasi, A. (2021). Vaccine development against COVID-19: Study from pre-clinical phases to clinical trials and global use. *Vaccines, 9*(8), 836.

AIFA (2021). *EMA receives application for marketing authorisation for Lagevrio (molnupiravir) for treating patients with COVID-19.* Retrieved from https://www.aifa.gov.it/en/web/guest/dettaglionotizia?p_p_id=it_gov_aifa_portlet_GestioneCookies&p_p_lifecycle=1&p_p_state=normal&p_p_mode=view&_it_gov_aifa_portlet_GestioneCookies_javax.portlet.action=deleteCookiesAIFA&p_auth=RsughA56

Alanagreh, L. A., Alzoughool, F., & Atoum, M. (2020). The human coronavirus disease COVID-19: Its origin, characteristics, and insights into potential drugs and its mechanisms. *Pathogens, 9*(5), 331.

Alberts, B., Johnson, A., Lewis, J., Raff, M., Roberts, K., & Walter, P. (2002). T cells and MHC proteins. In *Molecular biology of the cell,* (4th ed.). Garland Science.

Aleem, A., Samad, A. B. A., & Slenker, A.K. (2022). Emerging variants of SARS-CoV-2 and novel therapeutics against coronavirus (COVID-19). In *StatPearls [Internet].* StatPearls Publishing.

Ali, N. (2020). Role of vitamin D in preventing of COVID-19 infection, progression and severity. *Journal of Infection and Public Health, 13*(10), 1373–1380.

Alloush, T. K., Mansour, O., Alloush, A. T., Roushdy, T., Hamid, E., El-Shamy, M., & Shokri, H. M. (2022). Rhino-orbito-cerebral mucormycosis during the COVID-19 third wave in 2021: An Egyptian preliminary report from a single tertiary hospital. *Neurological Sciences, 43*(2), 799–809.

Al-Tawfiq, J. A., Alhumaid, S., Alshukairi, A. N., Temsah, M. H., Barry, M., Al Mutair, A., & Dhama, K. (2021). COVID-19 and mucormycosis superinfection: The perfect storm. *Infection, 49*(5), 833–853.

Amfani, A. H., & Ibrahim, G. (2017). English-hausa glossary of HIV, AIDS and ebola-related terms. Retrieved from http://oer.ui.edu.ng/sites/default/files/English-Hausa%20glossary%20of%20HIV%2C%20AIDS%20and%20Ebola-related%20terms.pdf

Araf, Y., Akter, F., Tang, Y. D., Fatemi, R., Parvez, M. S. A., Zheng, C., & Hossain, M. G. (2022). Omicron variant of SARS-CoV-2: Genomics, transmissibility, and responses to current COVID-19 vaccines. *Journal of Medical Virology, 94*(5), 1825–1832.

Aranjani, J. M., Manuel, A., Abdul Razack, H. I., & Mathew, S. T. (2021). COVID-19-associated mucormycosis: Evidence-based critical review of an emerging infection burden during the pandemic's second wave in India. *PLoS Neglected Tropical Diseases, 15*(11), Article e0009921.

Arastehfar, A., Carvalho, A., van de Veerdonk, F. L., Jenks, J. D., Koehler, P., Krause, R., & Hoenigl, M. (2020). COVID-19 associated pulmonary aspergillosis (CAPA): From immunology to treatment. *Journal of Fungi, 6*(2), 91.

Artis, W. M., Fountain, J. A., Delcher, H. K., & Jones, H. E. (1982). A mechanism of susceptibility to mucormycosis in diabetic ketoacidosis transferrin and iron availability. *Diabetes, 31*(12), 1109–1114.

Astuti, I. (2020). Severe acute respiratory syndrome coronavirus 2 (SARS-CoV-2): An overview of viral structure and host response. *Diabetes & Metabolic Syndrome: Clinical Research & Reviews, 14*(4), 407–412.

Ayenigbara, I. O. (2020). COVID-19: An international public health concern. *Central Asian Journal of Global Health, 9*(1), 1–11.

Banerjee, M., Pal, R., & Bhadada, S. K. (2022). Intercepting the deadly trinity of mucormycosis, diabetes and COVID-19 in India. *Postgraduate Medical Journal, 98*(e2), e108–e109.

Bayram, N., Ozsaygılı, C., Sav, H., Tekin, Y., Gundogan, M., Pangal, E., & Özcan, İ. (2021). Susceptibility of severe COVID-19 patients to rhino-orbital mucormycosis fungal infection in different clinical manifestations. *Japanese Journal of Ophthalmology, 65*(4), 515–525.

BBC. (2022). Covid vaccine: India becomes second country to cross two billion Covid jabs. Retrieved from https://www.bbc.com/news/world-asia-india-56345591

Beig Parikhani, A., Bazaz, M., Bamehr, H., Fereshteh, S., Amiri, S., Salehi-Vaziri, M., & Azadmanesh, K. (2021). The inclusive review on SARS-CoV-2 biology, epidemiology, diagnosis, and potential management options. *Current Microbiology, 78*(4), 1099–1114.

Belser, J. A., Rota, P. A., & Tumpey, T. M. (2013). Ocular tropism of respiratory viruses. *Microbiology and Molecular Biology Reviews, 77*(1), 144–156.

Bhatt, K., Agolli, A., Patel, M. H., Garimella, R., Devi, M., Garcia, E., & Sanchez-Gonzalez, M. (2021). High mortality co-infections of COVID-19 patients: Mucormycosis and other fungal infections. *Discoveries, 9*(1), 1–12.

Biancolella, M., Colona, V. L., Mehrian-Shai, R., Watt, J. L., Luzzatto, L., Novelli, G., & Reichardt, J. K. (2022). COVID-19 2022 update: Transition of the pandemic to the endemic phase. *Human Genomics, 16*(1), 1–12.

Bisht, D., & Dubey, S. (2020). *Clinical trials knowhows for novel vaccine development.* incom

Bloch, E. M., & Feldweg, A. M. (2021). COVID-19: Convalescent plasma and hyperimmune globulin. *UpToDate.*

Bonow, R. O., Fonarow, G. C., O'Gara, P. T., & Yancy, C. W. (2020). Association of coronavirus disease 2019 (COVID-19) with myocardial injury and mortality. *JAMA Cardiology, 5*(7), 751–753.

Bourouiba, L. (2020). Turbulent gas clouds and respiratory pathogen emissions: Potential implications for reducing transmission of COVID-19. *JAMA, 323*(18), 1837–1838.

Braun, F., Lütgehetmann, M., Pfefferle, S., Wong, M. N., Carsten, A., Lindenmeyer, M. T., & Huber, T. B. (2020). SARS-CoV-2 renal tropism associates with acute kidney injury. *The Lancet, 396*(10251), 597–598.

Broadbent, A. J., Santos, C. P., Anafu, A., Wimmer, E., Mueller, S., & Subbarao, K. (2016). Evaluation of the attenuation, immunogenicity, and efficacy of a live virus vaccine generated by codon-pair bias de-optimization of the 2009 pandemic H1N1 influenza virus, in ferrets. *Vaccine, 34*(4), 563–570.

Bruyninckx, H. (2020). *Air pollution goes down as Europe takes hard measures to combat coronavirus.* Retrieved from https://www.eea.europa.eu/highlights/air-pollution-goes-down-as

Cano, R. L. E., & Lopera, H. D. E. (2013). Introduction to T and B lymphocytes. In *Autoimmunity: From Bench to Bedside [Internet].* El Rosario University Press.

Cao, Y., Yang, R., Lee, I., Zhang, W., Sun, J., Wang, W., & Meng, X. (2021). Characterization of the SARS-CoV-2 E protein: Sequence, structure, viroporin, and inhibitors. *Protein Science, 30*(6), 1114–1130.

Cascella, M., Rajnik, M., Aleem, A., Dulebohn, S. C., & Di Napoli, R. (2022). *Features, evaluation, and treatment of coronavirus (COVID-19).* Statpearls [internet].

Castrejón-Pérez, A. D., Welsh, E. C., Miranda, I., Ocampo-Candiani, J., & Welsh, O. (2017). Cutaneous mucormycosis. *Anais Brasileiros De Dermatologia, 92*, 304–311.

Centers for Disease Control and Prevention [CDC]. (2020). *Notice on health protection of COVID-19 in public places issued by pneumonia mechanism*. Retrieved from www.nhc.gov.cn/jkj/s7916/202001/d9ae8301384a4239a8041d6f77da09b6.shtml.com

Centers for Disease Control and Prevention [CDC]. (2021a). *SARS-CoV-2 variant classifications and definitions*. Retrieved from https://stacks.cdc.gov/view/cdc/105817

Centers for Disease Control and Prevention [CDC]. (2021b). Science brief: SARS-CoV-2 and surface (fomite) transmission for indoor community environments. In *CDC COVID-19 science briefs [internet]*. Centers for Disease Control and Prevention (US).

Centers for Disease Control and Prevention [CDC]. (2021c). *SARS-CoV-2 transmission*. Retrieved from https://www.ncbi.nlm.nih.gov/books/NBK570442/

Centers for Disease Control and Prevention [CDC]. (2021d). *About mucormycosis*. Retrieved from https://www.cdc.gov/fungal/diseases/mucormycosis/definition.html#:~:text=Disseminated%20mucormycosis%20occurs%20when%20the,spleen%2C%20heart%2C%20and%20skin

Centers for Disease Control and Prevention [CDC]. (2022a). *SARS-CoV-2 variant classifications and definitions*. Retrieved from https://www.cdc.gov/coronavirus/2019-ncov/variants/variant-classifications.html

Centers for Disease Control and Prevention [CDC]. (2022b). *COVID-19 treatments and medications*. Retrieved from https://www.cdc.gov/coronavirus/2019-ncov/your-health/treatments-for-severe-illness.html?adlt=strict&redig=897B07AA89A648E2B142728D564C7384&toWww=1

Central Pollution Control Board [CPCB]. (2021). *Central control room for air quality management-all India*. Retrieved from https://app.cpcbccr.com/ccr/#/caaqm-dashboard-all/caaqm-landing. Accessed 17 June 2021.

Chakravarti, R., Singh, R., Ghosh, A., Dey, D., Sharma, P., Velayutham, R., & Ghosh, D. (2021). A review on potential of natural products in the management of COVID-19. *RSC Advances, 11*(27), 16711–16735.

Chamilos, G., Lewis, R. E., & Kontoyiannis, D. P. (2008). Delaying amphotericin B-based frontline therapy significantly increases mortality among patients with hematologic malignancy who have zygomycosis. *Clinical Infectious Diseases, 47*(4), 503–509.

Chan, J. F. W., Kok, K. H., Zhu, Z., Chu, H., To, K. K. W., Yuan, S., & Yuen, K. Y. (2020). Genomic characterization of the 2019 novel human-pathogenic coronavirus isolated from a patient with atypical pneumonia after visiting Wuhan. *Emerging Microbes & Infections, 9*(1), 221–236.

Chan, L., Chaudhary, K., Saha, A., Chauhan, K., Vaid, A., Zhao, S., & Nadkarni, G. N. (2021). AKI in hospitalized patients with COVID-19. *Journal of the American Society of Nephrology, 32*(1), 151–160.

Chauhan, G., Madou, M. J., Kalra, S., Chopra, V., Ghosh, D., & Martinez-Chapa, S. O. (2020). Nanotechnology for COVID-19: Therapeutics and vaccine research. *ACS Nano, 14*(7), 7760–7782.

Chavda, V. P., & Apostolopoulos, V. (2022). Is booster dose strategy sufficient for omicron variant of SARS-CoV-2? *Vaccines, 10*(3), 367.

Chen, N., Zhou, M., Dong, X., Qu, J., Gong, F., Han, Y., & Zhang, L. (2020). Epidemiological and clinical characteristics of 99 cases of 2019 novel coronavirus pneumonia in Wuhan, China: A descriptive study. *The Lancet, 395*(10223), 507–513.

Chen, L., Li, J., Pang, X., Shi, K., Chen, J., Wang, J., & Xu, M. (2021a). Impact of COVID-19 lockdown on air pollutants in a coastal area of the Yangtze river delta, China, measured by a low-cost sensor package. *Atmosphere, 12*(3), 345.

Chen, W., Tian, Y., Li, Z., Zhu, J., Wei, T., & Lei, J. (2021b). Potential interaction between SARS-CoV-2 and thyroid: A review. *Endocrinology, 162*(3), bqab004.

Chinnaswamy, S. (2021). Understanding the devastating second wave of COVID-19 pandemic in India. *American Journal of Human Biology, 33*(6), 1–12.

Chirani, M. R., Kowsari, E., Teymourian, T., & Ramakrishna, S. (2021). Environmental impact of increased soap consumption during COVID-19 pandemic: Biodegradable soap production and sustainable packaging. *Science of the Total Environment, 796*, Article 149013.

COMBATCOVID. (n.d.). What are monoclonal antibodies? Retrieved from https://combatcovid.hhs.gov/. Accessed on 28 Sept, 2022.

Cook, T. M., El-Boghdadly, K., McGuire, B., McNarry, A. F., Patel, A., & Higgs, A. (2020). Consensus guidelines for managing the airway in patients with COVID-19: Guidelines from the Difficult Airway Society, the Association of Anaesthetists the Intensive Care Society, the Faculty of Intensive Care Medicine and the Royal College of Anaesthetists. *Anaesthesia, 75*(6), 785–799.

Cornely, O. A., Alastruey-Izquierdo, A., Arenz, D., Chen, S. C., Dannaoui, E., Hochhegger, B., & Mucormycosis, E. C. M. M. (2019). Global guideline for the diagnosis and management of mucormycosis: An initiative of the European Confederation of Medical Mycology in cooperation with the Mycoses Study Group Education and Research Consortium. *The Lancet Infectious Diseases, 19*(12), e405–e421.

Corpus-Mendoza, A. N., Ruiz-Segoviano, H. S., Rodríguez-Contreras, S. F., Yañez-Dávila, D., & Hernández-Granados, A. (2021). Decrease of mobility, electricity demand, and NO_2 emissions on COVID-19 times and their feedback on prevention measures. *Science of the Total Environment, 760*, Article 143382.

Cosgrove, K., Vizcaino, M., & Wharton, C. (2021). COVID-19-related changes in perceived household food waste in the united states: A cross-sectional descriptive study. *International Journal of Environmental Research and Public Health, 18*(3), 1104.

Courtier-Orgogozo, V., & de Ribera, F. A. (2022). SARS-CoV-2 infection at the Huanan seafood market. *Environmental Research, 214*, Article 113702.

COVID-19 Treatment Guidelines Panel. (n.d.). Coronavirus disease 2019 (COVID-19) treatment guidelines. National Institutes of Health. Available at https://www.covid19treatmentguidelines.nih.gov/. Accessed: 28 Sept 2022.

COVID19 VACCINE TRACKER. (2022). *Learn more about vaccine clinical trials*. Retrieved from https://covid19.trackvaccines.org/trials-vaccine-testing/

Cowley, J. A. (2016). Nidoviruses of fish and crustaceans. In *Aquaculture virology* (pp. 443–472). Academic Press.

Crommelin, D. J., Anchordoquy, T. J., Volkin, D. B., Jiskoot, W., & Mastrobattista, E. (2021). Addressing the cold reality of mRNA vaccine stability. *Journal of Pharmaceutical Sciences, 110*(3), 997–1001.

Cui, Z. (2005). DNA vaccine. *Advances in Genetics, 54*, 257–289.

Dai, X., Xiong, Y., Li, N., & Jian, C. (2019). Vaccine types. In *Vaccines-the history and future*. IntechOpen.

Danchin, A., Ng, T. W., & Turinici, G. (2020). A new transmission route for the propagation of the SARS-CoV-2 coronavirus. *Biology, 10*(1), 10.

Darwish, R. M., AlMasri, M., & Al-Masri, M. M. (2022). Mucormycosis: The hidden and forgotten disease. *Journal of Applied Microbiology, 132*(6), 4042–4057.

Das, R. R., Jaiswal, N., Dev, N., Jaiswal, N., Naik, S. S., & Sankar, J. (2020). Efficacy and safety of anti-malarial drugs (chloroquine and hydroxy-chloroquine) in treatment of COVID-19 infection: A systematic review and meta-analysis. *Frontiers in Medicine, 7*, 482.

De Groot, R. J., Baker, S. C., Baric, R., Enjuanes, L., Gorbalenya, A. E., Holmes, K. V., & Lefkowitz, E. J. (2011). Family coronaviridae. incom.

Deng, Y., Liu, W., Liu, K., Fang, Y. Y., Shang, J., Zhou, L., & Liu, H. G. (2020). Clinical characteristics of fatal and recovered cases of coronavirus disease 2019 in Wuhan, China: A retrospective study. *Chinese Medical Journal, 133*(11), 1261–1267.

Deplanque, D., & Launay, O. (2021). Efficacy of COVID-19 vaccines: From clinical trials to real life. *Therapies, 76*(4), 277–283.

Desforges, M., Le Coupanec, A., Dubeau, P., Bourgouin, A., Lajoie, L., Dubé, M., & Talbot, P. J. (2019). Human coronaviruses and other respiratory viruses: Underestimated opportunistic pathogens of the central nervous system? *Viruses, 12*(1), 14.

Dhama, K., Patel, S. K., Sharun, K., Pathak, M., Tiwari, R., Yatoo, M. I., & Rodriguez-Morales, A. J. (2020). SARS-CoV-2 jumping the species barrier: Zoonotic lessons from SARS, MERS and recent advances to combat this pandemic virus. *Travel Medicine and Infectious Disease, 37*, Article 101830.

Dixit, S. B., Zirpe, K. G., Kulkarni, A. P., Chaudhry, D., Govil, D., Mehta, Y., & Mhatre, U. (2020). Current approaches to COVID-19: Therapy and prevention. *Indian Journal of Critical Care Medicine: Peer-reviewed, Official Publication of Indian Society of Critical Care Medicine, 24*(9), 838.

Docherty, A. B., Harrison, E. M., Green, C. A., Hardwick, H. E., Pius, R., Norman, L., & Semple, M. G. (2020). Features of 20,133 UK patients in hospital with covid-19 using the ISARIC WHO clinical characterisation protocol: Prospective observational cohort study. *BMJ*, 369.

Dockrell, D. H., Russell, C. D., McHugh, B., & Fraser, R. (2022). Does autonomous macrophage-driven inflammation promote alveolar damage in COVID-19? *European Respiratory Journal., 60*(6), 1521–1531.

Doerfler, W. (2021). Adenoviral vector DNA-and SARS-CoV-2 mRNA-based COVID-19 vaccines: Possible integration into the human genome-are adenoviral genes expressed in vector-based vaccines? *Virus Research, 302*, Article 198466.

Down to Earth. (2021). *As told to parliament (December 3, 2021): Diabetes prevalence to grow almost 70% by 2045*. Retrieved from https://www.downtoearth.org.in/news/agriculture/as-told-to-parliament-december-3-2021-diabetes-prevalence-to-grow-almost-70-by-2045-80537#:~:text=The%20estimated%20number%20of%20diabetes,Lok%20Sabha%20December%203%2C%202021

Eccles, R., & Wilkinson, J. E. (2015). Exposure to cold and acute upper respiratory tract infection. *Rhinology, 53*(2), 99–106.

Eide, D. J. (2011). The oxidative stress of zinc deficiency. *Metallomics, 3*(11), 1124–1129.

El-Sayed, A., & Kamel, M. (2021). Coronaviruses in humans and animals: The role of bats in viral evolution. *Environmental Science and Pollution Research, 28*(16), 19589–19600.

European Centre for Disease Prevention and Control [ECDPC]. (2020). *Questions and answers on COVID-19*. Retrieved from https://www.ecdc.europa.eu/en/covid-19/questions-answers

Evert, K., Dienemann, T., Brochhausen, C., Lunz, D., Lubnow, M., Ritzka, M., & Simon, M. (2021). Autopsy findings after long-term treatment of COVID-19 patients with microbiological correlation. *Virchows Archiv, 479*(1), 97–108.

Fan, Y., Zhao, K., Shi, Z. L., & Zhou, P. (2019). Bat coronaviruses in China. *Viruses, 11*(3), 210.

Fang, E., Liu, X., Li, M., Zhang, Z., Song, L., Zhu, B., & Li, Y. (2022). Advances in COVID-19 mRNA vaccine development. *Signal Transduction and Targeted Therapy, 7*(1), 1–31.

Federation, I. D. (2019). *IDF diabetes atlas* (9th ed.). Place n Pub.

Fehr, A. R., & Perlman, S. (2015). Coronaviruses: An overview of their replication and pathogenesis. *Coronaviruses*, 1–23.

Ferlicot, S., Jamme, M., Gaillard, F., Oniszczuk, J., Couturier, A., May, O., & AP-HP/Universities/Inserm COVID-19 research collaboration. (2021). The spectrum of kidney biopsies in hospitalized patients with COVID-19, acute kidney injury and/or proteinuria. *Nephrology Dialysis Transplantation, 36*(7), 1253–1262.

Fernandez, G. J., Ramírez-Mejía, J. M., & Urcuqui-Inchima, S. (2022). Vitamin D boosts immune response of macrophages through a regulatory network of microRNAs and mRNAs. *The Journal of Nutritional Biochemistry, 109*, Article 109105.

Feumba, G. S., Laplante, J., Lasek-Nesselquist, E., & George, K. S. (2022). Detection of SARS-CoV-2 variants by multiple diagnostic assays Second RESUBMISSION JCV-D-21-00675R2. *Journal of Clinical Virology, 150*, Article 105160.

Fiorino, G., Allocca, M., Furfaro, F., Gilardi, D., Zilli, A., Radice, S., & Danese, S. (2020). Inflammatory bowel disease care in the COVID-19 pandemic era: The Humanitas, Milan, experience. *Journal of Crohn's and Colitis, 14*(9), 1330–1333.

Fong, M. W., Gao, H., Wong, J. Y., Xiao, J., Shiu, E. Y., Ryu, S., & Cowling, B. J. (2020). Nonpharmaceutical measures for pandemic influenza in nonhealthcare settings—social distancing measures. *Emerging Infectious Diseases, 26*(5), 976.

Foxnews. (2021). *Omicron is better at evading vaccines, new COVID-19 transmissibility data confirms*. Retrieved from https://www.foxnews.com/world/omicron-vaccines-new-covid-transmissibility-data

Frederiksen, L. S. F., Zhang, Y., Foged, C., & Thakur, A. (2020). The long road toward COVID-19 herd immunity: Vaccine platform technologies and mass immunization strategies. *Frontiers in Immunology, 11*, 1817.

Freuling, C. M., Breithaupt, A., Müller, T., Sehl, J., Balkema-Buschmann, A., Rissmann, M., & Mettenleiter, T. C. (2020). Susceptibility of raccoon dogs for experimental SARS-CoV-2 infection. *Emerging Infectious Diseases, 26*(12), 2982.

Gangavarapu, K., Latiff, A. A., Mullen, J. L., Alkuzweny, M., Hufbauer, E., Tsueng, G. , Haag, E., Zeller, M., Aceves, C. M., Zaiets, K., Cano, M., Zhou, J., Qian, Z., Sattler, R., Matteson, N. L., Levy, J. I., Lee, R. T., Freitas, L., Maurer-Stroh, S., GISAID core and curation team, & Hughes, L.D. (2022). Outbreak. Info genomic reports: Scalable and dynamic surveillance of SARS-CoV-2 variants and mutations. *Research square*, rs.3.rs-1723829. https://doi.org/10.21203/rs.3.rs-1723829/v1

Garbee, D. D., Pierce, S. S., & Manning, J. (2017). Opportunistic fungal infections in critical care units. *Critical Care Nursing Clinics, 29*(1), 67–79.

Garg, S., Singh, P., & Aggarwal, P. (2022). Combating COVID-19 with proficiency and precision. *Indian Journal of Community Health, 34*(2), 144–146.

Gasmi, A., Peana, M., Noor, S., Lysiuk, R., Menzel, A., Gasmi Benahmed, A., & Bjørklund, G. (2021). Chloroquine and hydroxychloroquine in the treatment of COVID-19: The never-ending story. *Applied Microbiology and Biotechnology, 105*(4), 1333–1343.

Gautam, S. (2020). The influence of COVID-19 on air quality in India: A boon or inutile. *Bulletin of Environmental Contamination and Toxicology, 104*(6), 724–726.

Gavi, The Vaccine Alliance. (2021a). *From alpha to omicron: Everything you need to know about SARS-CoV-2 variants of concern*. Retrieved from https://www.gavi.org/vaccineswork/alpha-omicron-everything-you-need-know-about-coronavirus-variants-concern?gclid=Cj0KCQiA64GRBhCZARIsAHOLriJkelG39ML3thDKAwB7nCb05lB6zRBKJ0NCMgMiQkE7obOn8djTnbYa

Gavi, The Vaccine Alliance. (2021b). *Five things we know about the Delta variant (and two things we don't)*. Retrieved from https://www.gavi.org/vaccineswork/five-things-we-know-about-delta-coronavirus-variant-and-two-things-we-still-need

Gavriatopoulou, M., Korompoki, E., Fotiou, D., Ntanasis-Stathopoulos, I., Psaltopoulou, T., Kastritis, E., & Dimopoulos, M. A. (2020). Organ-specific manifestations of COVID-19 infection. *Clinical and Experimental Medicine, 20*(4), 493–506.

Ghosh, A., Magar, D. G., Thapa, S., Nayak, N., & Talwar, O. P. (2019). Histopathology of important fungal infections—a summary. *Journal of Pathology of Nepal, 9*(1), 1490–1496.

Ghosh, D., Bernstein, J. A., & Mersha, T. B. (2020). COVID-19 pandemic: The African paradox. *Journal of Global Health, 10*(2).

Ghosh, S. M. (n.d.). *Opportunities for European–India collaboration*. Retrieved from https://www.bvmw.de/fileadmin/suborganization/Ausland/Aussenwirtschaft/Dateien/BVMW_Medtech_Webinar_PPT_2_.pdf

Goel, I., Sharma, S., & Kashiramka, S. (2021). Effects of the COVID-19 pandemic in India: An analysis of policy and technological interventions. *Health Policy and Technology, 10*(1), 151–164.

Goher, S. S., Ali, F., & Amin, M. (2022). The Delta variant mutations in the receptor binding domain of SARS-CoV-2 show enhanced electrostatic interactions with the ACE2. *Medicine in Drug Discovery, 13*, Article 100114.

Guan, Y., Zheng, B. J., He, Y. Q., Liu, X. L., Zhuang, Z. X., Cheung, C. L., & Poon, L. L. (2003). Isolation and characterization of viruses related to the SARS coronavirus from animals in southern China. *Science, 302*(5643), 276–278.

Guan, W. J., Ni, Z. Y., Hu, Y., Liang, W. H., Ou, C. Q., He, J. X., & Zhong, N. S. (2020). Clinical characteristics of coronavirus disease 2019 in China. *New England Journal of Medicine, 382*(18), 1708–1720.

Guarner, J., & Brandt, M. E. (2011). Histopathologic diagnosis of fungal infections in the 21st century. *Clinical Microbiology Reviews, 24*(2), 247–280.

Guarneri, C., Rullo, E. V., Pavone, P., Berretta, M., Ceccarelli, M., Natale, A., & Nunnari, G. (2021). Silent COVID-19: What your skin can reveal. *The Lancet Infectious Diseases, 21*(1), 24–25.

Güner, H. R., Hasanoğlu, İ, & Aktaş, F. (2020). COVID-19: Prevention and control measures in community. *Turkish Journal of Medical Sciences, 50*(9), 571–577.

Guo, Z. D., & Wang, Z. Y., Zhang, K., Gao, Y. W., Lu, B., & Chen, W. (2020). Aerosol and surface distribution of severe acute respiratory syndrome 290 Coronavirus 2 in hospital wards, Wuhan, China, 2020. *Emerging Infectious Disease Journal, 26*(7), 1583.

Gupta, K. L., Joshi, K., Pereira, B. J., & Singh, K. (1987). Disseminated mucormycosis presenting with acute renal failure. *Postgraduate Medical Journal, 63*(738), 297.

Häder, D. P., Banaszak, A. T., Villafañe, V. E., Narvarte, M. A., González, R. A., & Helbling, E. W. (2020). Anthropogenic pollution of aquatic ecosystems: Emerging problems with global implications. *Science of the Total Environment, 713*, Article 136586.

Hammoud, D., & Myronchenko, S. I. (2021). Comparative characteristics of approaches for designing a vaccine against SARS-Co-2. incom.

Hasan, M. K., Shahriar, A., & Jim, K. U. (2019). Water pollution in Bangladesh and its impact on public health. *Heliyon, 5*(8), Article e02145.

Hatmal, M. M. M., Alshaer, W., Al-Hatamleh, M. A., Hatmal, M., Smadi, O., Taha, M. O., & Plebanski, M. (2020). Comprehensive structural and molecular comparison of spike proteins of SARS-CoV-2, SARS-CoV and MERS-CoV, and their interactions with ACE2. *Cells, 9*(12), 2638.

Helenglass, G., Elliott, J. A., & Lucie, N. P. (1981). An unusual presentation of opportunistic mucormycosis. *British Medical Journal, 282*(6258), 108.

Helms, J., Kremer, S., Merdji, H., Clere-Jehl, R., Schenck, M., Kummerlen, C., & Meziani, F. (2020). Neurologic features in severe SARS-CoV-2 infection. *New England Journal of Medicine, 382*(23), 2268–2270.

Henriques, M. (2020). Will Covid-19 have a lasting impact on the environment? *BBC News, 27*.

Hoffmann, M., Kleine-Weber, H., Schroeder, S., Krüger, N., Herrler, T., Erichsen, S., & Pöhlmann, S. (2020). SARS-CoV-2 cell entry depends on ACE2 and TMPRSS2 and is blocked by a clinically proven protease inhibitor. *Cell, 181*(2), 271–280.

Hoffmann, M., Krüger, N., Schulz, S., Cossmann, A., Rocha, C., Kempf, A., & Pöhlmann, S. (2022). The Omicron variant is highly resistant against antibody-mediated neutralization: Implications for control of the COVID-19 pandemic. *Cell, 185*(3), 447–456.

Hossain, M. G., Tang, Y. D., Akter, S., & Zheng, C. (2022). Roles of the polybasic furin cleavage site of spike protein in SARS-CoV-2 replication, pathogenesis, and host immune responses and vaccination. *Journal of Medical Virology, 94*(5), 1815–1820.

Hu, B., Guo, H., Zhou, P., & Shi, Z. L. (2021). Characteristics of SARS-CoV-2 and COVID-19. *Nature Reviews Microbiology, 19*(3), 141–154.

Huang, C., Wang, Y., Li, X., Ren, L., Zhao, J., Hu, Y., & Cao, B. (2020). Clinical features of patients infected with 2019 novel coronavirus in Wuhan, China. *The Lancet, 395*(10223), 497–506.

Huang, S., & Kamihira, M. (2013). Development of hybrid viral vectors for gene therapy. *Biotechnology Advances, 31*(2), 208–223.

Hughes, C., Narayanaswamy, A., Vellingiri, B., Venkatesan, D., Warren, S., & Waters, M. (2022). What we know about the life-threatening novel human coronavirus. incom.

Hui, D. S., & Zumla, A. (2019). Severe acute respiratory syndrome: Historical, epidemiologic, and clinical features. *Infectious Disease Clinics, 33*(4), 869–889.

Hussain, S., & Kareem, S. (2020). Expectations of Indian guests from hotel preparedness in COVID-19. *Journal of Tourism, Hospitality and Culinary Arts, 12*, 31–51.

Hussain, S., Baxi, H., Riad, A., Klugarová, J., Pokorná, A., Slezáková, S., & Klugar, M. (2021). COVID-19-associated mucormycosis (CAM): An updated evidence mapping. *International Journal of Environmental Research and Public Health, 18*(19), 10340.

Indian Express. (2020). *Lockdown effect: Ganga water fit for drinking after decades, say experts*. Retrieved from https://www.newindianexpress.com/nation/2020/may/08/lockdown-effect-ganga-water-fit-for-drinking-after-decades-say-experts-2140622.html

Indian Express. (2021).*Why WHO skipped two Greek letters to name new Covid-19 variant Omicron*. Retrieved from https://indianexpress.com/article/cities/pune/who-greek-letters-name-new-covid-19-variant-omicron-7645228/

Indian Express. (2022). *Centre signals third wave is ending, writes to states: Remove extra curbs*. Retrieved from https://indianexpress.com/article/india/govt-asks-states-to-review-amend-additional-covid-19-restrictions-7777000/

International Federation of Pharmaceutical Manufacturers & Associations [IFPMA]. (2019). *The complex journey of a vaccine—Part III*. Retrieved from https://www.ifpma.org/wp-content/uploads/2019/07/IFPMA-ComplexJourney-2019_Stage-5_Web_High-Res.pdf

Iqbal, N., Irfan, M., Jabeen, K., Kazmi, M. M., & Tariq, M. U. (2017). Chronic pulmonary mucormycosis: An emerging fungal infection in diabetes mellitus. *Journal of Thoracic Disease, 9*(2), E121.

Islam, M. S., Hasib, F. Y., Nath, C., Ara, J., Nu, M. S., Fazal, M. A., & Chowdhury, S. (2021). Coronavirus disease 2019 and its potential animal reservoirs: A review. *Health, 7*(2), 171–181.

Jackson, C. B., Farzan, M., Chen, B., & Choe, H. (2022). Mechanisms of SARS-CoV-2 entry into cells. *Nature Reviews Molecular Cell Biology, 23*(1), 3–20.

Jafari, A., Danesh Pouya, F., Niknam, Z., AbdollahpourAlitappeh, M., Rezaei-Tavirani, M., & Rasmi, Y. (2022). Current advances and challenges in COVID-19 vaccine development: From conventional vaccines to next-generation vaccine platforms. *Molecular Biology Reports*, 1–15.

Jager, K. J., Kramer, A., Chesnaye, N. C., Couchoud, C., Sánchez-Álvarez, J. E., Garneata, L., & Massy, Z. A. (2020). Results from the ERA-EDTA Registry indicate a high mortality due to COVID-19 in dialysis patients and kidney transplant recipients across Europe. *Kidney International, 98*(6), 1540–1548.

Jaiswal, N. K., & Saxena, S. K. (2020). Classical coronaviruses. In *Coronavirus disease 2019 (COVID-19)* (pp. 141–150). Springer.

Jayaweera, M., Perera, H., Gunawardana, B., & Manatunge, J. (2020). Transmission of COVID-19 virus by droplets and aerosols: A critical review on the unresolved dichotomy. *Environmental Research, 188*, Article 109819.

Jeon, K. Y. (2020). COVID-19 vaccines-safety first, alleged "greater good" last. *American Journal of Epidemiology Public Heal, 4*(4), 012–016.

Jeong, W., Keighley, C., Wolfe, R., Lee, W. L., Slavin, M. A., Kong, D. C. M., & Chen, S. A. (2019). The epidemiology and clinical manifestations of mucormycosis: A systematic review and meta-analysis of case reports. *Clinical Microbiology and Infection, 25*(1), 26–34.

Jeyanathan, M., Afkhami, S., Smaill, F., Miller, M. S., Lichty, B. D., & Xing, Z. (2020). Immunological considerations for COVID-19 vaccine strategies. *Nature Reviews Immunology, 20*(10), 615–632.

Jhun, H., Park, H. Y., Hisham, Y., Song, C. S., & Kim, S. (2021). SARS-CoV-2 Delta (B. 1.617. 2) variant: A unique T478K mutation in receptor binding motif (RBM) of spike gene. *Immune Network, 21*(5).

Ji, B., Zhao, Y., Wei, T., & Kang, P. (2021). Water science under the global epidemic of COVID-19: Bibliometric tracking on COVID-19 publication and further research needs. *Journal of Environmental Chemical Engineering, 9*(4), Article 105357.

Jiang, S., Shi, Z., Shu, Y., Song, J., Gao, G. F., Tan, W., & Guo, D. (2020). Lancet. https://doi.org/10.1016/S0140-6736(20):30419-0

Jin, X., Lian, J. S., Hu, J. H., Gao, J., Zheng, L., Zhang, Y. M., & Yang, Y. (2020). Epidemiological, clinical and virological characteristics of 74 cases of coronavirus-infected disease 2019 (COVID-19) with gastrointestinal symptoms. *Gut, 69*(6), 1002–1009.

John, T. M., Jacob, C. N., & Kontoyiannis, D. P. (2021). When Uncontrolled Diabetes Mellitus and Severe COVID-19 Converge: The Perfect Storm for Mucormycosis. *Journal of fungi (Basel, Switzerland), 7*(4), 298. https://doi.org/10.3390/jof7040298

Jose, A., Singh, S., Roychoudhury, A., Kholakiya, Y., Arya, S., & Roychoudhury, S. (2021). Current understanding in the pathophysiology of SARS-CoV-2-associated rhino-orbito-cerebral mucormycosis: A comprehensive review. *Journal of Maxillofacial and Oral Surgery, 20*(3), 373–380.

Joshi, S., Parkar, J., Ansari, A., Vora, A., Talwar, D., Tiwaskar, M., & Barkate, H. (2021). Role of favipiravir in the treatment of COVID-19. *International Journal of Infectious Diseases, 102*, 501–508.

Joy, J. B. (2020). *Reciprocating the pandemic—COVID-19*. Retrieved from https://pulse.nitk.ac.in/2020/04/13/2751/

Karotia, D., & Kumar, A. (2020). A perspective on India's fight against COVID-19. *Epidemiology International, 5*(1), 22–28. E-ISSN: 2455-7048

Kaur, S. P., & Gupta, V. (2020). COVID-19 vaccine: A comprehensive status report. *Virus Research, 288*, Article 198114.

Kazybay, B., Ahmad, A., Mu, C., Mengdesh, D., & Xie, Y. (2022). Omicron N501Y mutation among SARS-CoV-2 lineages: Insilico analysis of potent binding to tyrosine kinase and hypothetical repurposed medicine. *Travel Medicine and Infectious Disease, 45*, Article 102242.

Katella, K. (2022). Omicron, delta, alpha, and more: What to know about the coronavirus variants. *Yale Medicine*, 28th February, available at: https://www.yalemedicine.org/news/covid-19-variants-of-concern-omicron. Accessed 8 March 2022.

Khan, F. I., Kang, T., Ali, H., & Lai, D. (2021a). Remdesivir strongly binds to RNA-dependent RNA polymerase, membrane protein, and main protease of SARS-CoV-2: Indication from molecular modeling and simulations. *Frontiers in Pharmacology, 12*, 1616.

Khan, I., Shah, D., & Shah, S. S. (2021b). COVID-19 pandemic and its positive impacts on environment: An updated review. *International Journal of Environmental Science and Technology, 18*(2), 521–530.

Khan, J., Asoom, L. I. A., Khan, M., Chakrabartty, I., Dandoti, S., Rudrapal, M., & Zothantluanga, J. H. (2021c). Evolution of RNA viruses from SARS to SARS-CoV-2 and diagnostic techniques for COVID-19: A review. *Beni-Suef University Journal of Basic and Applied Sciences, 10*(1), 1–14.

Khanna, R. C., Cicinelli, M. V., Gilbert, S. S., Honavar, S. G., & Murthy, G. V. (2020). COVID-19 pandemic: Lessons learned and future directions. *Indian Journal of Ophthalmology, 68*(5), 703–710.

Khoo, B., Tan, T., Clarke, S. A., Mills, E. G., Patel, B., Modi, M., & Dhillo, W. S. (2021). Thyroid function before, during, and after COVID-19. *The Journal of Clinical Endocrinology & Metabolism, 106*(2), e803–e811.

Klemeš, J. J., Van Fan, Y., Tan, R. R., & Jiang, P. (2020). Minimising the present and future plastic waste, energy and environmental footprints related to COVID-19. *Renewable and Sustainable Energy Reviews, 127*, Article 109883.

Koehler, P., Cornely, O. A., Böttiger, B. W., Dusse, F., Eichenauer, D. A., Fuchs, F., & Shimabukuro-Vornhagen, A. (2020). COVID-19 associated pulmonary aspergillosis. *Mycoses, 63*(6), 528–534.

Kowalzik, F., Schreiner, D., Jensen, C., Teschner, D., Gehring, S., & Zepp, F. (2021). MRNA-based vaccines. *Vaccines, 9*(4), 390.

Krammer, F. (2020). SARS-CoV-2 vaccines in development. *Nature, 586*(7830), 516–527.

Kumar, A., Kubota, Y., Chernov, M., & Kasuya, H. (2020). Potential role of zinc supplementation in prophylaxis and treatment of COVID-19. *Medical Hypotheses, 144*, Article 109848.

Lai, M. M., & Cavanagh, D. (1997). The molecular biology of coronaviruses. *Advances in Virus Research, 48*, 1–100.

Lauer, S. A., Grantz, K. H., Bi, Q., Jones, F. K., Zheng, Q., Meredith, H. R., & Lessler, J. (2020). The incubation period of coronavirus disease 2019 (COVID-19) from publicly reported confirmed cases: Estimation and application. *Annals of Internal Medicine, 172*(9), 577–582.

Lazarevic, I., Pravica, V., Miljanovic, D., & Cupic, M. (2021). Immune evasion of SARS-CoV-2 emerging variants: What have we learnt so far? *Viruses, 13*(7), 1192.

Leecaster, M., Gesteland, P., Greene, T., Walton, N., Gundlapalli, A., Rolfs, R., & Samore, M. (2011). Modeling the variations in pediatric respiratory syncytial virus seasonal epidemics. *BMC Infectious Diseases, 11*(1), 1–9.

Leung, W. K., To, K. F., Chan, P. K., Chan, H. L., Wu, A. K., Lee, N., & Sung, J. J. (2003). Enteric involvement of severe acute respiratory syndrome-associated coronavirus infection. *Gastroenterology, 125*(4), 1011–1017.

Levi, M., Thachil, J., Iba, T., & Levy, J. H. (2020). Coagulation abnormalities and thrombosis in patients with COVID-19. *The Lancet Haematology, 7*(6), e438–e440.

Li, H., Liu, S. M., Yu, X. H., Tang, S. L., & Tang, C. K. (2020a). Coronavirus disease 2019 (COVID-19): Current status and future perspectives. *International Journal of Antimicrobial Agents, 55*(5), Article 105951.

Li, Q., Wu, J., Nie, J., Zhang, L., Hao, H., Liu, S., & Wang, Y. (2020b). The impact of mutations in SARS-CoV-2 spike on viral infectivity and antigenicity. *Cell, 182*(5), 1284–1294.

Li, X., Zai, J., Wang, X., & Li, Y. (2020c). Potential of large "first generation" human-to-human transmission of 2019-nCoV. *Journal of Medical Virology, 92*(4), 448–454.

Li, Y., Qian, H., Hang, J., Chen, X., Cheng, P., Ling, H., Wang, S., Liang, P., Li, J., Xiao, S., Wei, J., Liu, L., Cowling, B. J., & Kang, M. (2021). Probable airborne transmission of SARS-CoV-2 in a poorly ventilated restaurant. *Building and environment, 196*, 107788. https://doi.org/10.1016/j.buildenv.2021.107788

Libby, P. (2020). The heart in COVID-19: Primary target or secondary bystander? *Basic to Translational Science, 5*(5), 537–542.

Lim, S., Bae, J. H., Kwon, H. S., & Nauck, M. A. (2021). COVID-19 and diabetes mellitus: From pathophysiology to clinical management. *Nature Reviews Endocrinology, 17*(1), 11–30.

Lin, L., Jiang, X., Zhang, Z., Huang, S., Zhang, Z., Fang, Z., & Shan, H. (2020). Gastrointestinal symptoms of 95 cases with SARS-CoV-2 infection. *Gut, 69*(6), 997–1001.

Lippi, G., & Favaloro, E. J. (2020). D-dimer is associated with severity of coronavirus disease 2019: A pooled analysis. *Thrombosis and Haemostasis, 120*(05), 876–878.

Lippi, G., Plebani, M., & Henry, B. M. (2020). Thrombocytopenia is associated with severe coronavirus disease 2019 (COVID-19) infections: A meta-analysis. *Clinica Chimica Acta, 506*, 145–148.

Lippi, G., & Plebani, M. (2020). Procalcitonin in patients with severe coronavirus disease 2019 (COVID-19): A meta-analysis. *Clinica Chimica Acta; International Journal of Clinical Chemistry, 505*, 190.

Liu, L., Iketani, S., Guo, Y., Chan, J. F. W., Wang, M., Liu, L., & Ho, D. D. (2022a). Striking antibody evasion manifested by the Omicron variant of SARS-CoV-2. *Nature, 602*(7898), 676–681.

Liu, Y., Liu, J., Johnson, B. A., Xia, H., Ku, Z., Schindewolf, C., & Shi, P. Y. (2022b). Delta spike P681R mutation enhances SARS-CoV-2 fitness over Alpha variant. *Cell Reports, 39*(7), Article 110829.

Lokhandwala, S., & Gautam, P. (2020). Indirect impact of COVID-19 on environment: A brief study in Indian context. *Environmental Research, 188*, Article 109807.

Lotfi, M., Hamblin, M. R., & Rezaei, N. (2020). COVID-19: Transmission, prevention, and potential therapeutic opportunities. *Clinica Chimica Acta, 508*, 254–266.

Ludvigsson, J. F. (2020). Systematic review of COVID-19 in children shows milder cases and a better prognosis than adults. *Acta Paediatrica, 109*(6), 1088–1095.

Lundstrom, K. (2020). Application of viral vectors for vaccine development with a special emphasis on COVID-19. *Viruses, 12*(11), 1324.

Lurie, N., Saville, M., Hatchett, R., & Halton, J. (2020). Developing Covid-19 vaccines at pandemic speed. *New England Journal of Medicine, 382*(21), 1969–1973.

Ma, Y., Zhao, Y., Liu, J., He, X., Wang, B., Fu, S., & Luo, B. (2020). Effects of temperature variation and humidity on the death of COVID-19 in Wuhan, China. *Science of the Total Environment, 724*, 138226.

Mahalmani, V. M., Mahendru, D., Semwal, A., Kaur, S., Kaur, H., Sarma, P., & Medhi, B. (2020). COVID-19 pandemic: A review based on current evidence. *Indian Journal of Pharmacology, 52*(2), 117.

Maier, C. L., Truong, A. D., Auld, S. C., Polly, D. M., Tanksley, C. L., & Duncan, A. (2020). COVID-19-associated hyperviscosity: A link between inflammation and thrombophilia? *The Lancet, 395*(10239), 1758–1759.

Mao, L., Jin, H., Wang, M., Hu, Y., Chen, S., He, Q., & Hu, B. (2020). Neurologic manifestations of hospitalized patients with coronavirus disease 2019 in Wuhan, China. *JAMA Neurology, 77*(6), 683–690.

Mariano, G., Farthing, R. J., Lale-Farjat, S. L., & Bergeron, J. R. (2020). Structural characterization of SARS-CoV-2: Where we are, and where we need to be. *Frontiers in Molecular Biosciences, 344*.

Mave, V., Shaikh, A., Monteiro, J. M., Bogam, P., Pujari, B. S., & Gupte, N. (2021). Impact of national and regional lockdowns on growth of COVID-19 cases in COVID-Hotspot City of Pune in Western India: A real-world data analysis. *medRxiv* (pp. 1–31).

McBride, R., Van Zyl, M., & Fielding, B. C. (2014). The coronavirus nucleocapsid is a multifunctional protein. *Viruses, 6*(8), 2991–3018.

Mehta, R. K., & Angadi, S. (2020). Hospital solid waste management during COVID-19 pandemic in Nepal. *Journal of Chitwan Medical College, 10*(4), 100–102.

Meini, S., Pagotto, A., Longo, B., Vendramin, I., Pecori, D., & Tascini, C. (2020). Role of Lopinavir/Ritonavir in the treatment of Covid-19: A review of current evidence, guideline recommendations, and perspectives. *Journal of Clinical Medicine, 9*(7), 2050.

Menebo, M. M. (2020). Temperature and precipitation associate with Covid-19 new daily cases: A correlation study between weather and Covid-19 pandemic in Oslo, Norway. *Science of the Total Environment, 737*, 139659.

Milani, G. P., Macchi, M., & Guz-Mark, A. (2021). Vitamin C in the treatment of COVID-19. *Nutrients, 13*(4), 1172.

Miller, E. H., Namale, V. S., Kim, C., Dugue, R., Waldrop, G., Ciryam, P., & Thakur, K. T. (2020). Cerebrospinal analysis in patients with COVID-19. In *Open forum infectious diseases* (Vol. 7, No. 11, p. ofaa501). Oxford University Press.

Ministry of Health and Family Welfare [MoHFW]. (2022). Retrieved from https://www.mohfw.gov.in/

Ministry of Home Affairs [MHA]. (2020a). *Special trains to be operated for movement of stranded persons across the country, due to lockdown restrictions to fight COVID-19*. Retrieved from https://www.mha.gov.in/sites/default/files/press_0_0.pdf

Ministry of Home Affairs [MHA]. (2020b). *New guidelines to fight COVID-19 to be effective from 1st June 2020*. Retrieved from https://pib.gov.in/PressReleasePage.aspx?PRID=1627965

Ministry of Home Affairs [MHA]. (2020c). Retrieved from https://www.mha.gov.in/sites/default/files/MHAOrder_29062020.pdf

Ministry of Home Affairs [MHA]. (2020d). Retrieved from https://www.civilaviation.gov.in/sites/default/files/MHA_SOP_dtd_22_08_2020_on_international_travel_under_Vande_Bharat_and_Bubble_flights.pdf

Ministry of Home Affairs [MHA]. (2020e). Retrieved from https://www.mha.gov.in/sites/default/files/Unlock3_29072020.pdf

Ministry of Home Affairs [MHA]. (2020f). Retrieved from https://www.mha.gov.in/sites/default/files/MHAOrder_Unlock4_29082020.pdf

Ministry of Home Affairs [MHA]. (2020g). Retrieved from https://www.mha.gov.in/sites/default/files/MHAOrderDt_30092020.pdf

Mishra, S. P. (2020). A review on current status of Corona Virus Disease-2019 (COVID-19) pandemic in India. *International Journal of Research and Analytical Review, 7*(3), 621–630.

Miyah, Y., Benjelloun, M., Lairini, S., & Lahrichi, A. (2022). COVID-19 impact on public health, environment, human psychology, global socioeconomy, and education. *The Scientific World Journal*.

Mlcochova, P., Kemp, S. A., Dhar, M. S., Papa, G., Meng, B., Ferreira, I. A., & Gupta, R. K. (2021). SARS-CoV-2 B. 1.617.2 Delta variant replication and immune evasion. *Nature, 599*(7883), 114–119.

Mohta, A., Neogi, S., & Das, S. (2011). Gastrointestinal mucormycosis in an infant. *Indian Journal of Pathology and Microbiology, 54*(3), 664–664.

Moreno-Sánchez, R. D. P., & Maldonado, J. H. (2006). Surviving from garbage: The role of informal waste-pickers in a dynamic model of solid-waste management in developing countries. *Environment and Development Economics, 11*(3), 371–391.

Muhammad, S., Long, X., & Salman, M. (2020). COVID-19 pandemic and environmental pollution: A blessing in disguise? *Science of the Total Environment, 728*, Article 138820.

Muralidar, S., Ambi, S. V., Sekaran, S., & Krishnan, U. M. (2020). The emergence of COVID-19 as a global pandemic: Understanding the epidemiology, immune response and potential therapeutic targets of SARS-CoV-2. *Biochimie, 179*, 85–100.

Murphy, K., & Weaver, C. (2016). *Janeway's immunobiology*. Garland Science.

Muthu, V., Rudramurthy, S. M., Chakrabarti, A., & Agarwal, R. (2021). Epidemiology and pathophysiology of COVID-19-associated mucormycosis: India versus the rest of the world. *Mycopathologia, 186*(6), 739–754.

Nardo, A. D., Schneeweiss-Gleixner, M., Bakail, M., Dixon, E. D., Lax, S. F., & Trauner, M. (2021). Pathophysiological mechanisms of liver injury in COVID-19. *Liver International, 41*(1), 20–32.

Nasereddin, A., Golan Berman, H., Wolf, D. G., Oiknine-Djian, E., & Adar, S. (2022). Identification of SARS-CoV-2 variants of concern using amplicon next-generation sequencing. *Microbiology Spectrum*, e00736-22.

Nassar, M., Bakhrebah, M. A., Meo, S. A., Alsuabeyl, M. S., & Zaher, W. A. (2018). Middle East Respiratory Syndrome Coronavirus (MERS-CoV) infection: Epidemiology, pathogenesis and clinical characteristics. *European Review for Medical and Pharmacological Sciences, 22*(15), 4956–4961.

National Centre for Disease Control-Libya. (2022). Retrieved from https://www.humanitarianresponse.info/sites/www.humanitarianresponse.info/files/documents/files/ncdc_essential_kowledge_for_covid-19_january_2022.pdf

National Centre for Immunisation Research and Surveillance [NCIRS]. (2020). *Phases of clinical trials*. Retrieved from https://www.ncirs.org.au/phases-clinical-trials?fbclid=IwAR3ireqihW5qqOH_-mjR-L0uQpfnLgeByMvVHC4zX29njTOZjcXK4rG4V5s

National Institutes of Health [NIH]. (2022). *Coronavirus disease 2019 (COVID-19) treatment guidelines*. Retrieved from https://www.covid19treatmentguidelines.nih.gov/about-the-guidelines/whats-new/

Nayak, S., & Herzog, R. W. (2010). Progress and prospects: Immune responses to viral vectors. *Gene Therapy, 17*(3), 295–304.

News 18. (2022). *Omicron is like a 'Flash Flood'. So how much longer for India's third Covid wave to ebb?* Retrieved from https://www.news18.com/news/opinion/omicron-is-like-a-flash-flood-so-how-much-longer-for-indias-third-covid-wave-to-ebb-4689407.html

Ni, L., Ye, F., Cheng, M. L., Feng, Y., Deng, Y. Q., Zhao, H., & Dong, C. (2020). Detection of SARS-CoV-2-specific humoral and cellular immunity in COVID-19 convalescent individuals. *Immunity, 52*(6), 971–977.

Niranjan, P. S. (2020). Corona virus pandemic impact on global education: A blessing in disguise. *Sustainable Humanosphere, 16*(2), 68–72.

NPSMEDICINEWISE. (2022). *Molnupiravir (Lagevrio) for mild-to-moderate COVID-19*. Retrieved from https://www.nps.org.au/radar/articles/molnupiravir-lagevrio-for-mild-to-moderate-covid-19

Nundy, S., Ghosh, A., Mesloub, A., Albaqawy, G. A., & Alnaim, M. M. (2021). Impact of COVID-19 pandemic on socio-economic, energy-environment and transport sector globally and sustainable development goal (SDG). *Journal of Cleaner Production, 312*, Article 127705.

Occupational Safety and Health Administration. (2022). Retrieved from https://www.osha.gov/coronavirus/control-prevention?fbclid=IwAR1mB_tnbSOA0odr7HeJDN2QFjHSkoiv5KAc

Our World in Data. (2022). *Coronavirus (COVID-19) vaccinations*. Retrieved from https://ourworldindata.org/covid-vaccinations

Pal, M., Berhanu, G., Desalegn, C., & Kandi, V. (2020). Severe acute respiratory syndrome coronavirus-2 (SARS-CoV-2): An update. *Cureus, 12*(3).

Pampel, J. (2022). *SARS-CoV-2 life cycle: Stages and inhibition targets*. Retrieved from https://www.antibodies-online.com/resources/18/5410/sars-cov-2-life-cycle-stages-and-inhibition-targets/

Pan, L., Wang, J., Wang, X., Ji, J. S., Ye, D., Shen, J., & Wang, L. (2022). Prevention and control of coronavirus disease 2019 (COVID-19) in public places. *Environmental Pollution, 292*, Article 118273.

Pandey, A., Singh, N., Vemula, S. V., Couëtil, L., Katz, J. M., Donis, R., & Mittal, S. K. (2012). Impact of preexisting adenovirus vector immunity on immunogenicity and protection conferred with an adenovirus-based H5N1 influenza vaccine. *PLoS ONE, 7*(3), Article e33428.

Pandey, A. (2020). *UP officials seen with crowd amid "Janata Curfew". Then, a clarification. NDTV*. Retrieved from https://www.ndtv.com/india-news/up-officials-seen-with-crowd-amid-janata-curfew-then-a-clarification-2198993

Pandiar, D., Kumar, N. S., Anand, R., Kamboj, M., Narwal, A., & Shameena, P. M. (2021). Does COVID 19 generate a milieu for propagation of mucormycosis? *Medical Hypotheses, 152*, Article 110613.

Pardi, N., Hogan, M. J., Porter, F. W., & Weissman, D. (2018). MRNA vaccines—A new era in vaccinology. *Nature Reviews Drug Discovery, 17*(4), 261–279.

Parolin, M., Parisotto, M., Zanchetta, F., Sartorato, P., & De Menis, E. (2021). Coronaviruses and endocrine system: A systematic review on evidence and shadows. *Endocrine, Metabolic & Immune Disorders – Drug Targets (Formerly Current Drug Targets – Immune, Endocrine & Metabolic Disorders), 21*(7), 1242–1251.

Patel, A., Patel, S., Fulzele, P., Mohod, S., & Chhabra, K. G. (2020). Quarantine an effective mode for control of the spread of COVID19: A review. *Journal of Family Medicine and Primary Care, 9*(8), 3867.

Patel, A., Agarwal, R., Rudramurthy, S. M., Shevkani, M., Xess, I., Sharma, R., & Network, M. (2021). Multicenter epidemiologic study of coronavirus disease-associated mucormycosis, India. *Emerging Infectious Diseases, 27*(9), 2349.

Pathak, V. V., Rai, A., Shukla, S. K., Jangra, S., & Tiwari, S. (2021). Prospects of iron oxide nanomaterial for remediation of wastewater. In *Functionalized nanomaterials based devices for environmental applications* (pp. 261–273). Elsevier.

Pei, G., Zhang, Z., Peng, J., Liu, L., Zhang, C., Yu, C., & Xu, G. (2020). Renal involvement and early prognosis in patients with COVID-19 pneumonia. *Journal of the American Society of Nephrology, 31*(6), 1157–1165.

Perisetti, A., Gajendran, M., Boregowda, U., Bansal, P., & Goyal, H. (2020). COVID-19 and gastrointestinal endoscopies: Current insights and emergent strategies. *Digestive Endoscopy, 32*(5), 715–722.

Perricone, C., Bartoloni, E., Bursi, R., Cafaro, G., Guidelli, G. M., Shoenfeld, Y., & Gerli, R. (2020). COVID-19 as part of the hyperferritinemic syndromes: The role of iron depletion therapy. *Immunologic Research, 68*(4), 213–224.

Petersen, E., Koopmans, M., Go, U., Hamer, D. H., Petrosillo, N., Castelli, F., & Simonsen, L. (2020). Comparing SARS-CoV-2 with SARS-CoV and influenza pandemics. *The Lancet Infectious Diseases, 20*(9), e238–e244.

Petrikkos, G., Skiada, A., Lortholary, O., Roilides, E., Walsh, T. J., & Kontoyiannis, D. P. (2012). Epidemiology and clinical manifestations of mucormycosis. *Clinical Infectious Diseases, 54*(suppl_1), S23–S34.

Pippal, S. K., Kumar, D., & Ukawat, L. (2021). Management challenge of rhino-orbito-cerebral mucormycosis in Covid 19 era: A prospective observational study. *Indian Journal of Otolaryngology and Head & Neck Surgery*, 1–7.

Pisoschi, A. M., Iordache, F., Stanca, L., Gajaila, I., Ghimpeteanu, O. M., Geicu, O. I., & Serban, A. I. (2022). Antioxidant, anti-inflammatory, and immunomodulatory roles of nonvitamin antioxidants in anti-SARS-CoV-2 therapy. *Journal of Medicinal Chemistry., 65*(19), 12562–12593.

Pozo Laderas, J. C., Pontes Moreno, A., Pozo Salido, C., Robles Arista, J. C., & Linares Sicilia, M. J. (2015). Disseminated mucormycosis in immunocompetent patients: A disease that also exists. *Revista Iberoamericana De Micologia, 32*(2), 63–70.

Prakash, H., & Chakrabarti, A. (2019). Global epidemiology of mucormycosis. *Journal of Fungi, 5*(1), 26.

Prates-Syed, W. A., Chaves, L. C., Crema, K. P., Vuitika, L., Lira, A., Côrtes, N., & Cabral-Miranda, G. (2021). VLP-based COVID-19 vaccines: An adaptable technology against the threat of new variants. *Vaccines, 9*(12), 1409.

Pulla, P. (2020). Covid-19: India imposes lockdown for 21 days and cases rise. *BMJ, 368*, m1251.

Pulliam, J. R., van Schalkwyk, C., Govender, N., von Gottberg, A., Cohen, C., Groome, M. J., & Moultrie, H. (2022). Increased risk of SARS-CoV-2 reinfection associated with emergence of Omicron in South Africa. *Science, 376*(6593), eabn4947.

Qin, F., Xia, F., Chen, H., Cui, B., Feng, Y., Zhang, P., & Luo, M. (2021). A guide to nucleic acid vaccines in the prevention and treatment of infectious diseases and cancers: From basic principles to current applications. *Frontiers in Cell and Developmental Biology, 830.*

Qualls, N., Levitt, A., Kanade, N., Wright-Jegede, N., Dopson, S., Biggerstaff, M., & Reed, C. (2017). Community mitigation guidelines to prevent pandemic influenza—United States, 2017. *MMWR Recommendations and Reports, 66*(1), 1.

Quartz India. (2022). *India's third wave of covid is here.* Retrieved from https://qz.com/india/2108711/indias-third-wave-of-covid-is-fueled-by-the-omicron-variant/

Rahman, H. S., Aziz, M. S., Hussein, R. H., Othman, H. H., Omer, S. H. S., Khalid, E. S., & Abdullah, R. (2020). The transmission modes and sources of COVID-19: A systematic review. *International Journal of Surgery Open, 26*, 125–136.

Rai, A., & Fulekar, M. H. (2020a). Nanomaterials development and their environmental applications. In *Bioremediation technology* (pp. 293–310). CRC Press.

Rai, A., & Fulekar, M. H. (2020b). Wastewater treatment: Common effluent treatment plant—case study. In *Bioremediation technology* (pp. 267–284). CRC Press.

Rai, A., Fulekar, J., & Fulekar, M. H. (2020). Rhizospheric microorganisms for the remediation of contaminants for ecological restoration. In *Plant microbiome paradigm* (pp. 163–174). Springer.

Rai, A., Amari, A., Yadav, V. K., Ismail, M. A., Elboughdiri, N., Fulekar, M. H., & Basnet, A. (2022). A synergistic effect of *Moringa oleifera*-based coagulant and ultrafiltration for the wastewater treatment collected from final ETP. *Adsorption Science & Technology.*

Rana, R., Kant, R., Huirem, R. S., Bohra, D., & Ganguly, N. K. (2022). Omicron variant: Current insights and future directions. *Microbiological Research*, 127204.

Raptis, C. A., Hammer, M. M., Short, R. G., Shah, A., Bhalla, S., Bierhals, A. J., & Henry, T. S. (2020). Chest CT and Coronavirus disease (COVID-19): A critical review of the literature to date. *AJR. American Journal of Roentgenology, 215*(4), 839–842.

Rauf, A., Abu-Izneid, T., Olatunde, A., Ahmed Khalil, A., Alhumaydhi, F. A., Tufail, T., & Rengasamy, K. R. (2020). COVID-19 pandemic: Epidemiology, etiology, conventional and non-conventional therapies. *International Journal of Environmental Research and Public Health, 17*(21), 8155.

Redondo, N., Zaldívar-López, S., Garrido, J. J., & Montoya, M. (2021). SARS-CoV-2 accessory proteins in viral pathogenesis: Knowns and unknowns. *Frontiers in Immunology*, 2698.

Reid, G., Lynch III, J. P., Fishbein, M. C., & Clark, N. M. (2020). Mucormycosis. In *Seminars in respiratory and critical care medicine* (Vol. 41, No. 01, pp. 099–114). Thieme Medical Publishers.

Richardson, M. (2009). The ecology of the Zygomycetes and its impact on environmental exposure. *Clinical Microbiology and Infection, 15*, 2–9.

Riddell, S., Goldie, S., Hill, A., Eagles, D., & Drew, T. W. (2020). The effect of temperature on persistence of SARS-CoV-2 on common surfaces. *Virology Journal, 17*(1), 1–7.

Robert-Guroff, M. (2007). Replicating and non-replicating viral vectors for vaccine development. *Current Opinion in Biotechnology, 18*(6), 546–556.

Rocha, I. C. N., Hasan, M. M., Goyal, S., Patel, T., Jain, S., Ghosh, A., & Cedeño, T. D. D. (2021). COVID-19 and mucormycosis syndemic: Double health threat to a collapsing healthcare system in India. *Tropical Medicine & International Health, 26*(9), 1016–1018.

Roden, M. M., Zaoutis, T. E., Buchanan, W. L., Knudsen, T. A., Sarkisova, T. A., Schaufele, R. L., & Walsh, T. J. (2005). Epidemiology and outcome of zygomycosis: A review of 929 reported cases. *Clinical Infectious Diseases, 41*(5), 634–653.

Rodrigues, A. F., Soares, H. R., Guerreiro, M. R., Alves, P. M., & Coroadinha, A. S. (2015). Viral vaccines and their manufacturing cell substrates: New trends and designs in modern vaccinology. *Biotechnology Journal, 10*(9), 1329–1344.

Rodriguez-Morales, A. J., Cardona-Ospina, J. A., Gutiérrez-Ocampo, E., Villamizar-Peña, R., Holguin-Rivera, Y., Escalera-Antezana, J. P., & Sah, R. (2020). Clinical, laboratory and imaging features of COVID-19: A systematic review and meta-analysis. *Travel Medicine and Infectious Disease, 34*, Article 101623.

Rojas, M., Rodríguez, Y., Monsalve, D. M., Acosta-Ampudia, Y., Camacho, B., Gallo, J. E., & Anaya, J. M. (2020). Convalescent plasma in Covid-19: Possible mechanisms of action. *Autoimmunity Reviews, 19*(7), Article 102554.

Saadat, S., Rawtani, D., & Hussain, C. M. (2020). Environmental perspective of COVID-19. *Science of the Total Environment, 728*, Article 138870.

Saha, J., Barman, B., & Chouhan, P. (2020). Lockdown for COVID-19 and its impact on community mobility in India: An analysis of the COVID-19 Community Mobility Reports, 2020. *Children and Youth Services Review, 116*, Article 105160.

Sahu, R. K., Salem-Bekhit, M. M., Bhattacharjee, B., Almoshari, Y., Ikbal, A. M. A., Alshamrani, M., & Elbagory, I. (2021). Mucormycosis in Indian COVID-19 Patients: Insight into its pathogenesis, clinical manifestation, and management strategies. *Antibiotics, 10*(9), 1079.

Saidha, P. K., Kapoor, S., Das, P., Gupta, A., Kakkar, V., Kumar, A., & Arya, V. (2021). Mucormycosis of paranasal sinuses of odontogenic origin post COVID19 infection: A case series. *Indian Journal of Otolaryngology and Head & Neck Surgery*, 1–5.

Saitoh, A., Sato, K., Magara, Y., Osaki, K., Narita, K., Shioiri, K., & Saint, S. (2020). Improving hand hygiene adherence in healthcare workers before patient contact: A multimodal intervention in four tertiary care hospitals in Japan. *Journal of Hospital Medicine, 15*(5), 262–267.

Salahshoori, I., Mobaraki-Asl, N., Seyfaee, A., Mirzaei Nasirabad, N., Dehghan, Z., Faraji, M., & Hamrang, A. (2021). Overview of COVID-19 disease: Virology, epidemiology, prevention diagnosis, treatment, and vaccines. *Biologics, 1*(1), 2–40.

Salama, C., Han, J., Yau, L., Reiss, W. G., Kramer, B., Neidhart, J. D., & Mohan, S. V. (2021). Tocilizumab in patients hospitalized with Covid-19 pneumonia. *New England Journal of Medicine, 384*(1), 20–30.

Salehi-Vaziri, M., Fazlalipour, M., Seyed Khorrami, S. M., Azadmanesh, K., Pouriayevali, M. H., Jalali, T., & Maleki, A. (2022). The ins and outs of SARS-CoV-2 variants of concern (VOCs). *Archives of Virology*, 1–18.

Samarasekera, U. (2021). India grapples with second wave of COVID-19. *The Lancet Microbe, 2*(6), Article e238.

Sanchis-Gomar, F., Lavie, C. J., Mehra, M. R., Henry, B. M., & Lippi, G. (2020). Obesity and outcomes in COVID-19: When an epidemic and pandemic collide. In *Mayo clinic proceedings* (Vol. 95, No. 7, pp. 1445–1453). Elsevier.

Sangkham, S. (2020). Face mask and medical waste disposal during the novel COVID-19 pandemic in Asia. *Case Studies in Chemical and Environmental Engineering, 2*, Article 100052.

Satarker, S., & Nampoothiri, M. (2020). Structural proteins in severe acute respiratory syndrome coronavirus-2. *Archives of Medical Research, 51*(6), 482–491.

Scacchi, A., Catozzi, D., Boietti, E., Bert, F., & Siliquini, R. (2021). COVID-19 lockdown and self-perceived changes of food choice, waste, impulse buying and their determinants in Italy: QuarantEat, a cross-sectional study. *Foods, 10*(2), 306.

Scheppler, L., De Clercq, N., McGoldrick, M., & Dias, J. (2021). Regulatory harmonization and streamlining of clinical trial applications globally should lead to faster clinical development and earlier access to life-saving vaccines. *Vaccine, 39*(5), 790–796.

Schoeman, D., & Fielding, B. C. (2019). Coronavirus envelope protein: Current knowledge. *Virology Journal, 16*(1), 1–22.

Shannon, A., Selisko, B., Le, N. T. T., Huchting, J., Touret, F., Piorkowski, G., & Canard, B. (2020). Rapid incorporation of Favipiravir by the fast and permissive viral RNA polymerase complex results in SARS-CoV-2 lethal mutagenesis. *Nature Communications, 11*(1), 1–9.

Sharma, V., Rai, H., Gautam, D. N., Prajapati, P. K., & Sharma, R. (2022). Emerging evidence on Omicron (B. 1.1.529) SARS-CoV-2 variant. *Journal of Medical Virology, 94*(5), 1876–1885.

Sharun, K., Tiwari, R., Dhama, J., & Dhama, K. (2020). Dexamethasone to combat cytokine storm in COVID-19: Clinical trials and preliminary evidence. *International Journal of Surgery, 82*, 179.

Shereen, M. A., Khan, S., Kazmi, A., Bashir, N., & Siddique, R. (2020). COVID-19 infection: Emergence, transmission, and characteristics of human coronaviruses. *Journal of Advanced Research, 24*, 91–98.

Shil, P., Atre, N. M., & Tandale, B. V. (2022). Epidemiological findings for the first and second waves of COVID-19 pandemic in Maharashtra, India. *Spatial and Spatio-temporal Epidemiology, 41*, 100507.

Shin, J. S., Jung, E., Kim, M., Baric, R. S., & Go, Y. Y. (2018). Saracatinib inhibits middle east respiratory syndrome-coronavirus replication in vitro. *Viruses, 10*(6), 283.

Silva, A. L. P., Prata, J. C., Walker, T. R., Duarte, A. C., Ouyang, W., Barcelò, D., & Rocha-Santos, T. (2021). Increased plastic pollution due to COVID-19 pandemic: Challenges and recommendations. *Chemical Engineering Journal, 405*, Article 126683.

Silveira, M. M., Oliveira, T. L., Schuch, R. A., McBride, A. J. A., Dellagostin, O. A., & Hartwig, D. D. (2017). DNA vaccines against leptospirosis: A literature review. *Vaccine, 35*(42), 5559–5567.

Singh, R. (2020). Covid-19 and its impact on Indian villages: A study of four identified villages of district Prayagraj. *International Journal of Research in Economics and Social Sciences, 10*(10), 1–90.

Singh, K., & Mehta, S. (2016). The clinical development process for a novel preventive vaccine: An overview. *Journal of Postgraduate Medicine, 62*(1), 4.

Singh, V., & Mishra, V. (2021). Environmental impacts of coronavirus disease 2019 (COVID-19). *Bioresource Technology Reports, 15*, Article 100744.

Singh, A. K., Singh, R., Joshi, S. R., & Misra, A. (2021). Mucormycosis in COVID-19: A systematic review of cases reported worldwide and in India. *Diabetes & Metabolic Syndrome: Clinical Research & Reviews, 15*(4), Article 102146.

Singh, J., Swaroop, S., Sharma, P., & Mishra, V. (2022). Real-time assessment of the Ganga River during pandemic COVID-19 and predictive data modeling by machine learning. *International Journal of Environmental Science and Technology*, 1–24.

Skiada, A., Pavleas, I., & Drogari-Apiranthitou, M. (2020). Epidemiology and diagnosis of mucormycosis: An update. *Journal of Fungi, 6*(4), 265.

Skiada, A., Lass-Floerl, C., Klimko, N., Ibrahim, A., Roilides, E., & Petrikkos, G. (2018). Challenges in the diagnosis and treatment of mucormycosis. *Medical Mycology, 56*(suppl_1), S93–S101.

Sky news. (2021). *COVID-19: 'Delta is making other coronavirus variants extinct'—but what does it mean for the UK?* Retrieved from https://news.sky.com/story/covid-19-delta-is-making-other-coronavirus-variants-extinct-but-what-does-it-mean-for-the-uk-12431343

Słomka, A., Kowalewski, M., & Żekanowska, E. (2020). Coronavirus disease 2019 (COVID–19): A short review on hematological manifestations. *Pathogens, 9*(6), 493.

Smith, H. W., & Kirchner, J. A. (1958). Cerebral mucormycosis: A report of three cases. *AMA Archives of Otolaryngology, 68*(6), 715–726.

Sobti, R. C., Dhalla, N. S., Watanabe, M., & Sobti, A. (2021). *Delineating health and health system: Mechanistic insights into Covid 19 complications*. Springer.

Somani, M., Srivastava, A. N., Gummadivalli, S. K., & Sharma, A. (2020). Indirect implications of COVID-19 towards sustainable environment: An investigation in Indian context. *Bioresource Technology Reports, 11*, Article 100491.

Southeast AIDS Education and Training Center (SEAETC). (n.d.). *HIV and oral health the era of COVID-19*. Retrieved from https://www.seaetc.com/wp-content/uploads/2020/12/20201216-HIV-and-Oral-Health-in-the-Era-of-COVID-19.pdf

Spellberg, B. (2012). Gastrointestinal mucormycosis: An evolving disease. *Gastroenterology & Hepatology, 8*(2), 140.

Spellberg, B., Edwards, J., Jr., & Ibrahim, A. (2005). Novel perspectives on mucormycosis: Pathophysiology, presentation, and management. *Clinical Microbiology Reviews, 18*(3), 556–569.

Srivastava, S., Kumar, A., Bauddh, K., Gautam, A. S., & Kumar, S. (2020). 21-day lockdown in India dramatically reduced air pollution indices in Lucknow and New Delhi, India. *Bulletin of Environmental Contamination and Toxicology, 105*(1), 9–17.

Talon, J., Salvatore, M., O'Neill, R. E., Nakaya, Y., Zheng, H., Muster, T., & Palese, P. (2000). Influenza A and B viruses expressing altered NS1 proteins: A vaccine approach. *Proceedings of the National Academy of Sciences, 97*(8), 4309–4314.

Tan, T., Khoo, B., Mills, E. G., Phylactou, M., Patel, B., Eng, P. C., & Dhillo, W. S. (2020). Association between high serum total cortisol concentrations and mortality from COVID-19. *The Lancet Diabetes & Endocrinology, 8*(8), 659–660.

Tandon, A., & Pandey, L. (2021). COVID-19, steroids, and mucormycosis: What an ophthalmologist should know. *Indian Journal of Ophthalmology, 69*(7), 1970.

Tang, T., Bidon, M., Jaimes, J. A., Whittaker, G. R., & Daniel, S. (2020). Coronavirus membrane fusion mechanism offers a potential target for antiviral development. *Antiviral Research, 178*, Article 104792.

Team, E. (2020). The epidemiological characteristics of an outbreak of 2019 novel coronavirus diseases (COVID-19)—China, 2020. *China CDC Weekly, 2*(8), 113.

Tekes, G., & Thiel, H. J. (2016). Feline coronaviruses: Pathogenesis of feline infectious peritonitis. *Advances in Virus Research, 96*, 193–218.

Thakar, A., & Lal, D. (2021). "Black fungus": A perspective on the coronavirus disease 2019 (COVID-19)—associated rhino-orbital mucormycosis epidemic in India. In *International forum of allergy & rhinology* (Vol. 11, No. 8, p. 1278). Wiley-Blackwell.

The Hill. (2022). *Paxlovid may be causing REBOUND Covid? Reporter breaks down Pfizer's data*. Retrieved from https://thehill.com/hilltv/rising/3509292-rising-june-2-2022/

Therakathu, J., Prabhu, S., Irodi, A., Sudhakar, S. V., Yadav, V. K., & Rupa, V. (2018). Imaging features of rhinocerebral mucormycosis: A study of 43 patients. *The Egyptian Journal of Radiology and Nuclear Medicine, 49*(2), 447–452.

Thomas, S. (2020). The structure of the membrane protein of SARS-CoV-2 resembles the sugar transporter SemiSWEET. *Pathogens and Immunity, 5*(1), 342.

Times of India [ToI]. 2021. *Devastating 2nd wave, billion + vaccinations: 12 charts show how India fought Covid in 2021*. Retrieved from https://timesofindia.indiatimes.com/india/devastating-2nd-wave-billion-vaccinations-10-charts-show-how-india-fought-covid-in-2021/articleshow/88564472.cms

Times of India [ToI]. (n.d.). *Covid-19: Spread*. Retrieved from https://timesofindia.indiatimes.com/coronavirus/data/covid-19-spread?from=mdr

Tiwari, N. R., Phatak, S., Sharma, V. R., & Agarwal, S. K. (2021). COVID-19 and thrombotic microangiopathies. *Thrombosis Research, 202*, 191–198.

Tobías, A., Carnerero, C., Reche, C., Massagué, J., Via, M., Minguillón, M. C., & Querol, X. (2020). Changes in air quality during the lockdown in Barcelona (Spain) one month into the SARS-CoV-2 epidemic. *Science of the Total Environment, 726*, Article 138540.

Torjesen, I. (2021). Covid-19: Omicron may be more transmissible than other variants and partly resistant to existing vaccines, scientists fear. incom.

Toscano, G., Palmerini, F., Ravaglia, S., Ruiz, L., Invernizzi, P., Cuzzoni, M. G., & Micieli, G. (2020). Guillain-Barré syndrome associated with SARS-CoV-2. *New England Journal of Medicine, 382*(26), 2574–2576.

Tripathi, A., Tyagi, V. K., Vivekanand, V., Bose, P., & Suthar, S. (2020). Challenges, opportunities and progress in solid waste management during COVID-19 pandemic. *Case Studies in Chemical and Environmental Engineering, 2*, Article 100060.

Tsagkovits, A., Ioannidis, D., & Rokade, A. (2021). The microscope drape method to reduce aerosolisation during endoscopic sinus and skull base surgery in the COVID era: How I do it. *European Archives of Oto-Rhino-Laryngology, 278*(2), 573–576.

Tsoupras, A. (2022). The anti-inflammatory and antithrombotic properties of bioactives from orange, sanguine and clementine juices and from their remaining by-products. *Beverages, 8*(3), 39.

Umbert, I. J., & Su, W. P. (1989). Cutaneous mucormycosis. *Journal of the American Academy of Dermatology, 21*(6), 1232–1234.

United States Food and Drug Administration [USFDA]. (2021). *Why you should not use Ivermectin to treat or prevent COVID-19*. Retrieved from https://www.fda.gov/consumers/consumer-updates/why-you-should-not-use-ivermectin-treat-or-prevent-covid-19

United States Food and Drug Administration [USFDA]. (2022a). *Fact sheet for healthcare providers: Emergency use authorization for PAXLOVID™*. Retrieved from https://www.fda.gov/media/155050/download

United States Food and Drug Administration [USFDA]. (2022b). *Frequently asked questions on the emergency use authorization for Lagevrio (Molnupiravir) for treatment of COVID-19*. Retrieved from https://www.fda.gov/media/155056/download#:~:text=A%3A%20FDA%20has%20issued%20an,whom%20alternative%20COVID%2D%2019%20treatment

United States Food and Drug Administration [USFDA]. (2022c). *Recommendations for investigational COVID-19 convalescent plasma*. Retrieved from https://www.fda.gov/vaccines-blood-biologics/investigational-new-drug-applications-inds-cber-regulated-products/recommendations-investigational-covid-19-convalescent-plasma#:~:text=On%20August%2023%2C%202020%2C%20FDA,reissued%20this%20EUA%20with%20revisions

United States Food and Drug Administration [USFDA]. (2022d). *Fact sheet for healthcare providers: Emergency use authorization for bebtelovimab*. Retrieved from https://www.fda.gov/media/156152/download

ur Rehman, M. F., Fariha, C., Anwar, A., Shahzad, N., Ahmad, M., Mukhtar, S., & Haque, M. F. U. (2021). Novel coronavirus disease (COVID-19) pandemic: A recent mini review. *Computational and Structural Biotechnology Journal, 19*, 612–623.

Van Doremalen, N., Bushmaker, T., Morris, D. H., Holbrook, M. G., Gamble, A., Williamson, B. N., & Munster, V. J. (2020). Aerosol and surface stability of SARS-CoV-2 as compared with SARS-CoV-1. *New England Journal of Medicine, 382*(16), 1564–1567.

Van Fan, Y., Jiang, P., Hemzal, M., & Klemeš, J. J. (2021). An update of COVID-19 influence on waste management. *Science of the Total Environment, 754*, Article 142014.

Vanhorebeek, I., Latronico, N., & Van den Berghe, G. (2020). ICU-acquired weakness. *Intensive Care Medicine, 46*(4), 637–653.

Varghese, G. M., & John, R. (2020). COVID-19 in India: Moving from containment to mitigation. *The Indian Journal of Medical Research, 151*(2–3), 136.

Velavan, T. P., & Meyer, C. G. (2020). La epidemia de COVID-19. *Tropical Medicine & International Health, 25*(3), 278–280.

Viana, R., Moyo, S., Amoako, D. G., Tegally, H., Scheepers, C., Althaus, C. L., & de Oliveira, T. (2022). Rapid epidemic expansion of the SARS-CoV-2 Omicron variant in southern Africa. *Nature, 603*(7902), 679–686.

Viner, R. M., & Whittaker, E. (2020). Kawasaki-like disease: Emerging complication during the COVID-19 pandemic. *The Lancet, 395*(10239), 1741–1743.

Wallace, G. (2020). Airlines and TSA report 96% drop in air travel as pandemic continues. *CNN* (April 9, 2020) (online) https://edition.cnn.com/2020/04/09/politics/airline-passengers-decline/index.html

Wang, Y., Sun, J., Zhu, A., Zhao, J., & Zhao, J. (2018). Current understanding of middle east respiratory syndrome coronavirus infection in human and animal models. *Journal of Thoracic Disease, 10*(19), S2260.

Wang, C., Horby, P. W., Hayden, F. G., & Gao, G. F. (2020a). A novel coronavirus outbreak of global health concern. *The Lancet, 395*(10223), 470–473.

Wang, D., Hu, B., Hu, C., Zhu, F., Liu, X., Zhang, J., & Peng, Z. (2020b). Clinical characteristics of 138 hospitalized patients with 2019 novel coronavirus-infected pneumonia in Wuhan, China. *JAMA, 323(11)*, 1061–1069.

Wang, J., Peng, Y., Xu, H., Cui, Z., & Williams, R. O. (2020c). The COVID-19 vaccine race: Challenges and opportunities in vaccine formulation. *An Official Journal of the American Association of Pharmaceutical Scientists, 21*(6), 1–12.

Wang, N., Shang, J., Jiang, S., & Du, L. (2020d). Subunit vaccines against emerging pathogenic human coronaviruses. *Frontiers in Microbiology, 11*, 298.

Wang, X., Pan, Z., & Cheng, Z. (2020e). Association between 2019-nCoV transmission and N95 respirator use. *Journal of Hospital Infection, 105*(1), 104–105.

Watson, M., Gilmour, R., Menzies, R., Ferson, M., McIntyre, P., & New South Wales Pneumococcal Network. (2006). The association of respiratory viruses, temperature, and other climatic parameters with the incidence of invasive pneumococcal disease in Sydney, Australia. *Clinical Infectious Diseases, 42*(2), 211–215.

Wei, J. F., Huang, F. Y., Xiong, T. Y., Liu, Q., Chen, H., Wang, H., & Chen, M. (2020). Acute myocardial injury is common in patients with COVID-19 and impairs their prognosis. *Heart, 106*(15), 1154–1159.

Wessels, I., Rolles, B., & Rink, L. (2020). The potential impact of zinc supplementation on COVID-19 pathogenesis. *Frontiers in Immunology*, 1712.

Wilder-Smith, A., & Freedman, D. O. (2020). Isolation, quarantine, social distancing and community containment: Pivotal role for old-style public health measures in the novel coronavirus (2019-nCoV) outbreak. *Journal of Travel Medicine., 27*(2), 1–4.

Wold, W., & Toth, K. (2013). Adenovirus vectors for gene therapy, vaccination and cancer gene therapy. *Current Gene Therapy, 13*(6), 421–433.

Wolfel, R., Corman, V. M., Guggemos, W., Seilmaier, M., Zange, S., Muller, M. A., & Ehmann, R. (2020). 292 Virological assessment of hospitalized patients with COVID-2019. *Nature, 293*, 581.

World Health Organization [WHO]. (2020k). India situation report. Retrieved from https://www.who.int/india/emergencies/india-situation-report

World Health Organization [WHO]. (2020i). *Update on COVID-19 vaccines & immune response*. Retrieved from https://www.who.int/docs/default-source/coronaviruse/risk-comms-updates/update52_vaccines.pdf?sfvrsn=b11be994_4

World Health Organization [WHO]. (2020h). *Coronavirus update 52*. Retrieved from http://en.sbmu.ac.ir/uploads/210312_-_EPI_WIN_Update52_COVID-19_vaccines.pdf

World Health Organization [WHO]. (2020m). *WHO situation report*. Retrieved from https://covid19.who.int/region/searo/country/in

World Health Organization [WHO]. (2020l). *India-situation report_1113*. Retrieved from https://www.thehinducentre.com/resources/article38485334.ece/binary/India-Situation-Report_1-113.pdf

World Health Organization [WHO]. (2020a). COVID-19—China. Retrieved from https://www.who.int/emergencies/disease-outbreak-news/item/2020-DON229

World Health Organization [WHO]. (2020b). *Modes of transmission of virus causing COVID-19: Implications for IPC precaution recommendations*. Retrieved from https://www.who.int/news-room/commentaries/detail/modes-of-transmission-of-virus-causing-covid-19-implications-for-ipc-precaution-recommendations

World Health Organization [WHO]. (2020c). *Clinical management of severe acute respiratory infection (SARI) when COVID-19 disease is suspected*. Retrieved from https://www.who.int/docs/default-source/coronaviruse/clinical-management-of-novel-cov.pdf

World Health Organization [WHO]. (2020d). Naming the coronavirus disease (COVID-19) and the virus that causes it. *Brazilian Journal of Implantology and Health Sciences, 2*(3).

World Health Organization [WHO]. (2020e). Novel Coronavirus (2019-nCoV). Situation Report–22, Data as reported by 11 February 2020.

World Health Organization [WHO]. (2020f). *DRAFT landscape of COVID-19 candidate vaccines*, 4 March 2020.

World Health Organization [WHO]. (2020g). *Clinical management of severe acute respiratory infection (SARI) when COVID-19 disease is suspected: Interim guidance, 13 March 2020 (No. WHO/2019-nCoV/clinical/2020.4)*. World Health Organization.

World Health Organization [WHO]. (2020j). *Water, sanitation, hygiene, and waste management for the COVID-19 virus: Interim guidance, 23 April 2020 (No. WHO/2019-nCoV/IPC_WASH/2020.3)*. World Health Organization.

World Health Organization [WHO]. (2021c). *WHO-convened global study of origins of SARS-CoV-2: China Part*.

World Health Organization [WHO]. (2021e). *Coronavirus disease (COVID-19): Dexamethasone*. Retrieved from https://www.who.int/news-room/questions-and-answers/item/coronavirus-disease-covid-19-dexamethasone

World Health Organization [WHO]. (2021a). *Genomic sequencing of SARS-CoV-2: A guide to implementation for maximum impact on public health*. Retrieved 8 Jan 2021.

World Health Organization [WHO]. (2021b). *WHO calls for further studies, data on origin of SARS-CoV-2 virus, reiterates that all hypotheses remain open*. Retrieved 30 March 2021.

World Health Organization [WHO]. (2021d). *Tracking SARS-CoV-2 variants*. Retrieved from https://www.who.int/activities/tracking-SARS-CoV-2-variants

World Health Organization [WHO]. (2022d). *Coronavirus update 73 COVID-19 vaccines & immune response*. Retrieved from https://www.who.int/docs/default-source/coronaviruse/risk-comms-updates/update73_covid-19-vaccines-and-immune-response.pdf?sfvrsn=7902cc35_5

World Health Organization [WHO]. (2022a). *WHO coronavirus (COVID-19) dashboard*. Retrieved from https://covid19.who.int/?adgroupsurvey={adgroupsurvey}&gclid=CjwKCAjwhNWZBhB_EiwAPzlhNph8a3cTqCCl_hFWEwcLVGgqYKUJ_cXyBJ7igwYjSQ2l8fdK7D-6rhoCBSsQAvD_BwE

World Health Organization [WHO]. (2022b). *Interim statement on COVID-19 vaccination for children*. Retrieved from https://www.who.int/news/item/11-08-2022-interim-statement-on-covid-19-vaccination-for-children

Wu, C., Chen, X., Cai, Y., Zhou, X., Xu, S., Huang, H., & Song, Y. (2020). Risk factors associated with acute respiratory distress syndrome and death in patients with coronavirus disease 2019 pneumonia in Wuhan, China. *JAMA Internal Medicine, 180*(7), 934–943.

Wu, L., Zhou, L., Mo, M., Li, Y., Han, J., Li, J., & Xu, Z. (2021). The effect of the multiple mutations in Omicron RBD on its binding to human ACE2 receptor and immune evasion: An investigation of molecular dynamics simulations. incom.

Wu, C. R., Yin, W. C., Jiang, Y., & Xu, H. E. (2022). Structure genomics of SARS-CoV-2 and its Omicron variant: Drug design templates for COVID-19. *Acta Pharmacologica Sinica,* 1–13.

Xiao, X., Newman, C., Buesching, C. D., Macdonald, D. W., & Zhou, Z. M. (2021). Animal sales from Wuhan wet markets immediately prior to the COVID-19 pandemic. *Scientific Reports, 11*(1), 1–7.

Xu, R. H., He, J. F., Evans, M. R., Peng, G. W., Field, H. E., Yu, D. W., & Schnur, A. (2004). Epidemiologic clues to SARS origin in China. *Emerging Infectious Diseases, 10*(6), 1030.

Xu, Y., Yuen, P. W., & Lam, J. K. W. (2014). Intranasal DNA vaccine for protection against respiratory infectious diseases: The delivery perspectives. *Pharmaceutics, 6*(3), 378–415.

Xu, Y., Li, X., Zhu, B., Liang, H., Fang, C., Gong, Y., & Gong, S. (2020). Characteristics of pediatric SARS-CoV-2 infection and potential evidence for persistent fecal viral shedding. *Nature Medicine, 26*(4), 502–505.

Yang, Z. Y., Wyatt, L. S., Kong, W. P., Moodie, Z., Moss, B., & Nabel, G. J. (2003). Overcoming immunity to a viral vaccine by DNA priming before vector boosting. *Journal of Virology, 77*(1), 799–803.

Yang, L., Dai, J., Zhao, J., Wang, Y., Deng, P., & Wang, J. (2020a). Estimation of incubation period and serial interval of COVID-19: Analysis of 178 cases and 131 transmission chains in Hubei province, China. *Epidemiology & Infection, 148*.

Yang, X., Yu, Y., Xu, J., Shu, H., Liu, H., Wu, Y., & Shang, Y. (2020b). Clinical course and outcomes of critically ill patients with SARS-CoV-2 pneumonia in Wuhan, China: A single-centered, retrospective, observational study. *The Lancet Respiratory Medicine, 8*(5), 475–481.

Yang, Y., Peng, F., Wang, R., Guan, K., Jiang, T., Xu, G., & Chang, C. (2020c). The deadly coronaviruses: The 2003 SARS pandemic and the 2020 novel coronavirus epidemic in China. *Journal of Autoimmunity, 109*, Article 102434.

Ye, Z. W., Yuan, S., Yuen, K. S., Fung, S. Y., Chan, C. P., & Jin, D. Y. (2020). Zoonotic origins of human coronaviruses. *International Journal of Biological Sciences, 16*(10), 1686.

Zaim, S., Chong, J. H., Sankaranarayanan, V., & Harky, A. (2020). COVID-19 and multiorgan response. *Current Problems in Cardiology, 45*(8), Article 100618.

Zambrano-Monserrate, M. A., Ruano, M. A., & Sanchez-Alcalde, L. (2020). Indirect effects of COVID-19 on the environment. *Science of the Total Environment, 728*, Article 138813.

Zegarelli, D. J. (1993). Fungal infections of the oral cavity. *Otolaryngologic Clinics of North America, 26*(6), 1069–1089.

Zepeda-Cervantes, J., Ramírez-Jarquín, J. O., & Vaca, L. (2020). Interaction between virus-like particles (VLPs) and pattern recognition receptors (PRRs) from dendritic cells (DCs): Toward better engineering of VLPs. *Frontiers in Immunology, 11*, 1100.

Zhang, J., Xie, B., & Hashimoto, K. (2020). Current status of potential therapeutic candidates for the COVID-19 crisis. *Brain, Behavior, and Immunity, 87*, 59–73.

Zhang, S., Wang, C., Shi, L., & Xue, Q. (2021). Beware of steroid-induced avascular necrosis of the femoral head in the treatment of COVID-19—Experience and lessons from the SARS epidemic. *Drug Design, Development and Therapy, 15*, 983.

Zhao, Z., Li, H., Wu, X., Zhong, Y., Zhang, K., Zhang, Y. P., & Fu, Y. X. (2004). Moderate mutation rate in the SARS coronavirus genome and its implications. *BMC Evolutionary Biology, 4*(1), 1–9.

Zhao, C. L., Rapkiewicz, A., Maghsoodi-Deerwester, M., Gupta, M., Cao, W., Palaia, T., & Hanna, I. (2021). Pathological findings in the postmortem liver of patients with coronavirus disease 2019 (COVID-19). *Human Pathology, 109*, 59–68.

Zhou, F., Yu, T., Du, R., Fan, G., Liu, Y., Liu, Z., & Cao, B. (2020a). Clinical course and risk factors for mortality of adult inpatients with COVID-19 in Wuhan, China: A retrospective cohort study. *The Lancet, 395*(10229), 1054–1062.

Zhou, P., Yang, X. L., Wang, X. G., Hu, B., Zhang, L., Zhang, W., & Shi, Z. L. (2020b). A pneumonia outbreak associated with a new coronavirus of probable bat origin. *Nature, 579*(7798), 270–273.

Zhu, F. C., Li, Y. H., Guan, X. H., Hou, L. H., Wang, W. J., Li, J. X., & Chen, W. (2020). Safety, tolerability, and immunogenicity of a recombinant adenovirus type-5 vectored COVID-19 vaccine: A dose-escalation, open-label, non-randomised, first-in-human trial. *The Lancet, 395*(10240), 1845–1854.

Zirpe, K. G., Dixit, S., Kulkarni, A. P., Pandit, R. A., Ranganathan, P., Prasad, S., & Joshi, A. (2021). The Second- vs First-wave COVID-19: More of the same or a lot worse? A comparison of mortality between the two waves in patients admitted to intensive care units in nine hospitals in western Maharashtra. *Indian Journal of Critical Care Medicine: Peer-reviewed, 25*(12), 1343.

Zumla, A., Hui, D. S., Azhar, E. I., Memish, Z. A., & Maeurer, M. (2020). Reducing mortality from 2019-nCoV: Host-directed therapies should be an option. *The Lancet, 395*(10224), e35–e36.

Chapter 8
COVID-19 as an Occupational Disease

1 Introduction

World Health Organization (WHO) China office was informed on December 31, 2019, regarding pneumonia-like case having unspecified aetiology from seafood market of Wuhan city (Chew et al., 2020). Further, WHO issued a second statement regarding the series of incidents on January 9, 2020. It stated Chinese experts have validated a preliminary characterization of a new coronavirus (CoVs), isolated from the hospitalised patients experiencing pneumonia-like symptoms. On February 11, 2020, the WHO updated the naming of the virus from 2019-nCoV to SARS-CoV-2 (WHO, 2020a). Animals, birds and human ribonucleic acid (RNA) genome may encapsulate CoVs (Koh, 2020). Humans coronaviruses appear in seven different forms. The 229E, NL63, OC43 and KHU1 forms of CoVs are often linked to minor-to-moderate respiratory disorders like flu or common cold. MERS-CoV and SARS-CoV, to the contrary, are two other forms of CoVs that can pose serious respiratory symptoms which might result in death (Li et al., 2020). SARS-CoV-2, is the seventh and latest type of CoV circulating in nature (Andersen et al., 2020).

1.1 COVID-19: Exposure in the Workforce

Humanity has witnessed several disastrous epidemics and pandemics pertaining to respiratory disorders in the past. The severely affected group includes daily-wage labourers, migrants and other working classes who are generally overlooked in such situations resulting in occupational morbidity and mortality (Madhav et al., 2016; Occupational Safety and Health Administration [OSHA] 2021). Throughout the crisis, occupational health risks have historically been seriously underrated or ignored. Examples include the current COVID-19 pandemic as well as the previous influenza, SARS and MERS (Chou et al., 2020). During these public health crisis,

M. H. Fulekar et al., *Pandemic Over the Centuries*,
https://doi.org/10.1007/978-3-031-94270-9_8

a range of new and chronic respiratory disorders were observed, including lung function deficits in SARS and MERS patients and persistent reticular radiographic anomalies (You et al., 2020). The present pandemic may also pose long-term consequences to workers' wellbeing, adding to the daily toll of sickness and fatalities (Semple & Cherrie, 2020). Badges of valour worn by first responders during such disasters may conceal injuries, if not outright fatalities, some of which may have been avoidable and could also be avoided during emergencies in the long term (Pandey & Sharma, 2020).

It has become evident how severely the pandemic has affected the health sector (particularly low income workers) (WHO, 2021a). About 3 million Americans who are employed in the health care sector offer patients direct support as nurses or household services (Spetz et al., 2019). Huge number of individuals operate in housekeeping, laundering, or catering services, all of which have a significant risk of infection, offering essential indirect care delivery (WHO, 2020b). The majority of those at risk for contacting SARS-CoV-2 are the health care workers (HCW) (Himmelstein & Woolhandler, 2020). Besides, financial crisis has become an irrefutable reality, even among the biggest economies (Borio, 2020). Owing to such uneven consequences of the pandemic on occupational and socioeconomically deprived groups have started to draw interest, notably in editorials in various medical publications and also in media (Choo, 2020; Van Dorn et al., 2020). People who began to work just before the pandemic in low-wage positions suffered additional revenue losses as a result of reduced working time, even if they retained employment (Antipova, 2021). A whopping 80% of low-income workers receive hourly pay, and 43% of them work for small businesses with less than 25 employees, whose chances of surviving the current financial crisis are most certainly thin (Garfield et al., 2020). Many employees struggled to make ends meet, making it harder for them to pay for their own healthcare (Blackburn & Ventura, 2021).

These results highlight the importance of focusing more attention to lowering the risk for these professionals during the pandemic. (Quigley et al., 2022). In low- and middle-income nations, combating COVID-19 in workplaces has been challenging (LMICs), because of scarce resources. Strengthening overall organisational safety protocols, ensuring access to Employees' Compensation allowances, sick leaves, and providing suitable health coverage, especially at a reduced cost or free are therefore important in protecting the employees.

1.2 Occupational Health and Hygiene

There are numerous distinct chemical, physical, biological and psychological health concerns in the profession. The field of occupational hygiene includes identification and evaluation of workplace threats (Antonini & Anderson, 2015). Worker protection against occupational diseases unfortunately, cannot be achieved solely through occupational hygiene. For timely identification of occupational disorders, the involvement of health personnel in the manner of substitution and physical examination is crucial

(Piňosová et al., 2021). To deploy employees in positions that are appropriate for their skills and limits, management conducts a replacement health assessment of each employee (Dubois & Singh, 2009). Additionally, it is important to conduct routine medical check-up on employees to determine their state of health (Hakro & Jinshan, 2019). The body organs and systems that are most likely to be impacted by toxic materials at work should be the main focus of the evaluation (Tulchinsky & Varavikova, 2014).

1.3 Occupational Diseases

Factories (Amendment) Act of 1987, Sections 89 and 90 of the third Schedule define "Notifiable Diseases". The diseases that must be reported are listed as follows:

1. Lead poisoning, including poisoning by any preparation of compounds of lead as their sequelae
2. Lead terra-ethyl poisoning
3. Phosphorus poisoning or its sequelae
4. Mercury poisoning or its sequelae
5. Manganese poisoning or its sequelae
6. Arsenic poisoning or its sequelae
7. Poisoning by nitrous fumes
8. Carbon disulfide poisoning
9. Benzene poisoning, including poisoning by any of its ho amino derivatives or its sequelae
10. Chrome ulceration or its sequelae
11. Anthrax
12. Silicosis
13. Poisoning by halogens or halogen derivatives of the hydrocarbons of the aliphatic series.
14. Pathological manifestations due to: (a) Radium and (b) X-rays
15. Primary epitheliomatous cancer of the skin
16. Toxic anemia
17. Toxic jaundice due to poisonous substances
18. Air acne or dermatitis due to mineral oils and compounds containing a mineral oil base
19. Byssinosis
20. Asbestosis
21. Occupational or contact dermatitis caused by direct contact with chemicals and paints are of two types, i.e., primarily irritants and allergic sensitizers.
22. Noise-induced hearing loss (exposure to high noise levels)
23. Beryllium poisoning
24. Carbon monoxide poisoning
25. Miners pneumoconiosis

26. Phosgene poisoning
27. Occupational cancer
28. Isocyanates poisoning
29. Toxic nephritis

1.4 Coronavirus (COVID-19): Outbreak Risk to Other Diseases and Challenges

Coronavirus (COVID-19) outbreak is more prone to the patients suffering from hypertension, diabetes, cancer, tuberculosis and respiratory related diseases.

(a) COVID-19 infection in cancer patient
(b) COV1D-19 cardiovascular diseases
(c) COVID-19 infection in diabetic patient
(d) COVID-19 infection in respiratory related diseases

1.5 Modes of Entry

A virus could infect an individual at workplace if it comes into direct contact with a susceptible area on the body or inside. The three main modes of entry are inhalation, direct contact and ingestion (Phillips et al., 2015).

1.5.1 Inhalation

Inhalation is possibly the most prevalent route for viruses to enter the human body. A variety of chemicals and foreign particles are inhaled and deposited in the upper and lower respiratory tracts (Stetzenbach, 2009). Whenever a sensitive person comes near to an infected individual, or within 1–2 m, the probability of infection by aerosol inhalation is higher (Li, 2021). The mucous layer or the alveoli become encased in virus-filled particles of various sizes. Aerosols add another layer of complexity to the situation. The risk of infection by inhaled microorganisms depends on the size of the particle (Thomas, 2013). Fine virus-laden particles with a diameter of less than 2.5 microns can enter the lungs profoundly, while particles with a diameter between 10.0 and 2.5 microns tend to accumulate in the larger tracheal and bronchial tubes and throughout the upper respiratory system.

1.5.2 Direct Contact

The skin protects against most of the particles present in environment. Skin-to-skin contact occurs when a pathogen is transferred from a reserve to a susceptible host through direct contact, also known as droplet transmission (Centers for Disease Control and Prevention [CDC], 2012). Direct contact is achieved under following circumstances:

(a) Abrasion and roughness promote direct contact
(b) Penetration directly via lesions or contusions
(c) The skin is degreased, after which the degreasing agent is assimilated
(d) Degreasing the skin, allows additional particles to be absorbed, and local and general sensitivity.

1.5.3 Ingestion

Despite the fact that ingestion is a rare entry method, it could be harmful if occupational hygiene standards are poor. Pathogens are consumed alongside food in the workplace when housekeeping is subpar, staff members are careless regarding sanitation, and they do not clean their hands thoroughly with soap after visiting lavatory. The dangerous particle, however, is eliminated through the intestinal tract when it is not soluble in body fluid after being consumed with food or saliva. Virions could either be consumed and directly impact the intestines and stomach or they may infect oropharyngeal cells first and replicate which later travels to the intestinal tract. Given its hard, layered epithelial tissue and quick transport of ingested particles above its surface, the oesophagus seldom becomes infected. Mucus, which may include particular secretory antibodies (IgA), protects the digestive tract, but the directed peristaltic movement of gut contents creates several possibilities for virus particles to come in touch with vulnerable epithelial cells (Burrell et al., 2017).

2 Occupational Safety and Health Administration Classification Related to Risk of Worker Exposure to SARS-CoV-2

Occupational SARS-CoV-2 exposure threat for people during the pandemic may vary based on local infection situations wherein individuals and their workplaces are located. Exposure risk largely relies on the physical environment of an organisation, including nature of task performed, overall health of the employee, the ability of staff to use facial masks, and necessity of direct contact (within 6 ft.) with other individuals, like those confirmed to have or presumed to have COVID-19, and those who could be affected or could spread SARS-CoV-2 without their awareness (Fig. 1).

The possible occupational risk pyramid, developed by OSHA, divides job responsibilities into four groups, viz. very high, high, medium and lower risk.

Group 1 is a very high-risk category. Professionals who have an extremely high risk of coming into contact with SARS-CoV-2 during particular medical, forensic, or testing operations. Workers who fall within this category include: people in the

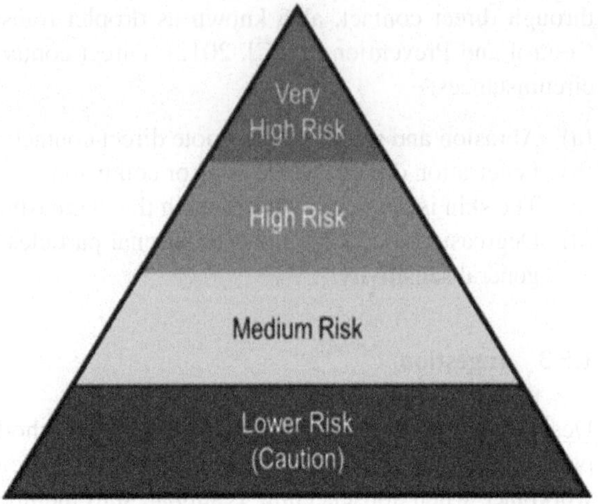

Fig. 1 Risk exposure pyramid. (*Source* OSHA, n.d.)

healthcare sector who execute respiratory induction techniques, bronchoscopies, intubation, certain dental treatments and evaluations, or intrusive sample collection upon patients who are suffering from confirmed or suspected COVID-19. Medical or laboratory personnel who collect or handle specimens from suspected or confirmed COVID-19 cases.

Group 2 is high-risk category wherein personnel coming into contact with confirmed or suspected SARS-CoV-2 infections. The following workers fall into this category: Hospital staff who frequently visit patients' rooms and administer services to vulnerable to suspected or confirmed COVID-19 patients. Ambulance drivers and accompanying staff who transfer patients with COVID-19. Personnel's in charge of handling the corpses who at the time had either confirmed or were suspected of COVID-19. People working under challenging environmental weather conditions as well as in different commercial, agricultural, and architectural sectors. People who work in retail stores, supermarkets or groceries, pharmacy, transportation, police departments and emergency service agencies, diners, and clubs routinely contact with general public.

Group 3 is the medium-risk category. Professionals that necessitate prolonged close contact with persons in communities with transmission. Workers who fall within this category include: Individuals who operate in a variety of industrial, manufacturing, agricultural, construction, or other essential infrastructure that need regular or ongoing contact with coworkers, particularly in confined spaces outdoors or in well-ventilated areas. Workers in major retailers, grocers or supermarkets, pharmacies, transportation, police departments and emergency response agencies, restaurants and bars are among those who frequently interact with the general public outside or in well-ventilated settings. People residing in similar shared residential buildings or transitory work camps.

Group 4 is the low-risk category. Professionals who do not require working closely with others. The public and other employees are rarely in contact with these workers during their work, e.g. remote workers (work-from-home), office workers who do not regularly interact closely with clients, coworkers, or the general public.

3 COVID-19: Hierarchy of Hazard Controls

SARS-COV-2 has been designated as the new occupational disease of the decade and is projected to be recognised globally ever since pandemic's beginning stages (Koh, 2020; Moen, 2020). The direct connection between occupational diseases and working life as well as their potential consequences contribute to their prominence. Those who must do essential tasks are probably more at risk of contracting SARS-CoV2 (Mutambudzi et al., 2021). The risk of contacting SARS-CoV-2 and other co-infections, as well as workplace violence, harassment, prejudice and bigotry, are all workplace worries. They also include a severe workload and extended PPE use (WHO, 2021b). According to some estimates, even though HCW make up less than 2% of the population in LMIC WHO, they account for 14–35% of COVID-19 infections (2020b). Workers are entitled to compensation due to occupational disorders under the law, and insurance companies recognised this right. Besides, a lot of developing nations lack surveillance mechanisms and frequently fail to report cases. It is perhaps plausible to strengthen the response by recognising that SARS-CoV-2 is an occupational hazard by consistently following the "hierarchy of controls" developed by OSHA experts in high-risk workplaces (Fig. 2).

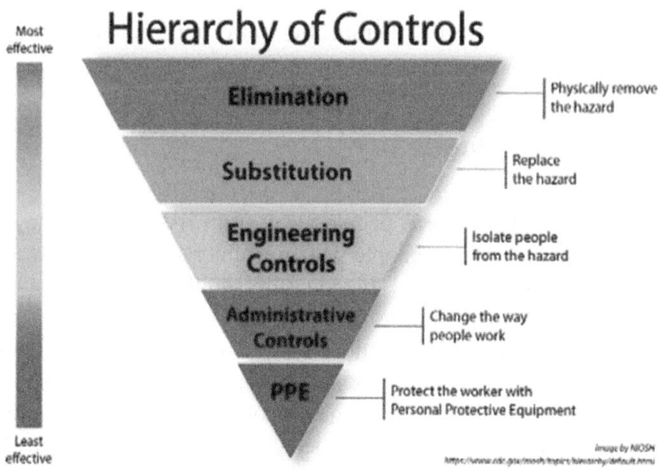

Fig. 2 Hazard hierarchy control. (*Source* NIOSH, 2022)

The hierarchy of controls is a model used in OSHA to assess the relative efficacy of various risk reduction techniques and to guide decision-making on how to put workable and efficient solutions into practice. The philosophy is represented as an upside-down pyramid, featuring 5 categories outlined in decreasing order of effectiveness. These include personal protective equipment (PPE), engineering controls, substitution, administrative controls, and controls (National Institute for Occupational Safety and Health [NIOSH], 2022). Given its broad applicability in assisting healthcare practitioners, administrators, experts and the general public effectively comprehend the relative efficacy of these approaches to prevent the dissemination of an airborne virus like SARS-CoV-2, the theory is increasingly being employed to truly understand COVID-19 risk reduction (NIOSH, 2020).

3.1 Elimination and Substitution

The best hazard mitigation techniques are elimination and substitution since they deal with both the hazard and any associated problems as well as alternatives that are less harmful. Earlier during the pandemic, stay-at-home directives were used in an effort to reduce the danger of infection within regions. However, in general, "lock-downs" are only practical for short periods of time whereas other safety measures are taken owing to the devastating social, psychological and economic effects. Isolation and quarantine are other techniques for lowering the risk of infection from infected or reported cases. Keeping away from unnecessary social assemblies is another effective strategy to lower the risk of SARS-CoV-2 exposure. Parallel to this, allowing remote work or shifting indoor operations outdoors substitutes less hazardous possibilities for the threats connected with in-person activities. COVID-19 vaccinations have proven to be an extremely effective therapeutic control approach wherever applicable.

3.2 Engineering Controls

Engineering controls merely keep people away from risks rather than eliminating them. Physical barriers between people who must interact closely are widely recognised type of engineering control. This measure should be applied in conjunction with certain other safeguards because they do not completely remove risks. Engineering controls that might further reduce the possibility of COVID-19 spread are still not widely used, such as enhancing ventilation and filtration, using upper-room disinfectant UV light, and increasing the quality of ambient air (Morawska & Milton, 2020; Morawska et al., 2021). Well-designed control activities can greatly improve safety of people who follow administrative controls or use PPE while also minimising danger regardless of their personal behaviour (Alle & Ibrahim, 2021). Contrary to business, it is not always feasible to alter interiors, ventilators, as well as other facilities to significantly reduce the likelihood of disease transmission. This is a major flaw in engineering controls as a method of community protection. Even yet, engineering measures are effective in preventing distant airborne transmission of contagious viruses like SARS-CoV-2 since they can lower indoor respiratory aerosol load.

3.3 *Administrative Controls*

Administrative controls are often used to promote social distance, decrease one-on-one contact, and minimise population size in designated areas in order to prevent COVID-19 infestation. These controls entail altering individual behaviour through policy or mandate in order to minimise hazard. Administrative measures to reduce danger associated with density include limiting shampered by political and social hostility. This highlights the major disadvantage of using administrative controls alone to stop the spread of disease. Administrative controls only work through agreement, therefore, even the most well-intentioned personnel might fall victim to adherence loopholes. Furthermore, strategies that rely on absolute compliance are futile and bound to failure, as examples from patient safety and risk management demonstrate.

3.4 *Personal Protective Equipment*

Personal protective equipment (PPE), including respirators, safety glasses, and many other personally used preventive items, is the final method for reducing exposure. Whilst PPE will always be lesser successful than measures at the top of the pyramid, its use has been an important tool to stop the spread of COVID-19. PPE's efficacy as a control has significant limitations because it depends on both an ample supply and appropriate, regular usage. Use of PPE is essential when threats are actually hard to avoid. Regulations also mandate appropriate controls in workplaces where PPE is employed to control hazards. Non-medical masks and cloth face coverings are not PPE, and using one does not make one immune to SARS-CoV-2 infection risk. Laymen are rarely trained in the proper use of PPE, and therefore are usually unprepared during the pandemic (Gandhi & Marr, 2021).

4 Occupational Safety and Health Administration Guidelines

4.1 Employers

Employers must strictly implement COVID-19 control measures at the earliest in order to mitigate the adversity caused by the pandemic on businesses, staff, consumers and the general public. Adapting strategies for businesses that have already made preparations for influenza pandemics in order to account for the specific workers exposed, origins of exposure, channels of dissemination, and other distinguishing elements of SARS-CoV-2 may be necessary when preparing for COVID-19. Employers that have not done should do it immediately in order to prepare themselves and their workforce for potentially dire pandemic conditions.

- Recognise complete responsibility for ensuring that all essential preventative and safety precautions are implemented to reduce workplace safety and health risks; refresher training on infection control and prevention (IPC) must be conducted.
- Adequate training on application, putting on, taking off, and disposal of personal protective equipment (PPE); information, guidance, and training on workplace safety and health must be extended to the employee.
- Provide healthcare or other employees caring for suspected or confirmed cases with appropriate IPC and PPE supplies.
- Familiarise staff with technical updates on COVID-19 and provide suitable methods to monitor, triage, test, and treat patients as well as share infection prevention and control information with patients and the public; as necessary, start providing with adequate security measures for personal safety; provide a blame-free environment for workers to report hazards.
- Respect the rights to compensation, rehabilitation, and curative services if COVID-19 infection results from occupational exposure. This would be regarded as an occupational exposure, and any illness that resulted would be regarded as an occupational disease. It would also make resources for mental health and counselling available, enable collaboration between management and employees or their representatives, and provide access to those resources.

4.2 Health Workers

- Use the supported procedures to evaluate, triage and treat patients with dignity and respect, kindness, and decency; follow proper occupational safety and health practises; minimize exposure oto health hazards; and engage in employer-provided occupational safety and health education.
- Keep patient information private.
- Promptly report suspected and confirmed cases in accordance with recognised public health reporting guidelines.

- Supply reliable information about infection control practices, and public health, even to those who are worried but do not show any symptoms or threat.
- Correctly put-on, use, remove, and discard personal protective equipment.
- Self-monitor for sickness symptoms and, if one arises, isolate oneself or notify supervisors.
- Alert management if they exhibit symptoms of excessive stress or mental health issues that call for support actions.
- Any circumstance that they have good reason to believe poses an immediate and serious risk to their life or health should be reported to the superiors.

5 Conclusion

The exposure of COVID-19 and its spread to people who are already ill with occupational distress, need to be addressed urgently with proper consideration given to their medical histories. In light of the newly prevalent occupational disorders among industrial workers and the exposure to COVID-19, appropriate preventative measures and treatment should be performed. The WHO has declared 2021 to be the International Year of Health and Care Workers in honour of their unrelenting commitment to the battle against COVID-19 pandemic (WHO, n.d.). Personal hygiene, PPE usage, social distancing are common recommendations for limiting the spread of COVID-19. Further, occupational and industrial data must be integrated into the public health monitoring and healthcare information system. Also, to strengthen the occupational health and safety of employees, WHO, ILO and other international organisations should emphasise on managing the challenges through regulations, guidelines and standards.

References

Allen, J. G., & Ibrahim, A. M. (2021). Indoor air changes and potential implications for SARS-CoV-2 transmission. *JAMA, 325*(20), 2112–2113.

Andersen, K. G., Rambaut, A., Lipkin, W. I., Holmes, E. C., & Garry, R. F. (2020). The proximal origin of SARS-CoV-2. *Nature Medicine, 26*(4), 450–452.

Antipova, A. (2021). Analysis of the COVID-19 impacts on employment and unemployment across the multi-dimensional social disadvantaged areas. *Social Sciences & Humanities Open, 4*(1), Article 100224. https://doi.org/10.1016/j.ssaho.2021.100224

Antonini, J. M., & Anderson, S. E. (2015). Occupational health and industrial hygiene. *Environmental Health Insights, 8*(Suppl 1), 97–98. https://doi.org/10.4137/EHI.S24583

Blackburn, R., & Ventura, M. (2021). Earning less than before COVID-19, the self-employed struggle to make ends meet. *LSE Business Review*.

Borio, C. (2020). The Covid-19 economic crisis: Dangerously unique. *Business Economics, 55*(4), 181–190. https://doi.org/10.1057/s11369-020-00184-2

Burrell, C. J., Howard, C. R., & Murphy, F. A. (2017). Pathogenesis of virus infections. *Fenner and White's Medical Virology*, 77–104. https://doi.org/10.1016/B978-0-12-375156-0.00007-2

Centers for Disease Control and Prevention [CDC]. (2012). *Lesson 1: Introduction to epidemiology*. Retrieved from https://www.cdc.gov/csels/dsepd/ss1978/lesson1/section10.html

Chew, N. W., Lee, G. K., Tan, B. Y., Jing, M., Goh, Y., Ngiam, N. J., & Sharma, V. K. (2020). A multinational, multicentre study on the psychological outcomes and associated physical symptoms amongst healthcare workers during COVID-19 outbreak. *Brain, Behavior, and Immunity, 88*, 559–565.

Choo, E. K. (2020). COVID-19 fault lines. *The Lancet, 395*(10233), 1333.

Chou, R., Dana, T., Buckley, D. I., Selph, S., Fu, R., & Totten, A. M. (2020). Epidemiology of and risk factors for coronavirus infection in health care workers: A living rapid review. *Annals of Internal Medicine, 173*(2), 120–136.

Dubois, C. A., & Singh, D. (2009). From staff-mix to skill-mix and beyond: Towards a systemic approach to health workforce management. *Human Resources for Health, 7*(1), 1–19.

Gandhi, M., & Marr, L. C. (2021). Uniting infectious disease and physical science principles on the importance of face masks for COVID-19. *Med, 2*(1), 29–32.

Garfield, R., Rae, M., Claxton, G., & Orgera, K. (2020). Double jeopardy: Low wage workers at risk for health and financial implications of COVID-19. *Kaiser Family Foundation, 29*.

Hakro, S., & Jinshan, L. (2019). Workplace employees' annual physical checkup and during hire on the job to increase health-care awareness perception to prevent disease risk: A work for policy-implementable option globally. *Safety and Health at Work, 10*(2), 132–140. https://doi.org/10.1016/j.shaw.2018.08.005

Himmelstein, D. U., & Woolhandler, S. (2020). Health insurance status and risk factors for poor outcomes with COVID-19 among US health care workers: A cross-sectional study. *Annals of Internal Medicine, 173*(5), 410–412.

Koh, D. (2020). Occupational risks for COVID-19 infection. *Occupational Medicine, 70*(1), 3–5.

Li, Y. (2021). Basic routes of transmission of respiratory pathogens—A new proposal for transmission categorization based on respiratory spray, inhalation, and touch. *Indoor Air, 31*(1), 3.

Li, Q., Guan, X., Wu, P., Wang, X., Zhou, L., Tong, Y., & Feng, Z. (2020). Early transmission dynamics in Wuhan, China, of novel coronavirus-infected pneumonia. *New England Journal of Medicine., 382*(13), 1199–1207.

Madhav, N., Oppenheim, B., Gallivan, M., Mulembakani, P., Rubin, E., & Wolfe, N. (2016). Chapter 17. Pandemics: Risks, impacts, and mitigation. *Disease Control Priorities, 9*.

Moen, B. E. (2020). COVID-19 should be recognized as an occupational disease worldwide. *Occupational Medicine,* 299.

Morawska, L., & Milton, D. K. (2020). It is time to address airborne transmission of coronavirus disease 2019 (COVID-19). *Clinical Infectious Diseases, 71*(9), 2311–2313.

Morawska, L., Allen, J., Bahnfleth, W., Bluyssen, P. M., Boerstra, A., Buonanno, G., & Yao, M. (2021). A paradigm shift to combat indoor respiratory infection. *Science, 372*(6543), 689–691.

Mutambudzi, M., Niedzwiedz, C., Macdonald, E. B., Leyland, A., Mair, F., Anderson, J., & Demou, E. (2021). Occupation and risk of severe COVID-19: Prospective cohort study of 120 075 UK Biobank participants. *Occupational and Environmental Medicine, 78*(5), 307–314.

National Institute for Occupational Safety and Health [NIOSH]. (2022). *Hierarchy of controls*. Retrieved from https://www.cdc.gov/niosh/topics/hierarchy/default.html

National Institute for Occupational Safety and Health [NIOSH]. (2020). *Guidance on preparing workplaces for COVID-19 (OSHA 3990-02 2020)*. Retrieved from https://www.osha.gov/sites/default/files/publications/OSHA3990.pdf

Occupational Safety and Health Administration [OSHA]. (2021). Occupational exposure to COVID-19; Emergency Temporary Standard. Federal Register/Vol. 86, No. 116.

Occupational Safety and Health Administration [OSHA]. (n.d.). *Hazard recognition*. Retrieved from https://www.osha.gov/coronavirus/hazards

Pandey, S. K., & Sharma, V. (2020). A tribute to frontline corona warriors—Doctors who sacrificed their life while saving patients during the ongoing COVID-19 pandemic. *Indian Journal of Ophthalmology, 68*(5), 939.

Phillips, L., Johnson, M., Deener, K., & Bonanni, C. (2015). EPA's exposure assessment toolbox (EPA-Expo-Box). *Journal of Environmental Informatics, 25*(2).

Piňosová, M., Andrejiova, M., Badida, M., & Moravec, M. (2021). Occupational disease as the bane of workers' lives: A chronological review of the literature and study of its development in Slovakia. Part 1. *International Journal of Environmental Research and Public Health, 18*(11), 5910. https://doi.org/10.3390/ijerph18115910

Quigley, D. D., Qureshi, N., Gahlon, G., & Gidengil, C. (2022). Worker and employer experiences with COVID-19 and the California workers' compensation system: A review of the literature. *American Journal of Industrial Medicine, 65*(3), 203–213. https://doi.org/10.1002/ajim.23326

Semple, S., & Cherrie, J. W. (2020). Covid-19: Protecting worker health. *Annals of Work Exposures and Health, 64*(5), 461–464.

Spetz, J., Stone, R. I., Chapman, S. A., & Bryant, N. (2019). Home and community-based workforce for patients with serious illness requires support to meet growing needs. *Health Affairs, 38*(6), 902–909.

Stetzenbach, L. D. (2009). Airborne infectious microorganisms. *Encyclopedia of Microbiology*, 175–182. https://doi.org/10.1016/B978-012373944-5.00177-2

Thomas, R. J. (2013). Particle size and pathogenicity in the respiratory tract. *Virulence, 4*(8), 847–858.

Tulchinsky, T. H., & Varavikova, E. A. (2014). Environmental and occupational health. *The New Public Health*, 471–533. https://doi.org/10.1016/B978-0-12-415766-8.00009-4

Van Dorn, A., Cooney, R. E., & Sabin, M. L. (2020). COVID-19 exacerbating inequalities in the US. *The Lancet, 395*(10232), 1243–1244.

World Health Organisation [WHO]. (2020a). *Listings of WHO's response to COVID-19*. Retrieved from https://www.who.int/news/item/29-06-2020-covidtimeline

World Health Organisation [WHO]. (2020b). *Keep health workers safe to keep patients safe*. Retrieved from https://www.who.int/news/item/17-09-2020-keep-health-workers-safe-to-keep-patients-safe-who

World Health Organization [WHO]. (2021a). *The impact of COVID-19 on health and care workers: A closer look at deaths (No. WHO/HWF/WorkingPaper/2021.1)*. World Health Organization.

World Health Organization [WHO]. (2021b). *COVID-19: Occupational health and safety for health workers: Interim guidance, 2 February 2021 (No. WHO/2019-nCoV/HCW_advice/2021.1)*. World Health Organization.

World Health Organization [WHO]. (n.d.). *Year of health and care workers 2021*. Retrieved from https://www.who.int/campaigns/annual-theme/year-of-health-and-care-workers-2021

You, J., Zhang, L., Ni-Jia-Ti, M. Y., Zhang, J., Hu, F., Chen, L., Dong, Y., Yang, K., Zhang, B., & Zhang, S. (2020). Anormal pulmonary function and residual CT abnormalities in rehabilitating COVID-19 patients after discharge. *The Journal of Infection, 81*(2), e150–e152. https://doi.org/10.1016/j.jinf.2020.06.003

Chapter 9
Migrants During Lockdown and Green Building Perspectives

1 Introduction

In India, the industries are divided into three categories, viz. large-scale (LSI), middle-scale (MSI), and small-scale (SSI) (Rai & Fulekar, 2020). LSIs and, to a lesser extent, MSIs provide housing amenities and employment within the industrial grounds; SSIs are generally characterised as a developing sector. They generally hire temporary workers and do not provide housing; as a result, workers must find their own lodging in an industrial setting close enough (Shelly et al., 2020). The COVID-19 pandemic has currently spread to many countries, including India. Due to lockdown because of Corona virus, there was closure of industries, restrictions on commercial activity, and delay in transportation (Fairlie, 2020). The lockdown situation in industrial and commercial settings made it difficult for temporary workers to stay in their rented accommodations and even made it difficult for them to make a living. Consequently, with few resources and limited transit systems, people began to migrate to their own states during the lockdown (International Labour Organization [ILO], 2020). Out of 1.21 billion people in India, labourers and migrants make up 37.68% of the workforce (Kumar & Choudhury, 2021). Now, lockdown has ended, industrial activity has begun using low-skilled labour, triggering an economic crisis. The migrants have returned to their job states in order to survive. Without a policy, plan, and programme for migrant workers' housing, whether on a rental basis or through the supply of housing facilities, this will lead to consequences. The Indian government intends to offer these migratory labourers shelter and employment opportunities in their respective states. In future, migrant workers should not depend on a socioeconomic emergency for their livelihood, therefore, housing for migrant workers will be designed for and provided together with other amenities (National Human Rights Commission [NHRC], 2020). In order to build housing for migrant workers who have been displaced by COVID-19, a model inexpensive house that conforms with green building principles and includes built-in infrastructure for environmental sustainability has been recommended. The scope of this will entail developing sustainable and cheap housing for migrants or

M. H. Fulekar et al., *Pandemic Over the Centuries*,
https://doi.org/10.1007/978-3-031-94270-9_9

unstructured groups based on a model home that will be created using the idea of green building.

A sustainable model home can be designed and built in India as part of the green building initiative in order to spread the idea of green building at a reasonable price using possibly accessible material resources (Khoshnava et al., 2020). The use of solar energy (renewable energy) and conversion into electricity for domestic use, as well as water reuse and recycling utilising modern methods such as bioremediation and nano-bioremediation, are all in conformity with the sustainable environment (Allen et al., 2015). The synthesis of nanomaterials for manufacturing nano tiles and nano catalyst for waste water treatment will be done using solid kitchen wastes (Abdelbasir et al., 2020). In order to meet the model home's need for self-sufficiency, eco-gardening, and eco-medicinal plants will be developed. Such sustainable model home design and construction will serve as a model for building housing for migrant workers who were displaced during COVID-19.

2 Covid-19: Lockdown Phases in India

Lockdown was formally announced by the Indian government in stages as follows.

2.1 Phase 1 (25th March to 14th April)

On March 24, 2020, a nationwide lockdown was declared (Ministry of Home Affairs [MHA], 2020a). This resulted in an industrial shutdown, limited commercial operation, halted domestic and international flights, travel restrictions, and closure of all government and private services. People were also confined to their homes (except medical condition). There was no mobility of individuals between states or within states (Shukla et al., 2021). During the lockdown, the Police were keeping a close watch on how the administration was enforcing the Law and Order (Datta et al., 2020). The services for the Covid-19 patients were being provided in the hospitals by the medical personnel and staffs.

2.2 Phase 2 (15th April to 3rd May, 2020)

The government maintained stringent rules and declared containment zones to further extend the lockdown; in these zones, only essential services were allowed. The authorities carried out "extensive contact tracing house-to-house surveillance and other clinical measures". On 16 April, presence of infection hotspots, orange zones, which indicated moderate infection, and green zones indicating no infection (MHA, 2020b). The government announced several relaxations on April 20, 2020, including the opening of agricultural enterprises, including dairy, aquaculture, and plantations, as well as stores selling farming goods. In addition, the government declared the restart of public services with directives to maintain social distance.

2.3 Phase 3 (4th to 17th May, 2020)

The state-wide lockdown was further extended for 2 weeks, effective from May 4 to May 17, 2020, depending upon the severe situation, and the rising number of coronavirus infections (MHA, 2020c). All state districts were categorised into the same three zones based on the virus's spread and comparable relaxations were implemented. On May 17, the National Disaster Management Authority further extended the lockdown until May 31 (MHA, 2020d).

2.4 Phase 4 (18th to 31st May, 2020)

The lockdown was extended with additional relaxations by the Ministry of Home Affairs and National Disaster Management Authority on May 17, 2020, until May 31, 2020. States were granted further power to define green, orange, and red zones, as well as to create an implementation roadmap, like previous amendments. Containment and buffer zones were created within red zones. The power to define containment and buffer zones was assigned to the local bodies.

On May 30, the government announced that the current lockdown would last till June 30 only for the containment zones, but it will be lifted moving ahead. Starting on June 8, services would be gradually reintroduced. The designation of it was "Unlock 1.0". Subsequently, the authorities made it clear that the country was no longer under lockdown and that "unlock" had already started (MHA, 2020e). Unlock 2.0, the second phase of unlocking, from July 1 to July 31, 2020 was made public with lesser restrictions.

3 Lockdown Impact

Migrant labourers and/or workers were unemployed and/or underpaid due to a nationwide lockdown to combat COVID-19 (Guha et al., 2021). Amid the pandemic, stranded workers were battling to return home with little to no savings. The severely affected were daily wagers. Migrant workers were left without a job or a place to stay in major cities like Mumbai, Delhi, and other cities due to the lockdown (Jesline et al., 2021). Due to the lockout, hundreds of people retreated important Indian cities in search of employment. Masses of migrant workers had to face problems like business and workplace closures, financial loss, food scarcity, and fears for the future (Irudaya

Rajan et al., 2020). The COVID-19 outbreak caused job losses for almost 87% of contract workers in urban areas. This was most common among temporary workers. In general, metropolitan India had a bigger loss of jobs than rural areas (Statista, 2022). Despite government's plan that assured those in need will receive higher rations as a response of the lockdown, the delivery system was ineffective, putting numerous migrants and their families without enough food. Given that they had no job and no income, multitudes of migrant workers were seen travelling hundreds of kilometres on foot or cycles to return to their villages (Times of India [ToI], 2020). The migrant labourers opted to travel in big groups, notwithstanding the government's provision of special trains and buses. They skipped the line to get on the government-run transportation because they were famished and anxious to get home (Hindustan Times, 2020a). They also believed that if they returned to their native place, they could resume agriculture and work in minor occupations underneath the MNREGA (The Hindu, 2020a). Many migrant labourers had to work for days without much access to food or water; few were detained for breaking the lockdown, and some of them eventually died from strain or accidents (Hindustan Times, 2020b, 2020c). As of May 10, 2020, there had been more than 350 deaths attributed to the lockdown, because of starvation, suicide (Panigrahi et al., 2021), fatigue, accidents, police abuse (Agoramoorthy & Hsu, 2021) and delayed medical attention. Most of the reported deaths during the lockdown were of labourers and migrants. Despite the Central Government's consent, numerous state governments-imposed restrictions on food delivery services (The Economic Times, 2020). From January to March 2020, the GDP growth rate dropped from 8.2% in 2018 to 3.1% (Statista, 2021).

However, Government of India report submitted to Supreme Court of India states that as of 7 April, 2020 state governments were running 22,567 relief camps for stranded migrant workers, with Kerala operating 15,541 camps (68%), Maharashtra 1,135 camps, Tamil Nadu 178 camps, and other states operating fewer camps. Camps were run by non-governmental organization were 3,909 (The Hindu, 2020b).

4 Green Building for Migrant Workers

India has the highest proportion of urban residents who live in slums, approximately 35% of the total population in 2018 (World Bank, 2022). The constraints of accommodating this population growth necessitate the implementation of fresh intervention strategies in urban and housing management with a focus on sustainable growth. Decision to concentrate its engagement on slums stems primarily from the fact that these areas and this situation are not the exception but rather the pattern in Indian cities (Nolan, 2015). Low-income people cannot get affordable houses from private builders, which drives them to the unregistered land market where prices are low (Garnham et al., 2022). Due to financial limitations and a lack instances, the existing urban planning and management criteria are completely irrelevant to the true demands of most of the people and are unable to address this changing configuration with framework that helps urban planning. In order to provide suitable

Fig. 1 Plight of migrant workers in India. (*Source* Indian Express)

housing policies and provide low cost houses or alternatives through social housing initiatives, the government becomes important. Government should investigate the possibility of providing houses to marginalized section of the society which also takes into consideration the concept of sustainability. Therefore, green building can be attributed as solution and maintain or improve the lifestyle of marginalized section together while maintaining the environment in which it is located (Fig. 1).

4.1 Green Building: Definition

Green building, as defined by the United States Environmental Protection Agency [USEPA] (2016), is the practice of enhancing the efficiency in which buildings and their locations utilise resources, like energy, water and materials, whilst also limiting the effects of building structures on the environment and human health over the course of the building's complete life cycle, from planning and design to use and reuse. The construction industry, amongst many other industrial and production sectors, is one that ought to incorporate sustainability and has served as the main driving force behind the sustainable building and green building initiatives. Green buildings outperform conventional (non-green) structures, according to the study (Dwaikat & Ali, 2018). Amongst frequently cited efficacies/benefits of green construction are energy conservation, water treatment and regeneration, improved indoor air quality, improved health and productivity, higher property value, and better light and ventilation (Patel & Chugan, 2013).

4.2 Components of Green Building

The UEPA states that following are the seven key criteria used to evaluate green buildings in order to determine their sustainability.

4.2.1 Energy Efficiency and Renewable Energy

Energy efficiency is the ratio between energy input and energy output. Measuring energy efficiency, especially on a large scale, is difficult to achieve throughout time and across different countries or industries due to methodological issues (Schipper & Haas, 1997). According to Hartley (1990), renewable energy technologies generate energy that is useful by transforming natural phenomena into practical energy forms. These technologies generate energy from sources such as gravitational forces (tides), the heat of the earth's core (geothermal), and the intrinsic energy of sun and its direct and indirect effects on the planet. Energy-efficient structures use less energy, function more affordably, and provide more convenience, reducing both household and business expenses (Vatalis et al., 2013). Energy savings 40 to 50% are achieved by green buildings in India that have been certified by the Indian Green Building Council [IGBC] (World Green Building Council, n.d.). A particularly efficient green building plan is to use renewable energy and the potential for on-site production in conjunction to including energy-saving fixtures and appliances. Solar energy panels should be installed close to green buildings to capture solar energy, store it, and then transform into electrical energy. The window and door frames are also made of material that can absorb solar energy throughout the day and, with the aid of an energy conversion system, transform it into electricity. With the development of infrastructure, LED lamps, which consume less energy and provide more light, can also be used. Another significant energy-consuming process is heating of water, particularly in residential structures. Energy usage can be decreased without impacting users by installing sensors to manage the heating mechanism and increasing insulation surrounding water heaters. Roofs have a significant capacity to reduce energy demand, yet they are frequently underutilised in houses. Over the course of entire building envelope, roofs receive the most solar irradiation. Green roofs employ flora as a cover to improve building's ability to cool. If available to building inhabitants, green roofs provide additional advantages like improve air quality and connect to nature (Costanzo et al., 2016).

4.2.2 Water Efficiency

Water efficiency has certainly grown with energy efficiency in the hierarchy of green construction components. Water conservation can be practised at any stage of a project, from planning to completion. Utilising more advanced techniques and technology to provide the same or better services while using less water is known as water efficiency (U.S. Green Building Council [USGBC], n.d.). In India, IGBC-certified green buildings use 20–30% less water than equivalent traditional construction (World Green Building Council, n.d.). Reducing operational costs and the need to create new sources and extend water infrastructure are all benefits of increasing water efficiency. Additionally, it decreases freshwater demands from finite sources, preserving additional water for future use and enhancing environmental water quality and aquatic ecology (Vatalis et al., 2013). Reduction in the utilisation of drinking water for non-consumption use can be seen as the first step in improving water efficiency at home. For water saving, the appropriate flow plugs and low-flow plumbing fixtures should be employed. The water flow is moderated by such machinery. Urinals that are water-efficient utilise less than 2.8 L per flush. Reservoirs that will collect and preserve runoff from roofs, either above or below ground should be constructed. In the Green Rating, water efficiency includes water harvesting, which is also vital for water efficiency (Shamanth, 2020) (Fig. 2).

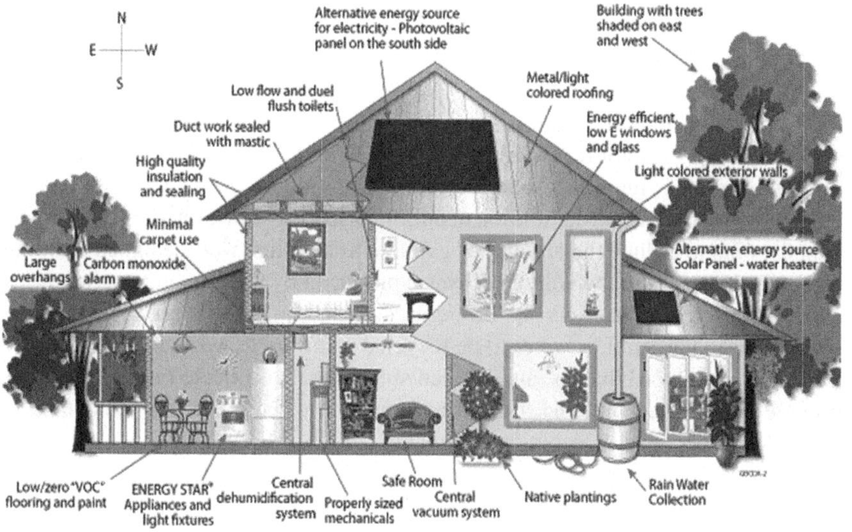

Fig. 2 Components of green buildings. (*Source* The Constructor)

4.2.3 Environmentally Preferred Building Materials and Specification

Endeavour to develop a high-performance structure must include the use of long-lasting, aesthetically pleasing, and ecologically friendly construction materials. The well-being of the residents and a sense of belonging with the abundance of the natural environment are enhanced using organic and healthy resources. Resource conservation and the use of non-toxic materials are the main ways to choose eco-friendly materials with minimal negative effects on the environment (Energy.gov., n.d.). Sustainable building materials can be derived from conventional sources like earth and stone, from process improvements or inputs like industrial waste that have been found by life-cycle analysis to be the least harmful to the environment, or from established manufacturing processes (Rousseau, 2009). Timbers from forests that have been certified to a third-party forest standard, quickly bio-renewable resources like bamboo and straw, shielding concrete forms, dimension stone, recycled stone, recycled metal and other non-toxic, reusable, renewable, and/or recyclable products (e.g., sheep wool, panels made from paper flakes, compressed earth block, adobe, baked earth, rammed earth, clay, etc.) are examples of "green" building materials. To reduce the energy required for their shipping, materials should be sourced and produced close to the construction site (Ji, 2016). Project specifications can assist in environmental processes, eco-friendly materials and environmental services (Vatalis et al., 2013).

4.2.4 Waste Reduction

Waste reduction initiatives are important to slow or stop per-person waste generation rate from escalation (Twardowska et al., 2004). The goal of green architecture is to minimise, repurpose, and recycle waste produced during the development, repair, reconstruction and demolition of buildings. By providing on-site solutions like compost bins to reduce the amount of garbage to landfills, well-designed buildings also assist in reducing the volume of waste produced by the inhabitants. Single-use and non-recyclable materials should be used minimum. Mulch pruning should be encouraged in garden. For water to be recycled and reused, waste water from bathrooms, showers, wash basins, and kitchen sinks must be collected in a storage tank through a connected pipeline, treated and purified to be reused for gardening, in toilets, and other similar activities (Shamanth, 2020).

4.2.5 Toxics Reduction

Toxin reduction within a structure can occur in many ways. Many conventionally harmful things can be replaced with more environmental friendly alternatives, including glues and coatings used on building materials and the chemicals to produce flame-retardant indoor furnishings. Organic repellents can replace pesticides and insecticides. For regular home functions, phosphate-free or biodegradable products should be utilised. Items made of plastic and cans that may cause cancer, be replaced with earthen or timber. Many construction materials release air pollution. The approach concentrates on three categories of indoor air pollutants, viz. carcinogen, irritant and smell, while also considering the decrease of harmful emissions from building materials. Few human carcinogens are known to cause serious illnesses, and is advisable to avoid their use. Since there are many eye and airway irritants and come in many different chemicals, there are significant probabilities that they can be replaced. In practice, we should avoid the odours. Currently, sealants, glues, adhesives, paints, lacquers and wall/floor coatings are all made of these components (Andersen et al., 1982).

4.2.6 Indoor Air Quality

Indoor air quality aims to eliminate VOCs and other air pollutants like microbiological contamination. Green structures depend on ventilation system that is appropriately built to supply enough ventilation of clean ambient air or re-circulating filtered. According to Levin (1991), there are seven main ventilation factors to consider when controlling indoor air quality: (1) dilution by outdoor air ventilation, (2) air intake locations, (3) building exhaust locations, (4) air cleaning and filtration, (5) space air distribution, (6) heat recovery and (7) microbial control. Choosing building supplies and interior finishing goods with zero or low VOC emissions is essential during the design and construction phase. Most of the cleaning and maintaining agents and construction materials release gases, some of which are harmful. Controlling dampness, which promotes the formation of moulds, the presence of bacteria and viruses, dust mites and other creatures, as well as microbiological problems, is also important to indoor air quality. However, proper ventilation is also required to remove humidity from indoor sources, like cooking, bathing, cleaning and other activities. A properly insulated and securely sealed enclosure will prevent water damage. Since sustainable materials like bamboo absorb more carbon dioxide than traditional wood, and are ideal for air quality.

4.2.7 Smart Growth and Sustainable Development

Sustainable development has been described in numerous ways, but the most cited definition is from 'Our Common Future', generally known as Brundtland Report (WCED, 1987): Development that is sustainable satisfies existing demands without jeopardising the capacity of future generations to satisfy their own needs. Smart growth is a concept for urban planning and transportation that concentrates development in small, walkable urban centres to prevent sprawl. Communities are conserving resources by reinvesting in current infrastructure and repurposing historic structures while developing, preserving natural lands and important environmental regions, protecting water and air quality, and reusing already built land. Communities are offering their residents and tourists the choice of walking, cycling, using public transport, or driving as they go about their business by planning neighbourhoods with businesses, offices, schools, churches, parks, and other services close to residences. These cities are making lively places to live, work, and play by using smart growth strategies that improve neighbourhoods and involve locals in development decisions. These cities with high quality of life boosts their economic competitiveness, fosters commercial prospects and broadens the local tax base (USEPA, 2006).

5 Conclusion

Green buildings not only minimise or eliminate negative environmental consequences by using less water, energy or natural resources, but can also have a positive environmental impact in many circumstances by generating their own energy or boosting biodiversity. These buildings can significantly improve living standard for the underprivileged especially impacted by COVID-19. In order to provide these underprivileged with shelter and job prospects in their respective states, governments and other organisations must take an active role.

References

Abdelbasir, S. M., McCourt, K. M., Lee, C. M., & Vanegas, D. C. (2020). Waste-derived nanoparticles: Synthesis approaches, environmental applications, and sustainability considerations. *Frontiers in Chemistry, 8*, 782.

Agoramoorthy, G., & Hsu, M. J. (2021). How the coronavirus lockdown impacts the impoverished in India. *Journal of Racial and Ethnic Health Disparities, 8*(1), 1–6.

Allen, J. G., MacNaughton, P., Laurent, J. G. C., Flanigan, S. S., Eitland, E. S., & Spengler, J. D. (2015). Green buildings and health. *Current Environmental Health Reports, 2*(3), 250–258.

Andersen, I., Seedorff, L., & Skov, A. (1982). A strategy for reduction of toxic indoor emissions. *Environment International, 8*, 11–16.

Costanzo, V., Evola, G., & Marletta, L. (2016). Energy savings in buildings or UHI mitigation? Comparison between green roofs and cool roofs. *Energy and Buildings, 114*, 247–255.

Datta, T., Ray, S., & Dawa, M. (2020). *Re-imagining the role of the police in COVID-19 times*. Retrieved from https://www.unicef.org/india/stories/re-imagining-role-police-covid-19-times

Dwaikat, L. N., & Ali, K. N. (2018). The economic benefits of a green building—evidence from Malaysia. *Journal of Building Engineering, 18*, 448–453.

Energy.gov. (n.d.). *Materials*. Retrieved from https://www.energy.gov/sites/prod/files/2013/12/f5/sustainable_guide_ch6.pdf

Fairlie, R. (2020). The impact of COVID-19 on small business owners: Evidence from the first three months after widespread social-distancing restrictions. *Journal of Economics & Management Strategy, 29*(4), 727–740.

Garnham, L., Rolfe, S., Anderson, I., Seaman, P., Godwin, J., & Donaldson, C. (2022). Intervening in the cycle of poverty, poor housing and poor health: The role of housing providers in enhancing tenants' mental wellbeing. *Journal of Housing and the Built Environment, 37*(1), 1–21.

Guha, P., Islam, B., & Hussain, M. A. (2021). COVID-19 lockdown and penalty of joblessness on income and remittances: A study of inter-state migrant labourers from Assam, India. *Journal of Public Affairs, 21*(4), e2470.

Hartley, D. L. (1990). Perspectives on renewable energy and the environment. In J. W. Tester, D. O. Wood, N. A. Ferrari (Eds.), *Energy and the environment in the 21st century*. MIT

Hindustan Times. (2020a). *Across the country, migrants still forced to walk thousands of miles*. Retrieved from https://www.hindustantimes.com/india-news/similar-scenes-in-several-states-despite-new-trains/story-YL9qNcF315SA5o645aj1DP.html

Hindustan Times. (2020b). *55 arrested after police, migrants clash in Gujarat's Surat*. Retrieved from https://www.hindustantimes.com/india-news/55-arrested-after-police-migrants-clash-in-surat/story-KC2XFIGcbwtAPeeMyeTkXO.html

Hindustan Times. (2020c). *198 migrant workers killed in road accidents during lockdown: Report*. Retrieved from https://www.hindustantimes.com/india-news/198-migrant-workers-killed-in-road-accidents-during-lockdown-report/story-hTWzAWMYn0kyycKw1dyKqL.html

International Labour Organization [ILO]. (2020). *The impact of the COVID-19 pandemic on jobs and incomes in G20 economies*. Retrieved from https://www.ilo.org/wcmsp5/groups/public/-dgreports/-cabinet/documents/publication/wcms_756331.pdf

Irudaya Rajan, S., Sivakumar, P., & Srinivasan, A. (2020). The COVID-19 pandemic and internal labour migration in India: A 'crisis of mobility'. *The Indian Journal of Labour Economics, 63*(4), 1021–1039.

Jesline, J., Romate, J., Rajkumar, E., & George, A. J. (2021). The plight of migrants during COVID-19 and the impact of circular migration in India: A systematic review. *Humanities and Social Sciences Communications, 8*(1), 1–12.

Ji, S. (2016). *Green building materials and their common use in everyday life*. incom.

Khoshnava, S. M., Rostami, R., Mohamad Zin, R., Štreimikienė, D., Mardani, A., & Ismail, M. (2020). The role of green building materials in reducing environmental and human health impacts. *International Journal of Environmental Research and Public Health, 17*(7), 2589.

Kumar, S., & Choudhury, S. (2021). Migrant workers and human rights: A critical study on India's COVID-19 lockdown policy. *Social Sciences & Humanities Open, 3*(1), Article 100130.

Levin, H. (1991). Critical building design factors for indoor air quality and climate. Current status and predicted trends. *Indoor Air, 1*, 79–92.

Ministry of Home Affairs [MHA]. (2020a). *Government of India issues orders prescribing lockdown for containment of COVID-19 epidemic in the country*. Retrieved from https://pib.gov.in/PressReleseDetail.aspx?PRID=1607997

Ministry of Home Affairs [MHA]. (2020b). *New guidelines on measures for containment of COVID 19*. Retrieved from https://www.mha.gov.in/sites/default/files/MHA%20Order%20Dt.%201.5.2020%20to%20extend%20Lockdown%20period%20for%202%20weeks%20w.e.f.%204.5.2020%20with%20new%20guidelines.pdf

Ministry of Home Affairs [MHA]. (2020c). *Extension of lockdown for a further period of two weeks with effect from May 4, 2020*. Retrieved from https://www.mha.gov.in/sites/default/files/press%20release_0.pdf

Ministry of Home Affairs [MHA]. (2020d). *Extension of lockdown up to May 31, 2020*. Retrieved from https://www.mha.gov.in/sites/default/files/17052020Lockdown_1.pdf

Ministry of Home Affairs [MHA]. (2020e). *New guidelines to fight COVID-19 to be effective from 1st June 2020*. Retrieved from https://www.mha.gov.in/sites/default/files/PR_UNLOCK1Guidelines_30052020_1.pdf

National Human Rights Commission [NHRC]. (2020). *A study on social security and health rights of migrant workers in India*. Retrieved from https://nhrc.nic.in/sites/default/files/Approved_Health%20and%20social%20security%20ISMW_KDS-NHRC.pdf

Nolan, L. B. (2015). Slum definitions in urban India: Implications for the measurement of health inequalities. *Population and Development Review, 41*(1), 59–84.

Panigrahi, M., Pattnaik, J. I., Padhy, S. K., Menon, V., Patra, S., Rina, K., & Patro, B. (2021). COVID-19 and suicides in India: A pilot study of reports in the media and scientific literature. *Asian Journal of Psychiatry, 57*, Article 102560.

Patel, C., & Chugan, P. K. (2013). Measuring awareness and preferences of real estate developers for green buildings over conventional buildings. In J. Aagja, A. K. Awasthi, & S. Jain (Eds.), *Consumer behaviour and emerging practices in marketing* (pp. 332–341). Institute of Management, Nirma University. Himalaya Publishing House.

Rai, A., & Fulekar, M. H. (2020). Wastewater treatment: Common effluent treatment plant—case study. In *Bioremediation technology* (pp. 267–284). CRC Press.

Rousseau, D. (2009). Environmental friendly building materials. *Sustainable Built Environment, 1*, 162–185.

Schipper, L., & Haas, R. (1997). Cross-country comparisons of indicators of energy use, energy efficiency and CO_2 emissions. *Energy Policy, 25*, 7–9.

Shamanth, M. K. (2020). *Water technologies for green building*. Retrieved from https://www.constrofacilitator.com/water-technologies-for-green-building/

Shelly, R., Sharma, T., & Bawa, S. S. (2020). Role of micro, small and medium enterprises in Indian economy. *International Journal of Economics and Financial Issues, 10*(5), 84.

Shukla, M., Pandey, R., Singh, T., Riddleston, L., Hutchinson, T., Kumari, V., & Lau, J. Y. (2021). The effect of COVID-19 and related lockdown phases on young peoples' worries and emotions: Novel data from India. *Frontiers in Public Health, 9*, Article 645183.

Statista. (2021). *Indian GDP takes second dip*. Retrieved from https://www.statista.com/chart/18245/india-quarterly-gdp-growth/

Statista. (2022). *Impact of the coronavirus (COVID-19) on rural and urban employment across India in 2020, by type*. Retrieved from https://www.statista.com/statistics/1142485/india-coronavirus-impact-on-rural-and-urban-employment-by-type/

The Economic Times. (2020). *Zomato, Swiggy ordered to shut down in several states*. Retrieved from https://economictimes.indiatimes.com/internet/zomato-swiggy-ordered-to-shut-down-in-several-states/articleshow/74828652.cms

The Hindu. (2020a). *MGNREGA has evolved a rights-based framework for migrant workers*. Retrieved from https://www.thehindu.com/news/national/other-states/mgnrega-has-evolved-a-rights-based-framework-for-migrant-workers/article31778484.ece

The Hindu. (2020b). *Coronavirus | Centre files report on migrant workers*. Retrieved from https://www.thehindu.com/news/national/coronavirus-centre-files-report-on-migrant-workers/article31283896.ece

Times of India [ToI]. (2020). *Suddenly jobless, migrant workers on long walk home*. Retrieved from https://timesofindia.indiatimes.com/india/suddenly-jobless-migrant-workers-on-long-walk-home/articleshow/74820123.cms

Twardowska, I., Allen, H. E., Kettrup, A. F., & Lacy, W. J. (2004). *Solid waste: Assessment*. Gulf Professional Publishing.

U.S. Green Building Council [USGBC]. (n.d.). *Green Building 101: How does water efficiency impact a building?* Retrieved from https://www.usgbc.org/articles/green-building-101-how-does-water-efficiency-impact-building

United States Environmental Protection Agency [USEPA]. (2006). *Life cycle assessment: Principles and practice*. Retrieved from https://nepis.epa.gov/Exe/ZyNET.exe/P1000L86.TXT?ZyActionD=ZyDocument&Client=EPA&Index=2006+Thru+2010&Docs=&Query=&Time=&EndTime=&SearchMethod=1&TocRestrict=n&Toc=&TocEntry=&QField=&QFieldYear=&QFieldMonth=&QFieldDay=&IntQFieldOp=0&ExtQFieldOp=0&XmlQuery=&File=D%3A%5Czyfiles%5CIndex%20Data%5C06thru10%5CTxt%5C00000002%5CP1000L86.txt&User=ANONYMOUS&Password=anonymous&SortMethod=h%7C-&MaximumDocuments=1&FuzzyDegree=0&ImageQuality=r75g8/r75g8/x150y150g16/i425&Display=hpfr&DefSeekPage=x&SearchBack=ZyActionL&Back=ZyActionS&BackDesc=Results%20page&MaximumPages=1&ZyEntry=1&SeekPage=x&ZyPURL

United States Environmental Protection Agency [USEPA]. (2016). *Green building*. Retrieved from https://archive.epa.gov/greenbuilding/web/html/

Vatalis, K. I., Manoliadis, O., Charalampides, G., Platias, S., & Savvidis, S. J. P. E. (2013). Sustainability components affecting decisions for green building projects. *Procedia Economics and Finance, 5*, 747–756.

World Commission on Environment and Development, WCED. (1987). *Our common future* (p. 43). Oxford University Press.

World Bank. (2022). *Population living in slums (% of urban population)—India.* Retrieved from https://data.worldbank.org/indicator/EN.POP.SLUM.UR.ZS?end=2018&locations=IN&start=1990&view=chart

World Green Building Council. (n.d.). *About green buildings.* Retrieved from https://www.worldgbc.org/benefits-green-buildings

Glossary

Aerosols An aerosol is a micro sized particulate or liquid suspension in atmosphere.

Ageusia Complete or partial loss of taste.

Anosmia Partial or total loss of scent perception.

Antibody tests Examinations of the blood that check for certain antibodies. These are also referred to as serology tests.

Antibody A pathogen-neutralizing protein produced by plasma cells, a part of the immune response. In reaction to a fresh infection, antibodies are produced that are exclusive to microorganism.

Antigen Any material that can trigger a particular immune reaction against it or is capable of being identified by immune system.

Antipyretic therapy A fever-reducing treatment.

Anti-vaxxer Individuals who are against immunisation.

Antiviral medicines Drugs that prevent viruses from growing as well as from contracting healthy tissue.

Asymptomatic An individual who, even though has contracted a disease-causing microorganism, shows no evidence of being ill.

Booster An additional dose of a vaccination administered after the initial course has indeed been completed.

Cardiovascular system A group of organs that includes the heart (cardio) and blood vessels (vascular) which is responsible for pumping and circulating blood around the body.

Case fatality rate (CFR), or the fatality ratio The percentage of persons that died as a result of an infection out of the entire people infected.

Clinical presentation The visible signs and symptoms of an illness that may be utilised to reach a medical assessment.

Clinical trials Clinical trials are human research studies which try to evaluate a medicinal, surgical, or behavioural intervention.

Close contact Whenever anyone spends a sufficient amount of time in a confined room with an infected patient or gets near enough to come into contact with infectious body fluids such as droplets.

M. H. Fulekar et al., *Pandemic Over the Centuries*,
https://doi.org/10.1007/978-3-031-94270-9

Cohort study Study that entails examining a collection of individuals across a timeframe.

Communicable period The time span across which an infectious agent can spread directly or indirectly from one individual to the other.

Community spread The spread of an infectious illness between a group of individuals that is unrelated to interaction with such an individual travelling from a reported infected region or with an infected individual.

Comorbidity The presence of one or more illnesses, ailments, health conditions, or pathological changes in an individual that occur simultaneously as a particular health condition.

Confirmed case An individual who fits all of the official case definition parameters for an illness or whose situation is verified by at a minimum one examination procedure.

Congregate settings Public venues which may become congested and also where sick persons can come into touch. These include shopping malls, cinema halls, and convenience stores.

Conjugate vaccine A vaccination that combines two components to boost its efficacy.

Contact tracing The procedure of locating persons who might have been subjected to a communicable diseases carrier in the interest of making sure they've been informed of the possible risks as well as to follow up with them.

Contagiousness The fundamental reproduction number of an infectious agent represents their infectiousness.

Containment Is a society's endeavour to identify and isolate sick members in order to minimize the risk whilst leaving the healthy population separately.

COVID-19 The condition brought on by SARS-CoV-2. The name is derived from CO-rona-VI-rus D-isease and the year it was detected (2019).

Curfew A prohibition on leaving one's house between certain hours.

Diagnostic test A laboratory study or diagnostic approach being used to validate or exclude out such a diagnosis.

Diluent Liquid which is employed to dilute a vaccination to the known concentration prior to application. Sterile water happens to be the most utilized diluent.

Donning and doffing The very arduous skill of correctly wearing and discarding PPE, including the disposal of contaminated equipment.

Double masking Layering one mask above another, often a non-medical mask above a surgical mask, is what double masking entails. This is often done to aid with fitting and purification.

Droplet transmission When germs or viruses migrate inside microscopic droplets of respiratory tract fluids. These droplets are created by coughing, sneezing, singing, or talking which may be inhaled or rest on a person's mouth, nose, or eyes.

Dyspnea Breathing difficulties or a subjective experience of respiratory discomfort.

Early detection Diagnosis of a particular condition at a preliminary phase in its natural cycle.

Effective reproduction number The average number of secondary cases caused by an infected individual in a community including both resistant and vulnerable individuals towards the active infection.

Endemic The prevalence of a condition in a certain area or group.

Epidemic peak Throughout epidemic, the period when the incidence of new infections is at its peak.

Epidemic plateau The point in an epidemic when the incidence of new cases becomes constant.

Epidemic threshold A specified critical incidence that, if surpassed, results in an epidemic proclamation or the activation of predefined public health actions.

Epidemic The accelerated transmission of a disease to a large number of individuals within a brief amount of time, exceeding what is anticipated.

Epidemiology The investigation of the incidence and consequences of health-related conditions or occurrences in specific groups.

Expectoration The mucus discharged from the respiratory system by coughing or cleansing the throat through the mouth.

Extubation The withdrawal of a hollow tube placed into the trachea through the mouth or nose to keep the airway open.

Epidemiologists Experts and scientists in public health who examine the trends, etiology, and safeguards against illness and suffering amongst individuals.

False negative A false negative occurs when a test done on a sample acquired from a patient with the illness being tested yields a negative result.

False positive A false positive occurs when a test done on a sample acquired from a person who does not have the illness being tested yields a positive result.

Flatten the curve Implement measures to decrease the rise in the amount of cases of a communicable diseases in order to prevent overburdening the health-care system during the epidemic.

Fomite A surface or item that can harbour an infectious pathogen and serve as a route of spread.

Frontline workers Professionals that offer a critical public service or important service.

Haemoptysis Coughing up blood that originates in the lower respiratory system.

Herd immunity The resilience of a community against the spreading of an active infection depending on the infection tolerance of a significant percentage of the group's individual members.

Hot zone A section of a hospital wherein patients with an active infection associated with an epidemic are allowed to treat.

Hygiene Sanitation measures which inhibit the disease from spreading and protect wellbeing.

Hypercytokinemia A disproportionate production of cytokines as a result of immune system over stimulation.

Hypoxemia A reduction in the oxygen concentration of arterial blood.

Immune response: The immune system's function which fights foreign material.

Immunity The capacity of an individual to fight against any infection/disorder when subjected to it.

Immunization The mechanism through which an individual develops resistance to an illness or pathogens.

Immunocompromised/immunodeficient This term refers to someone whose immune system is unable to withstand or battle against infections in addition to most individuals.

Immunogenicity A substance's potential to stimulate an immunological response.

Immunological memory The ability of an organism's immune system to identify and respond to an antigen with which it has previously come into contact.

Incidence The number of new cases of a disease or condition in a population at risk over a given period, usually one year.

Incubation period The time between an infectious agent infecting a vulnerable host and the beginning of symptoms and indications of an active infection.

Incubation The proliferation and spread of a pathogen in a vulnerable host prior to the manifestation of communicable disease signs and symptoms.

Infectivity An infectious agent's capacity to enter, live, and reproduce in a host.

Infodemic A cascade of information about a disaster, whether factual or not, that spreads quickly and uncontrolled.

Intensive care Care that necessitates the health care team's continuous assessment of an individual.

Isolation A method of preventing the transmission of a communicable disease that entails segregating an affected individual from non-infected persons throughout the disease's communicable phase.

Lockdown An emergency plan intended to restrict population mobility and assemblies for public safety or wellness concerns.

Lockdown lifting The procedure of lifting travel and gathering limitations placed on a population for public health or safety grounds.

Monoclonal antibodies Proteins created in a lab using clones of a single B cell that attach to a particular target, replicating the innate immune reaction.

Morbidity Disease severity commonly expressed as cases per 10,000 people each year.

Mortality The number of people who die from the disease annually expressed in units of 100,000 people.

Mortality rate The frequency of fatalities seen within a given population over a certain time frame.

Mutation An alteration in an organism's DNA sequence.

Myalgia Muscle pain

Myocarditis Heart muscle inflammation.

Nonpharmaceutical Interventions (NPIs) Besides getting vaccinated and taking medication, people and communities may take activities to help reduce the spread of diseases.

N95 mask The designation "N95" refers to the National Institute for Occupational Safety and Health (NIOSH) standard in the United States; "N" indicates that the item is not oil resistant, and "95" indicates that it can filter 95% of particles 0.3 μm and bigger when used appropriately.

Non-medical mask A fabric attachment worn over the mouth, nose, and chin to avoid pathogens particle transmission.

Outbreak A localised rise in the occurrence of a disease.

Oxygen therapy Procedure that provides more oxygen to breathe in. Supplemental oxygen is another name for it.

Pandemic fatigue The progressive loss of desire to follow health precautions which is impacted by the sentiments, experiences, and judgements of the individuals impacted by a pandemic.

Pandemic An epidemic that occurs globally or across a big geographic region, crosses international boundaries, and typically affects a substantial population.

Pathogenicity The pathogen's capacity to cause illness.

Person under investigation (PUI) When a doctor believes a patient has the coronavirus. Yet, no tests have verified the illness.

Personal protective equipment or PPE Device/item required to reduce a person's exposure to potentially hazardous compounds that might cause disease or injury—gloves, full body suits, protective glasses, etc.

Physical distance A distance established by public health officials between persons in order to avoid the transmission of an infectious illness.

Plank the curve Perform aggressive efforts to significantly reduce the rise in the amount of instances of contagious diseases in order to prevent overloading the health-care system during the epidemic's peak.

Plasma The blood's liquid component. It is primarily water and comprises electrolytes as well as important proteins such as antibodies.

Positivity rate The proportion of tests that yielded a positive result.

Preclinical study A preliminary research of a novel medication performed in vitro (on cells or tissues) or in vivo (in animals) prior to clinical testing in humans.

Preservatives If the vaccination will be administered to more than one person, preservatives keep the vaccine from getting contaminated once the vial has been unsealed. 2-phenoxyethanol is the preservative that is most frequently utilized.

Pre-symptomatic Individuals have tested positive for infection but have not yet shown any symptoms or indications.

Primary case The initial instance of an infectious illness in a population.

Probable case An individual who has substantial disease symptoms, and whose diagnosis has not been supported by a laboratory test.

Quarantine An action taken to stop the transmission of contagious diseases whereby a healthy individual who may have come into contact with an infected person is confined while the ailment is still in its incubation stage.

Reservoir Environment in which a pathogen develops and reproduces.

Resilience The capacity to evolve and adjust in regardless of challenging conditions while effectively handling a highly stressful scenario.

Secondary case A person who contracts a disease after coming into touch with the primary patient.

Self-isolation A method of preventing the transmission of a contagious diseases in which an affected individual decides to avoid contact with non-infected persons throughout the disease's communicable phase.

Sero-surveillance Epidemiological research that employs serological testing.

Serological test A blood serum test that determines the existence of antibodies against a particular infectious pathogen.

Seroprevalence The number of persons in a particular group that have antibodies specific to an active infection throughout a time frame specified.

Side effect A unintended consequence of a therapy that varies from the desired treatment response.

Silent hypoxia An extremely low blood oxygen level without any obvious symptoms.

Social distancing Increasing physical isolation between persons to aid in the prevention of the transmission of a communicable disease.

Stabilizers Stabilizers protect the vaccine components from attaching to the vaccination container and avoid chemical reactions within the vaccine from happening.

Stay-at-home order A formal rule mandating residents of a community to stay at home unless conducting necessary tasks.

Superspreading A single infectious individual infecting an abnormally large population relative to how many persons the typical infectious individual infects.

Surfactants Surfactants keep the vaccine's components uniformly mixed. They stop the components in the vaccine from aggregating and settling.

Surge capacity The capacity to swiftly mobilise extra resources to address the heightened demands caused by a mass casualty disaster.

Susceptible host An individual who is susceptible to an infectious pathogen.

Suspect case An individual who has been subjected to an infectious vector or who exhibits certain disease-related symptoms.

Trace contacts Identify persons who might have been contacted to someone with a contagious disease and follow up with them in order to verify they are mindful of their possible risk.

Transmission The spreading of an active infection from one individual to the other.

Vaccination The process of administering a vaccine to create resistance against a particular illness or pathogen.

Vaccine bank An inventory of vaccinations or vaccine ingredients.

Vaccine hesitancy Reluctance or refusal to get a vaccination regardless of the availability of vaccination services.

Vaccine A medication that increases a person's immunity to a specific illness.

Vector A living or insect that spreads an infectious pathogen from a diseased host to a vulnerable host.

Ventilator Medical equipment that helps people who are having trouble breathing using artificial ventilation.

Viral load The amount of a virus present in an infected individual.

Viral shedding The process through which viral particles from an affected person's body are released into the surroundings.

Viremia Viral contamination in the blood.

Virion The whole viral particle, which is a virus's extracellular infective form and is made up of RNA or DNA and is encased in a protein shell and occasionally an envelope.

Virulence A way to evaluate how serious an illness is that is brought on by a particular pathogen.

Virus strain A subspecies of a viral species that exhibits certain genetic traits.

Virus variant A virus that has undergone mutations to become different from the original viral strain.

Virus A protein-encased infectious agent consisting of DNA or RNA that can only spread and reproduce by using the machinery of other cells.

Voluntary quarantine An infection control strategy in which a healthy individual who may have encountered an infected person decides to isolate themself.

Warm zone Yellow zone A section of a hospital where suspicious and potential cases of an infectious illness impacted in an epidemic are managed.

World Health Organization Health-related directing and coordinating agency within the United Nations organisation. It has the responsibility for offering global health direction, guiding the health research program, establishing standards and guidelines, communicating evidence-based alternative approaches, assisting nations with technical assistance, and reviewing and evaluating global patterns of health.

Zoonosis An infectious illness that can be transferred directly, indirectly, or by a carrier from animals to people.

Zoonotic disease A sickness induced by an infectious pathogen that can be transmitted from humans to other animals.

Index

A
Accessory protein, 120, 131
Acute Respiratory Distress Syndrome (ARDS), 133
Administrative controls, 205
Alert phase, 5
Alpha (B.1.1.7), 117, 123, 127
αCoVs, 121
Amphixenoses, 47
Anthrapozoonoses, 47
Antigenic drift, 83
Antigenic shift, 83
Antonine plague, 25, 32, 64
Approval and licensure, 157
Arogya Setu, 170
Asian flu, 30, 84
Asiatic cholera, 27, 96
Asymptomatic, 131
Athenian plague, 24, 32

B
Bacterial zoonoses, 45
Baricitinib, 154
Bebtelovimab, 150, 154
Beta (B.1.351), 117, 124, 127
βCoVs, 121
Biowarfare weapon, 78
Black death, 24, 32, 68, 70
Black fungus, 137
Booster doses, 175
Bubo, 68
Bubonic plague, 24, 69, 70, 75, 86

C
Cardiovascular manifestation, 134
Chemoprophylaxis, 56
Chlamydial zoonoses, 46
Climate change, 13, 44, 48, 57
Clinical trials, 156
Community participation, 75
Continuum of pandemic phases, 4
Convalescent plasma, 153
Coronaviridae, 32, 115
Coronavirus disease 2019 (COVID-19), 115
Covaxin, 165, 175
Covid-19 Associated Mucormycosis (CAM), 137
Covishield, 165, 175
Cross-species transmission, 12
Cutaneous mucormycosis, 140
Cyclozoonoses, 46
Cytokine storm, 86, 134

D
Delta (B.1.617.2), 117, 125, 127
δCoVs, 121
Dermal manifestations, 136
Dexamethasone, 137, 154
Dilution effect, 52
Direct contact, 129
Disinfection, 146
Disseminated mucormycosis, 140
DNA vaccines, 161

E
Elimination and substitution, 204
El Tor form, 26, 99, 100
Energy efficiency, 216
Engineering controls, 204
Entry of virus, 130

M. H. Fulekar et al., *Pandemic Over the Centuries*,
https://doi.org/10.1007/978-3-031-94270-9

Envelope protein, 120, 131
Epidemic, 1, 2
Epidemiologic surveillance, 12
Exoanthropic, 48
Extensive contact tracing, 212

F

Faecal-oral transmission, 129
Favipiravir, 152
Fifth cholera pandemic, 98
First-line therapy, 144
First wave, 166, 169
Fiscal crisis, 1
Flagellant, 66
Fomites transmission, 129
Food security, 14
Fourth cholera pandemic, 98

G

γCoVs, 121
Gamma (P.1), 117, 124, 127
Gastrointestinal manifestation, 134
Gastrointestinal mucormycosis, 140
Genomic sequence technologies, 12
Global Influenza Programme (GIP), 90
Global Influenza Strategy (GIS), 91
Global Influenza Surveillance and Response System (GISRS), 90
Global Task Force on Cholera Control (GTFCC), 102
Green building, 215
Green building materials, 218
Greenhouse Gas (GHG), 48, 167
Green zones, 212

H

H1N1 swine flu, 32, 84
Haematological manifestation, 135
Hazard hierarchy control, 203
Health Care Workers (HCW), 198
Health education, 75
Hemagglutinin (HA), 83
Herd immunity, 11, 106
Home isolation, 148
Host resistance, 56
House-to-house surveillance, 172
Human Immune Deficiency virus infection/Acquired Immune Deficiency Syndrome (HIV/AIDS), 31
Hydroxychloroquine, 152
Hygeia, 68

I

Immunization, 56
Inactivated virus vaccines, 163
Incidence proportion, 3
Indoor air quality, 219
Ingestion, 201
Inhalation, 200
Insufflation, 75
Interpandemic era, 4
Ivermectin, 153

J

Janata curfew, 170
Justinian plague, 25, 32, 64

L

Lagevrio, 152
Life cycle of SARS-CoV-2, 130
Live-attenuated virus vaccines, 163
Lockdown, 170
Long-Acting Antibody (LAAB), 150
Lopinavir/Ritonavir, 153

M

Membrane protein, 120
Metazoonoses, 46
Middle East Respiratory Syndrome Coronavirus (MERS-CoV), 1, 115
Migrants, 211
Monoclonal (mAb) Antibodies, 150
mRNA-based vaccine, 160
Mucormycosis, 137, 138
Mycotic zoonoses, 45

N

Neuraminidase (NA), 29, 83
Neurological manifestation, 136
Non-replicating viral vectors, 161
Non-structural polyprotein, 120
Nucleocapsid protein, 120, 131

O

Occupational diseases, 198
Omicron (B.1.1.529), 117, 126
Oral Cholera Vaccine (OCV), 103
Oral rehydration therapy, 105
Orange zones, 212
Orthozoonoses, 46
O1 serogroups, 26, 100, 101

O139 serogroups, 26, 100
Our common future, 219

P
Pandemic, 1, 2
Pandemic Influenza Preparedness (PIP), 90
Pandemic phase, 5
Pandemic Severity Assessment Framework (PSAF), 4
Pandemic transition phase, 7
Parasitic zoonoses, 45
Personal prophylactic measures, 74
Personal protective measures, 148
Pfizer, 34, 124, 126, 149, 165
Pherozoonoses, 46
Plague doctors, 67
Pneumonic, 70, 71
Poverty, 15
Precautionary doses, 175
Precipitation, 49
Preclinical trials, 155
Presymptomatic, 131
Probability of exceedance, 2
Probiotics, 107
Protein subunit vaccine, 158
Protozoal zoonoses, 45
Pulmonary manifestation, 133
Pulmonary mucormycosis, 139

Q
Quarantine, 148

R
Red zones, 213
Remdesivir, 151
Renal manifestations, 136
Replicating viral vectors, 161
Replication, 130
Reservoir neutralization, 54
Respiratory transmission, 128
Rhino-Orbital Cerebral Mucormycosis (ROCM), 139
Rickettsial zoonoses, 45
Risk exposure pyramid, 202
Rodent control, 73
Russian flu, 29

S
Salvage therapy, 144
Saprozoonoses, 46
Sea level rise, 49
Secondary Attack Rate (SAR), 4
Second cholera pandemic, 96
Second wave, 124, 137, 172
Septicaemic, 63
Seventh pandemic of cholera, 99
Severe Acute Respiratory Syndrome Coronavirus (SARS-CoV), 32
Severe Acute Respiratory Syndrome Coronavirus 2 (SARS-CoV-2), 115
Sixth cholera pandemic, 98
Smart growth, 219
Social distancing, 145
Social inequalities, 15
Social mobilization, 103
Spikeprotein, 120, 130
Spillover incidents, 13
Sputnik V, 175
Structural proteins, 131
Surgical transmission, 129
Surveillance, 12
Sustainable development, 219
Synanthropic, 48

T
The unlocks, 171
Third cholera pandemic, 97
Third wave, 175
Tocilizumab, 153
Toxics reduction, 218
Toxin Co-regulated Pilus (TCP), 100
Translation of the viral replication machinery, 130
Translation of the viral structure proteins, 130
Type II transmembrane serine protease (TMPRSS2), 130

V
Vaccination, 155
Variant of High Consequence (VOHC), 128
Variant of Interest (VOI), 127
Variants Being Monitored (VBM), 123
Variants of Concern (VOCs), 123
Vector control, 74
Viral zoonoses, 45
Virion assembly, 130
Virus-Like Particles (VLPs), 159
Virus release, 130

W
Waste reduction, 218
Water efficiency, 217
Water, Sanitation and Hygiene (WASH), 101
Whole virus vaccines, 162
World Health Assembly (WHA), 119

Z
Zooanthroponoses, 47
Zoonosis, 13

The manufacturer's authorised representative in the EU is Springer Nature Customer Service Centre GmbH, Europaplatz 3, 69115 Heidelberg, Germany. If you have any concerns regarding our products, please contact ProductSafety@springernature.com

Printed and bound by CPI Group (UK) Ltd, Croydon, CR0 4YY
30/09/2026
02225589-0008